The Convolution Transform

The Convolution Transform

Isidore Isaac Hirschman
and
David Vernon Widder

DOVER PUBLICATIONS, INC.
Mineola, New York

Bibliographical Note

This Dover edition, first published in 2005, is an unabridged republication of the work first published by Princeton University Press, Princeton, New Jersey, in 1955.

Library of Congress Cataloging-in-Publication Data

Hirschman, I. I. (Isidore Isaac), 1922–
The convolution transform / Isidore Isaac Hirschman and David Vernon Widder.
p. cm.
Originally published: Princeton : Princeton University Press, 1955, in series: Princeton mathematical series ; 20.
Includes bibliographical references and index.
ISBN 0-486-44175-X (pbk.)
1. Convolutions (Mathematics) I. Widder, D. V. (David Vernon), 1898– II. Title.

QA601.H58 2005
515'.723—dc22

2004063427

Manufactured in the United States of America
Dover Publications, Inc., 31 East 2nd Street, Mineola, N.Y. 11501

Preface

The operation of convolution applied to sequences or functions is basic in analysis. It arises when two power series or two Laplace (or Fourier) integrals are multiplied together. Also most of the classical integral transforms involve integrals which define convolutions. For the present authors the convolution transform came as a natural generalization of the Laplace transform. It was early recognized that the now familiar real inversion of the latter is essentially accomplished by a *particular* linear differential operator of infinite order (in which translations are allowed). When one studies *general* operators of the same nature one encounters immediately general convolution transforms as the objects which they invert. This relation between differential operators and integral transforms is the basic theme of the present study.

The book may be read easily by anyone who has a working knowledge of real and complex variable theory. For such a reader it should be complete in itself, except that certain fundamentals from *The Laplace Transform* (number 6 in this series) are assumed. However, it is by no means necessary to have read that treatise completely in order to understand this one. Indeed some of those earlier results can now be better understood as special cases of the newer developments.

In conclusion we wish to thank the editors of the Princeton Mathematical Series for including this book in the series.

I. I. Hirschman
D. V. Widder

Contents

Chapter I

INTRODUCTION

Chapter II

THE FINITE KERNELS

Chapter III

THE NON-FINITE KERNELS

CHAPTER IV

VARIATION DIMINISHING TRANSFORMS

CHAPTER V

ASYMPTOTIC BEHAVIOUR OF KERNELS

CHAPTER VI

REAL INVERSION THEORY

CHAPTER VII

REPRESENTATION THEORY

CHAPTER VIII

THE WEIERSTRASS TRANSFORM

CHAPTER IX

COMPLEX INVERSION THEORY

CHAPTER X

MISCELLANEOUS TOPICS

The Convolution Transform

CHAPTER I

Introduction

1. INTRODUCTION

1.1. In this preliminary chapter is presented a heuristic introduction to the material which is to be given detailed treatment in later chapters. The method here is to illustrate rather than to prove. As illustrations we use four special examples of convolution transforms which taken together will show clearly the variety of properties which such transforms may have. The first of these examples involves an exponential kernel and is, in a sense, trivial. However, from another point of view, this kernel is the atomic material from which all others are constructed, so that its use for introductory purposes is perhaps mandatory. The last two examples reduce to the Laplace and Stieltjes transforms. Since we regard the fundamental properties of these as known, any new results about the convolution transform can be checked against the corresponding known ones for these two classic transforms.

2. CONVOLUTIONS

2.1. When two Laurent series

$$A(z) = \sum_{k=-\infty}^{\infty} a_k z^k, \qquad B(z) = \sum_{k=-\infty}^{\infty} b_k z^k$$

are multiplied together formally a new series of the same type results

$$A(z)B(z) = \sum_{k=-\infty}^{\infty} c_k z^k$$

where the new coefficients c_k are related to the old ones as follows:

$$c_n = \sum_{k=-\infty}^{\infty} a_{n-k} b_k \qquad n = 0, \pm 1, \pm 2, \cdots$$

The sequence $\{c_n\}_{-\infty}^{\infty}$ is called the *convolution* of the sequences $\{a_n\}_{-\infty}^{\infty}$ and $\{b_n\}_{-\infty}^{\infty}$. We arrive at the continuous analogue of this operation when we multiply together two bilateral Laplace integrals,

$$A(s) = \int_{-\infty}^{\infty} e^{-st} a(t)\, dt, \qquad B(s) = \int_{-\infty}^{\infty} e^{-st} b(t)\, dt.$$

The result is an integral of the same form,

$$A(s)B(s) = \int_{-\infty}^{\infty} e^{-st}c(t)\,dt,$$

where

$$c(x) = \int_{-\infty}^{\infty} a(x-t)b(t)\,dt. \tag{1}$$

This combination of functions occurs so frequently that it may be regarded as one of the fundamental operations of analysis. The function $c(x)$ is called the *convolution* of $a(x)$ and $b(x)$, and the integral (1) is commonly abbreviated as $a(x) * b(x)$ or as $a * b$.

If we take one of the functions, say $a(x)$, as fixed equation (1) may be considered as an integral equation with $c(x)$ the given function and $b(x)$ the unknown. Alternatively, the equation may be thought of as an integral transform. Our usual notation will be

$$f(x) = \int_{-\infty}^{\infty} G(x-t)\varphi(t)\,dt, \tag{2}$$

and this will be called the *convolution transform with kernel* $G(x)$ *of the function* $\varphi(x)$ *into* $f(x)$. Let us list the following four examples which will serve as the illustrations mentioned in § 1.

A. Choose the kernel as

$$\begin{aligned} g(x) &= e^x & -\infty < x < 0 \\ &= 0 & 0 \leqslant x < \infty. \end{aligned}$$

Then

$$g * \varphi = \int_{-\infty}^{0} e^t\varphi(x-t)\,dt = e^x \int_x^{\infty} e^{-t}\varphi(t)\,dt.$$

B. If $G(x) = (1/2)e^{-|x|}$, then

$$G * \varphi = \tfrac{1}{2} \int_{-\infty}^{\infty} e^{-|x-t|}\,\varphi(t)\,dt.$$

C. If $G(x) = (2\pi)^{-1}\operatorname{sech}(t/2)$, then (2) becomes

$$f(x) = \frac{1}{2\pi} \int_{-\infty}^{\infty} \operatorname{sech}\frac{x-t}{2}\,\varphi(t)\,dt. \tag{3}$$

However, if we make an exponential change of variable, replacing e^x and e^t by x and t, respectively, this becomes

$$F(x) = \int_0^{\infty} \frac{\Phi(t)}{x+t}\,dt, \tag{4}$$

where

$$f(x) = e^{x/2}F(e^x), \qquad \varphi(x) = \pi e^{x/2}\Phi(e^x).$$

This is the familiar Stieltjes transform.

D. If

$$G(x) = e^{-e^x}e^x, \qquad f(x) = F(e^x)e^x, \qquad \varphi(x) = \Phi(e^{-x}),$$

then (2) becomes the Laplace transform

$$F(x) = \int_0^\infty e^{-xt}\Phi(t)\,dt. \tag{5}$$

Thus we see that both the Stieltjes transform and the *unilateral* Laplace transform are special cases of the convolution transform, as predicted in § 1.

3. OPERATIONAL CALCULUS

3.1. Very useful as a guide to the following theory is a rudimentary knowledge of operational calculus. This is a technique which treats an operational symbol such as "D," for differentiation, as if it were a number throughout a calculation and at the last step gives the symbol its original meaning. We shall not be concerned with the justification of this process, which of course lies in the fact that there is a one-to-one correspondence between the laws of combination of the operation on the one hand and the fundamental operations of arithmetic on the other. We take rather the point of view that a suitable meaning for a new operation may be deduced by algebraic procedures on old ones and then adopted as a *definition*; that new results may be conjectured by operational calculus and then *proved*.

As a first illustration let us deduce a meaning for the operation e^{aD}. We expand this exponential in power series as if D were a number and then interpret D^k as a derivative of order k, giving the Maclaurin series:

$$e^{aD} = \sum_{k=0}^{\infty} \frac{a^k D^k}{k!}$$

$$e^{aD}f(x) = \sum_{k=0}^{\infty} \frac{a^k}{k!} f^{(k)}(x) = f(x+a). \tag{1}$$

Finally, we *define* $e^{aD}f(x)$ as $f(x+a)$ and observe that in spite of the background of the definition $e^{aD}f(x)$ is well defined even if $f(x)$ is not differentiable.

As a second application of the operational calculus let us solve the differential system

$$(1 - D)f(x) = \varphi(x) \tag{2}$$

$$f(x) = o(e^x) \qquad x \to +\infty, \tag{3}$$

where $\varphi(x) \in C \cdot B$ (is continuous and bounded) for all x. We observe first that if a solution exists, it is unique. For, the corresponding homogeneous equation

$$(4) \qquad (1 - D)f(x) = 0$$

has the general solution Ae^x, and the latter solution can satisfy (3) only if the constant A is zero. That is, the system (4) (3) has only the identically vanishing solution.

The symbolic solution of (2) is

$$f(x) = (1 - D)^{-1}\varphi(x),$$

so that an interpretation of the operator $(1 - D)^{-1}$ is needed. This may be supplied by equation (1) and the familiar Laplace integral

$$(5) \qquad \frac{1}{1-s} = \int_{-\infty}^{\infty} e^{-st}g(t)\,dt = \int_{-\infty}^{0} e^{-st}e^{t}\,dt \qquad -\infty < s < 1,$$

where $g(t)$ is the kernel of Example A. Replacing s by D in (5) and using (1) for the interpretation of e^{-tD}, we obtain

$$(1 - D)^{-1}\varphi(x) = \int_{-\infty}^{\infty} g(t)e^{-tD}\varphi(x)\,dt = g * \varphi.$$

That is, the operational solution of the system (2) (3) is

$$(6) \qquad f(x) = g * \varphi = e^{x}\int_{x}^{\infty} e^{-t}\varphi(t)\,dt.$$

We can now show directly that this is the actual solution. Since $\varphi(x) \in B$ the integral (6) converges, so that the boundary condition (3) is clearly satisfied. Since $\varphi(x) \in C$, formal differentiation of equation (6) shows that $f(x)$ satisfies (2).

In spite of the trivial nature of this result, let us record it as a theorem, for we shall need to refer to it in an inductive proof of Chapter II.

THEOREM 3.1. *If $\varphi(x) \in C \cdot B$ in $(-\infty, \infty)$, and if*

$$f(x) = g * \varphi = e^{x}\int_{x}^{\infty} e^{-t}\varphi(t)\,dt,$$

then $f(x)$ is the unique solution of the system (2) (3).

In a similar way we could show that the unique solution of the system

$$(7) \qquad (1 - D^2)f(x) = \varphi(x)$$

$$(8) \qquad f(x) = o(e^{x}) \qquad x \to +\infty$$

$$(9) \qquad f(x) = o(e^{-x}) \qquad x \to -\infty$$

is $f(x) = \varphi * G$, where G is the kernel of Example B.

The role of the Laplace transform in these examples should be noticed. If a function $F(s)$ can be expressed as a Laplace integral, then an interpretation of the operator $F(D)$ is immediately available by use of (1).

4. GREEN'S FUNCTIONS

4.1. It is a familiar fact that the Green's function of a non-homogeneous differential system enables one to solve the system explicitly. As a simple illustration let us consider the differential system, (2) (3) of § 3. Formally, the Green's function $G(x, t)$ of the system is a function of x and of a parameter t such that

$$(1) \qquad (1 - D)G(x, t) = \delta(x - t) \qquad -\infty < x, t < \infty$$

$$(2) \qquad G(x, t) = o(e^x) \qquad x \to +\infty.$$

Here, as previously, D stands for differentiation with respect to x, the parameter t being held fixed. The function $\delta(x)$ is Dirac's symbolic function with the following properties:

$$\delta(x) = \begin{cases} 0 & x \neq 0 \\ +\infty & x = 0 \end{cases}$$

$$\int_{-h}^{h} \delta(x)\, dx = 1 \qquad h > 0.$$

In terms of $G(x, t)$ the solution of the given system (2) (3) of § 3 is

$$(3) \qquad f(x) = \int_{-\infty}^{\infty} G(x, t)\varphi(t)\, dt,$$

at least formally. For, differentiation under the integral sign gives

$$(1 - D)f(x) = \int_{-\infty}^{\infty} \delta(x - t)\varphi(t)\, dt = \int_{-\infty}^{\infty} \varphi(x - t)\delta(t)\, dt = \varphi(x).$$

Also $f(x)$ satisfies the desired boundary condition by (2).

The above heuristic procedure may serve as a guide. We take as our actual definition of $G(x, t)$ a function of x which satisfies the equation

$$(1 - D)G(x, t) = 0$$

in each of the intervals $(-\infty, t)$ (t, ∞), has a unit jump at $x = t$,

$$(4) \qquad G(t-, t) - G(t+, t) = 1$$

and satisfies (2). Let us compute $G(x, t)$. It must evidently have the form

$$\begin{aligned} G(x, t) &= A(t)e^x \qquad & -\infty < x < t, \\ &= B(t)e^x \qquad & t < x < \infty. \end{aligned}$$

By (2), $B(t) = 0$. By (4), $A(t) = 1$. That is, $G(x, t) = g(x - t)$, where $g(t)$ is the kernel of Example A of § 2. That the function (3), deduced formally as a solution of the given system, is indeed the solution now follows by Theorem 3.1 provided that $\varphi(x) \in C \cdot B$.

In a similar way we could show that the Green's function of the system (7) (8) (9) of § 3 is the kernel of Example B. In a later chapter we shall see that the kernels of Examples C and D may also be regarded as Green's functions of differential systems of infinite order.

5. OPERATIONAL CALCULUS CONTINUED

5.1. We may apply the operational procedure of § 3 to obtain an inversion formula for the convolution transform

$$f(x) = \int_{-\infty}^{\infty} G(x-t)\varphi(t)\,dt. \tag{1}$$

Let

$$\int_{-\infty}^{\infty} G(t)e^{-st}\,dt = \frac{1}{E(s)} \tag{2}$$

be the bilateral Laplace transform of $G(t)$. We have

$$\frac{1}{E(D)}\varphi(x) = \left[\int_{-\infty}^{\infty} G(t)e^{-Dt}\,dt\right]\varphi(x) = \int_{-\infty}^{\infty} G(t)\varphi(x-t)\,dt, \tag{3}$$

$$= f(x).$$

Multiplying by $E(D)$ we obtain

$$E(D)f(x) = \varphi(x), \tag{4}$$

our desired inversion formula. The function $E(s)$ defined by (2) will be called the *inversion function* corresponding to the kernel $G(t)$.

Let us compute the bilateral Laplace transforms of the kernels considered in § 2. In these formulas $s = \sigma + i\tau$ is a complex variable.

A. $$\int_{-\infty}^{\infty} g(t)e^{-st}\,dt = \frac{1}{1-s} \qquad -\infty < \sigma < 1.$$

B. $$\frac{1}{2}\int_{-\infty}^{\infty} e^{-|t|}\,e^{-st}\,dt = \frac{1}{1-s^2} \qquad -1 < \sigma < 1.$$

C. $$\frac{1}{2\pi}\int_{-\infty}^{\infty} e^{-st}\operatorname{sech}\left(\frac{t}{2}\right)dt = \frac{1}{\pi}\int_{-\infty}^{\infty}\frac{e^{-st}}{e^{t/2}+e^{-t/2}}\,dt$$

$$= \frac{1}{\pi}\int_0^{\infty}\frac{x^{s-1/2}}{1+x}\,dx \qquad x = e^{-t}$$

$$= \frac{1}{\pi}\Gamma\left(\frac{1}{2}+s\right)\Gamma\left(\frac{1}{2}-s\right)$$

$$= 1/\cos \pi s \qquad -\frac{1}{2} < \sigma < \frac{1}{2}.$$

D. $$\int_{-\infty}^{\infty} e^{-e^t}e^t e^{-st}\,dt = \int_0^{\infty} x^{-s}e^{x}\,dx \qquad x = e^t$$

$$= \Gamma(1-s) \qquad -\infty < \sigma < 1.$$

Here we have used several familiar formulas from the theory of the Gamma function. Compare, for example, E. C. Titchmarsh [1939; 106].

A. If

$$f(x) = \int_{-\infty}^{\infty} g(x-t)\varphi(t)\,dt,$$

then formula (4) implies that

$$(1-D)f(x) = \varphi(x).$$

We have already verified this.

B. If

$$f(x) = \frac{1}{2}\int_{-\infty}^{\infty} e^{-|x-t|}\,\varphi(t)\,dt,$$

then formula (4) implies that

$$(1-D)\,(1+D)f(x) = \varphi(x).$$

We have

$$f(x) = \frac{1}{2}\int_{-\infty}^{x} e^{-x+t}\varphi(t)\,dt + \frac{1}{2}\int_{x}^{\infty} e^{x-t}\varphi(t)\,dt,$$

from which we obtain

$$(1+D)\,\frac{1}{2}\int_{-\infty}^{x} e^{-x+t}\varphi(t)\,dt = \frac{1}{2}\,\varphi(x),$$

$$(1+D)\,\frac{1}{2}\int_{x}^{\infty} e^{x-t}\varphi(t)\,dt = -\,\frac{1}{2}\,\varphi(x) + \int_{x}^{\infty} e^{x-t}\varphi(t)\,dt,$$

$$(1+D)f(x) = \int_{x}^{\infty} e^{x-t}\varphi(t)\,dt.$$

Referring to Example A we see that

$$(1-D)\,(1+D)f(x) = \varphi(x).$$

C. If

$$f(x) = (2\pi)^{-1}\int_{-\infty}^{\infty} \operatorname{sech}\left(\frac{x-t}{2}\right)\varphi(t)\,dt, \tag{5}$$

then the formula (4) implies that

$$\cos \pi D\, f(x) = \varphi(x).$$

This formula is incomplete however since no definite meaning attaches to $\cos \pi D$. We have, see E. C. Titchmarsh [1939; 114],

$$\cos \pi x = \lim_{n\to\infty} e^{b_n x} \prod_{k=-n}^{n} \left(1 - \frac{2x}{2k+1}\right) e^{2x/(2k+1)}$$

where the $\{b_n\}$ are real and $\lim\limits_{n\to\infty} b_n = 0$, and this suggests

$$\lim_{n\to\infty} e^{b_n D} \prod_{k=-n}^{n} \left(1 - \frac{2D}{2k+1}\right) e^{2D/(2k+1)} f(x) = \varphi(x) \tag{6}$$

as a possible inversion formula for (5).

D. If

$$f(x) = \int_{-\infty}^{\infty} e^{-e^{(x-t)}} e^{(x-t)} \varphi(t)\, dt, \tag{7}$$

then the formula (4) implies that

$$\frac{1}{\Gamma(1-D)} f(x) = \varphi(x),$$

and since, see Titchmarsh [1939; 257],

$$\frac{1}{\Gamma(1-x)} = \lim_{n\to\infty} e^{b_n x} e^{-\gamma x} \prod_{k=1}^{n} \left(1 - \frac{x}{k}\right) e^{x/k},$$

we may conjecture that

$$\lim_{n\to\infty} e^{b_n D} e^{-\gamma D} \prod_{k=1}^{n} \left(1 - \frac{D}{k}\right) e^{D/k} f(x) = \varphi(x). \tag{8}$$

Other expansions for $\cos \pi x$ and $1/\Gamma(1-x)$ would lead to different "definitions" of $\cos \pi D$ and $1/\Gamma(1-D)$. The product definitions given here are characteristic of our theory.

If suitable choices are made for b_n then the formulas (6) and (8) become, after a change of variables, well known operational inversion formulas for the Stieltjes and Laplace transforms. Let us verify this in detail for the Laplace transform. We have shown in § 2 that if in

$$f(x) = \int_{-\infty}^{\infty} e^{-e^{(x-t)}} e^{x-t} \varphi(t)\, dt$$

we put

$$f(x) = e^x F(e^x), \qquad \varphi(t) = \Phi(e^{-t}), \qquad x = \log y, \qquad t = -\log u, \tag{9}$$

then we obtain

$$F(y) = \int_0^{\infty} e^{-uy} \Phi(u)\, du.$$

If we define

$$b_n = \log n - \sum_{k=1}^{n} \frac{1}{k} + \gamma \qquad n = 1, 2, \cdots,$$

then $\lim_{n\to\infty} b_n = 0$ and

$$\lim_{n\to\infty} P_n(D) f(x) = \varphi(x), \tag{10}$$

where

$$P_n(D) = e^{(\log n) D} \prod_{k=1}^{n} \left(1 - \frac{D}{k}\right).$$

We have

$$\left(1-\frac{D}{1}\right)f(x)=\left(1-\frac{D}{1}\right)e^{x}F(e^{x})=-e^{2x}F'(e^{x}),$$

$$\left(1-\frac{D}{2}\right)\left(1-\frac{D}{1}\right)f(x)=-\left(1-\frac{D}{2}\right)e^{2x}F'(e^{x})=+\frac{1}{2}e^{3x}F''(e^{x}),$$

$$\vdots$$

$$\prod_{k=1}^{n}\left(1-\frac{D}{k}\right)f(x)=\frac{(-1)^n}{n!}e^{(n+1)x}F^{(n)}(e^{x}),$$

$$P_n(D)f(x)=\frac{(-1)^n}{n!}e^{(n+1)(x+\log n)}F^{(n)}(e^{x+\log n}).$$

Making use of (9) we see that (10) is equivalent to the familiar inversion formula

$$\lim_{n\to\infty}\frac{(-1)^n}{n!}\left(\frac{n}{y}\right)^{n+1}F^{(n)}\left(\frac{n}{y}\right)=\Phi(y),$$

see D. V. Widder [1946; 288].

For a similar discussion of the Stieltjes transform see § 9 of Chapter III.

6. THE GENERATION OF KERNELS

6.1. Let b, $\{a_k\}_{k=1}^{\infty}$ be real numbers such that

$$\sum_{1}^{\infty}\frac{1}{a_k^2}<\infty. \tag{1}$$

We define

$$E(s)=e^{bs}\prod_{k=1}^{\infty}\left(1-\frac{s}{a_k}\right)e^{s/a_k}. \tag{2}$$

The condition (1) insures that the infinite product (2) is convergent. See E. C. Titchmarsh [1939; 250]. If there exists a function $G(t)$ such that

$$\int_{-\infty}^{\infty}e^{-st}G(t)\,dt=\frac{1}{E(s)} \tag{3}$$

then the considerations of the preceding section suggest strongly that the convolution transform

$$f(x)=\int_{-\infty}^{\infty}G(x-t)\varphi(t)\,dt \tag{4}$$

is inverted by the formula

$$\lim_{n\to\infty} P_n(D)f(x) = \varphi(x), \tag{5}$$

where

$$P_n(D) = e^{(b-b_n)} \prod_{k=1}^{n} \left(1 - \frac{D}{a_k}\right) e^{D/a_k}.$$

Here the b_n are real numbers such that $\lim_{n\to\infty} b_n = 0$.

The complex inversion formula for the bilateral Laplace transform asserts that if the transform

$$\int_{-\infty}^{\infty} e^{-st} a(t)\, dt = A(s)$$

converges absolutely in the strip $\sigma_1 < \sigma < \sigma_2$ then (under certain restrictions)

$$a(t) = \frac{1}{2\pi i} \int_{d-i\infty}^{d+i\infty} e^{st} A(s)\, ds \qquad \sigma_1 < d < \sigma_2, \quad -\infty < t < \infty.$$

We therefore set

$$G(t) = (2\pi i)^{-1} \int_{-i\infty}^{i\infty} [E(s)]^{-1} e^{st}\, ds. \tag{6}$$

We shall ultimately prove that $G(t)$ defined by (6) satisfies (3) and that the convolution transform (4) is indeed inverted by the operational inversion formula (5).

In point of fact we shall treat a slightly more general class of kernels. Let a_k, $k = 1, 2, \cdots, b, c$ be real and such that

$$\sum_k a_k^{-2} < \infty, \qquad c \geqq 0, \tag{7}$$

and let

$$G(t) = (2\pi i)^{-1} \int_{-i\infty}^{i\infty} \left[e^{-cs^2+bs} \prod_k \left(1 - \frac{s}{a_k}\right) e^{s/a_k} \right]^{-1} e^{st}\, ds. \tag{8}$$

It is to the study of the kernels (8) and their associated convolution transforms that the present book is devoted.

7. VARIATION DIMINISHING CONVOLUTIONS

7.1. It is natural to ask why when our operational procedures apply, at least formally, to every convolution transform we have limited ourselves to the kernels 6.1(8). The reasons for this lie somewhat deeper than the operational calculus, and depend upon the following result.

Let $G(t) \in L(-\infty, \infty)$,

$$\int_{-\infty}^{\infty} G(t)\,dt = 1,$$

and let

$$f(x) = \int_{-\infty}^{\infty} G(x - t)\varphi(t)\,dt,$$

where $\varphi(t)$ is bounded and continuous. The kernel $G(t)$ will be said to be *variation diminishing* if the number of changes of sign of $f(x)$ for $-\infty < x < \infty$ never exceeds the number of changes of sign of $\varphi(t)$ for $-\infty < t < \infty$. It has been shown by Schoenberg [1947, 1948b] that $G(t)$ is variation diminishing if and only if it is of the form 6.1(8). See Chapter IV for a proof in a form applicable to our theory.

As an example we may verify that

$$g(t) = (2\pi i)^{-1}\int_{-i\infty}^{i\infty} (1 - s)^{-1}e^{st}\,ds$$

is variation diminishing. We have from § 3 that

$$\varphi(x) = (1 - D)f(x),$$

$$= -e^{x}De^{-x}f(x).$$

If $f(x)$ has n changes of sign, then there exist points

$$-\infty < x_1 < x_2 < \cdots < x_n < \infty$$

such that

$$f(x_i) = 0 \qquad\qquad i = 1, \cdots, n,$$

and such that $f(x)$ is not identically zero in any of the intervals

$$[-\infty, x_1], [x_1, x_2], \cdots, [x_{n-1}, x_n], [x_n, +\infty].$$

Applying Rolle's theorem we see that $\varphi(x)$ has at least one change of sign in each of the intervals

$$[x_1, x_2], \cdots, [x_{n-1}, x_n], [x_n, +\infty].$$

Thus $\varphi(x)$ has at least n changes of sign. Note that in the above argument use was made of the fact that $\lim\limits_{x\to+\infty} f(x)e^{-x} = 0$.

Thus the kernels 6.1(8) are characterized by an important intrinsic property. Since this property of being variation diminishing plays an essential role in our theory our choice of kernels is seen to be dictated not by convenience but by intrinsic mathematical structure.

8. OUTLINE OF PROGRAM

8.1. In this section we shall very briefly describe and illustrate the program of the following chapters. Chapters II through V are devoted to the study of the kernels $G(t)$ defined by 6.1(8). We list a few typical properties. It is shown that $G(t)$ is a frequency function with mean b and variance $2c + \sum_k a_k^{-2}$; that is

$$\begin{gathered} G(t) \geqq 0, \\ \int_{-\infty}^{\infty} G(t)\, dt = 1, \\ \int_{-\infty}^{\infty} G(t)(t-b)\, dt = 0, \\ \int_{-\infty}^{\infty} G(t)(t-b)^2\, dt = 2c + \sum_k a_k^{-2}. \end{gathered} \tag{1}$$

We also prove, as we indicated before, that $G(t)$ is variation diminishing, a fact which has many interesting consequences, among them that $G^{(n)}(t)$ has, for each value of n for which it is defined, exactly n changes of sign, and that $-\log G(t)$ is convex.

In Chapters VI and VII we study the convolution transform with kernel $G(t)$ defined by the formula 6.1(8) with $c = 0$. These kernels must be subdivided into three classes:

$G(t) \in$ *class I if there are positive and negative roots* a_k;

$G(t) \in$ *class II if there are only positive roots* a_k *and if* $\sum_k a_k^{-1} = \infty\cdot$

$G(t) \in$ *class III if there are only positive roots* a_k *and if* $\sum_k a_k^{-1} < \infty$.

If $G(t)$ satisfies the restriction that $c = 0$ then $G(t)$ or $G(-t)$ belongs to one of these classes. The following results may serve as examples.

The kernel of Example C belongs to class I. It is well known that if the Stieltjes transform

$$F(y) = \int_{0+}^{\infty} \frac{1}{y+u} \Phi(u)\, du$$

converges (conditionally) for any value of $y > 0$ then it converges for all $y > 0$, see D. V. Widder [1946; 326]. After an exponential change variable we find that if

$$f(x) = (2\pi)^{-1} \int_{-\infty}^{\infty} \operatorname{sech}\left[\tfrac{1}{2}(x-t)\right]\varphi(t)\, dt$$

converges (conditionally) for any value of x, $-\infty < x < \infty$, then it converges for all x. This convergence behaviour is characteristic of class I kernels.

The kernel of Example D belongs to class II. If the Laplace transform

$$F(y) = \int_{0+}^{\infty} e^{-yu}\Phi(u)\,du$$

converges (conditionally) for any value of $y > 0$ then it converges for all larger values of y, from which it follows that if

$$f(x) = \int_{-\infty}^{\infty} e^{-e^{(x-t)}}\, e^{(x-t)}\varphi(t)\,dt$$

converges for any value of x, $-\infty < x < \infty$, then it converges for all larger values of x. See D. V. Widder [1946; 37]. This behaviour is typical of class II kernels. The kernels of Examples A and B belong to classes III and I, respectively.

In Chapter VIII a study is made of the transforms with kernels of the form

$$\int_{-\infty}^{\infty} e^{-st}G(t)\,dt = 1/e^{cs^2}. \tag{3}$$

In this case $G(t)$ may be computed explicitly,

$$G(t) = (4\pi c)^{-1/2}e^{-t^2/4c}.$$

The theory of the convolution transforms with these kernels is closely bound up with the heat equation

$$\frac{\partial u}{\partial h} = \frac{\partial^2 u}{\partial x^2}.$$

For instance it will be shown that necessary and sufficient conditions for $u(x, h)$ to be of the form

$$u(x, h) = (4\pi h)^{-\frac{1}{2}} \int_{-\infty}^{\infty} e^{-(x-t)^2/4h}\varphi(t)\,dt \qquad -\infty < x < \infty;\ h > 0$$

where $|\varphi(t)| \leq M$ are that:

a. $|u(x, h)| \leq M \qquad -\infty < x < \infty; \quad h > 0,$

b. $\dfrac{\partial u}{\partial h} = \dfrac{\partial^2 u}{\partial x^2} \qquad -\infty < x < \infty; \quad h > 0.$

Chapter IX is devoted to complex inversion formulas. It has long been known that if

$$F(w) = \int_{0}^{\infty} (w + u)^{-1}\Phi(u)\,du$$

then

$$\Phi(u) = \lim_{\epsilon\to 0+} (2\pi i)^{-1}\,[F(-u - i\epsilon) - F(-u + i\epsilon)],$$

or, what is substantially the same,

$$\Phi(u) = \lim_{\epsilon\to 0+} (2\pi i)^{-1}[e^{i\epsilon/2}F(-ue^{i\epsilon}) - e^{-i\epsilon/2}F(-ue^{-i\epsilon})].$$

See D. V. Widder [1946; 340]. If in this formula we make the change of variable described in Example C of § 2 we find that if

$$f(x) = (2\pi)^{-1} \int_{-\infty}^{\infty} \operatorname{sech} \tfrac{1}{2}(x - t)\varphi(t)\, dt,$$

then

$$\varphi(t) = \lim_{p \to 1-} \tfrac{1}{2}[f(t + ip\pi) + f(t - ip\pi)].$$

We shall see how to associate a similar formula with each kernel $G(t)$ of the form

$$G(t) = (2\pi i)^{-1} \int_{-i\infty}^{i\infty} \left[\prod_{1}^{\infty} \left(1 - \frac{s^2}{a_k^{\,2}}\right) \right]^{-1} e^{st}\, ds,$$

where

$$\lim_{k \to \infty} k/a_k = \Omega, \qquad 0 < \Omega < \infty.$$

Chapter X contains a number of shorter topics.

9. SUMMARY

9.1. In this chapter we have attempted to show a few of the principal ideas of our general theory by illustrating them with the simplest possible examples. We have chosen only a few of the many applications with the hope that the reader may obtain the essential flavor of the results before entering into the detailed study necessary for a deeper understanding.

CHAPTER II

The Finite Kernels

1. INTRODUCTION

1.1. In this chapter we shall be concerned with the problem of inverting the convolution transform

$$f(x) = \int_{-\infty}^{\infty} G(x-t)\varphi(t)\,dt, \tag{1}$$

where $G(x)$ belongs to a class of *finite kernels*. These are the functions $G(x)$ which are such that

$$\frac{1}{E(s)} = \int_{-\infty}^{\infty} e^{-st}G(t)\,dt, \tag{2}$$

where $E(s)$ is a polynomial of degree n with real zeros. The term "finite" is intended to correspond to the finite degree of the polynomial $E(s)$. In a later chapter we consider kernels $G(t)$ satisfying (2) with $E(s)$ an entire function.

If

$$E(s) = \left(1 - \frac{s}{a_1}\right)\left(1 - \frac{s}{a_2}\right)\cdots\left(1 - \frac{s}{a_n}\right),$$

we shall show that (1) is inverted by the linear differential operator $E(D)$, where D stands for differentiation with respect to x. That is,

$$E(D)f(x) = \varphi(x).$$

Since the kernels (2) turn out to be frequency functions in the sense of the theory of statistics, we begin our study with a brief discussion of the pertinent portion of that theory.

2. DISTRIBUTION FUNCTIONS

2.1. Before introducing the frequency functions mentioned in the previous section let us define the related distribution functions.

DEFINITION 2.1. A function $\alpha(t)$ defined in $-\infty < t < \infty$, is a distribution function if and only if*

1. $\alpha(t) \in \uparrow \qquad -\infty < t < \infty$
2. $\alpha(-\infty) = 0$
3. $\alpha(+\infty) = 1.$

* The logical symbols used, such as $\in\uparrow$, are familiar. They are explained, for example, in *Advanced Calculus*, by D. V. Widder [1947a; 5]. Dates in brackets refer to the bibliography, and the number following the semicolon is a page number.

Several examples follow:

$$\alpha(t) = \begin{cases} 0 & -\infty < t < 0 \\ 1/2 & t = 0 \\ 1 & 0 < t < \infty, \end{cases}$$

$$\beta(t) = \begin{cases} e^t & -\infty < t \leqq 0 \\ 1 & 0 \leqq t < \infty, \end{cases}$$

$$\gamma(t) = \frac{1}{2} + \frac{1}{\pi} \tan^{-1} t.$$

A distribution function is *normalized* if

$$\alpha(t) = \frac{\alpha(t+) + \alpha(t-)}{2} \qquad -\infty < t < \infty. \tag{1}$$

All three examples above are normalized. If $\alpha(t)$ were defined differently at the origin it would not be normalized.

In the theory of statistics it is found useful at times to think of $\alpha(t)$ as defining a distribution of unit mass on the infinite t-axis in such a way that there are $\alpha(b) - \alpha(a)$ units of mass in the interval (a, b). Hence the term.

Two distribution functions are *equivalent* if and only if they are equal at all points of continuity. From (1) it is clear that equivalent normalized distribution functions are equal at all points.

2.2. We now define a statistical term, the *mean* of a distribution function.

DEFINITION 2.1. The mean of a distribution function $\alpha(t)$, denoted by m_α, is the integral

$$m_\alpha = \int_{-\infty}^{\infty} t\, d\alpha(t) \tag{1}$$

when that integral converges.

For the examples of § 2.1, $m_\alpha = 0$, $m_\beta = -1$, and m_γ does not exist since the integral

$$\int_{-\infty}^{\infty} t\, d\gamma(t) = \frac{1}{\pi} \int_{-\infty}^{\infty} \frac{t}{1 + t^2}\, dt$$

diverges.

If $\alpha(t)$ is thought of as a distribution of mass, as explained in the previous section, then m_α locates the center of gravity of the distribution.

For example, if $\alpha(t)$ is a step-function with jumps m_k at points t_k, $k = 1, 2, \cdots, n$, then (1) becomes the familiar formula

$$m_\alpha = \sum_{k=1}^{n} m_k t_k = \frac{\sum_{k=1}^{n} m_k t_k}{\sum_{k=1}^{n} m_k}.$$

2.3. Another useful concept from statistical theory is *variance*.

Definition 2.3. The variance of a distribution function $\alpha(t)$, denoted by V_α, is the integral

$$V_\alpha = \int_{-\infty}^{\infty} (t - m_\alpha)^2 \, d\alpha(t)$$

when m_α is defined and the integral converges.

For the examples of § 2.1, $V_\alpha = 0$,

$$V_\beta = \int_{-\infty}^{0} (t+1)^2 e^t \, dt = 1,$$

and V_γ is not defined.

Using the foregoing physical interpretation of $\alpha(t)$, we see that V_α is the moment of inertia of the mass distribution about the center of gravity. It can consequently be considered as a measure of how far the mass is from its mean. Physically it is clear that V_α can vanish only if all the mass is concentrated at a point, as in the first example above. The fact is easily proved analytically.

3. FREQUENCY FUNCTIONS

3.1. A special case of a distribution function is one which is absolutely continuous. Its derivative is then called a *frequency function*.

Definition 3.1. A function* $\varphi(t)$ defined in $-\infty < t < \infty$ is a frequency function if and only if the function

$$\alpha(t) = \int_{-\infty}^{t} \varphi(u) \, du \qquad -\infty < t < \infty$$

is a distribution function.

Thus any non-negative function $\phi(u)$ for which

$$\int_{-\infty}^{\infty} \varphi(u) \, du = 1$$

* It will be understood that all functions considered are integrable in every sub-interval of the interval of definition.

is a frequency function. Using the examples of § 2.1, we see that $\varphi(t) = \beta'(t)$, $t \neq 0$, $\varphi(0) = 1$, is a frequency function. The same is true for $\gamma'(t) = \pi^{-1}(1 + t^2)^{-1}$. But there is no distribution function corresponding to $\alpha(t)$.

The mean and variance of frequency functions are defined in the obvious way:

$$m_\varphi = \int_{-\infty}^{\infty} t\varphi(t)\,dt,$$

$$V_\varphi = \int_{-\infty}^{\infty} (t - m_\varphi)^2\varphi(t)\,dt.$$

Thus, if $\varphi(t) = a\pi^{-1/2}e^{-a^2x^2}$, $a > 0$, then $m_\varphi = 0$ and

$$V_\varphi = \frac{a}{\sqrt{\pi}}\int_{-\infty}^{\infty} t^2e^{-a^2t^2}\,dt = \frac{1}{2a^2}. \tag{1}$$

In particular, if $a = 1/\sqrt{2}$, this function $\varphi(t)$ is called the *normal* frequency function.

4. CHARACTERISTIC FUNCTIONS

4.1. Distribution and frequency functions are often studied most easily by use of their Fourier or bilateral Laplace transforms. We shall refer to the transformed functions as *characteristic functions* in either case.

DEFINITION 4.1. The characteristic function, $\chi_\alpha(s)$, of a distribution function $\alpha(t)$ is

$$\chi_\alpha(s) = \int_{-\infty}^{\infty} e^{-st}\,d\alpha(t). \tag{1}$$

The characteristic function, $\chi_\varphi(s)$, of a frequency function $\varphi(t)$ is

$$\chi_\varphi(s) = \int_{-\infty}^{\infty} e^{-st}\varphi(t)\,dt. \tag{2}$$

Here, as elsewhere, s is a complex variable, $s = \sigma + i\tau$. Notice that $\chi_\alpha(-i\tau)$ and $\chi_\varphi(-i\tau)$ are Fourier transforms, and it is these that are usually called characteristic functions in statistical works. No confusion will arise by giving the functions (1) and (2) this name also.

Using the examples of § 2.1 we have

$$\chi_\alpha(s) = 1 \qquad -\infty < \sigma < \infty$$

$$\chi_\beta(s) = \frac{1}{1 - s} \qquad -\infty < \sigma < 1,$$

the transforms converging in the intervals indicated. Observe that the integrals (1) and (2) must always converge on the imaginary axis $\sigma = 0$ and may converge in a larger vertical strip. At the origin a characteristic function must equal unity.

As a final example choose $\varphi(t)$ the last function of § 3.1. Then

$$\chi_\varphi(s) = \frac{a}{\sqrt{\pi}} \int_{-\infty}^{\infty} e^{-st} e^{-a^2t^2}\, dt = e^{s^2/4a^2} \qquad -\infty < \sigma < \infty. \tag{3}$$

By differentiating this function twice with respect to s and setting $s = 0$ we may clearly obtain anew V_φ, (1) § 3.

4.2. Operations with characteristic functions in place of distribution functions can be useful only if it is possible to pass conveniently from the former to the latter. This is indeed the case, by use of the known inversion formulas for the bilateral Laplace transform. For easy reference we quote the pertinent results here; see D. V. Widder [1946; 241].

THEOREM A. *If $\alpha(t)$ is a normalized function of bounded variation in every finite interval, and if the integral*

$$f(s) = \int_{-\infty}^{\infty} e^{-st}\, d\alpha(t)$$

converges in the strip $k < \sigma < l$, then for all t

$$\lim_{T\to\infty} \frac{1}{2\pi i} \int_{c-iT}^{c+iT} \frac{f(s)}{s} e^{st}\, ds = \begin{cases} \alpha(t) - \alpha(-\infty) & c > 0,\ k < c < l \\ \alpha(t) - \alpha(\infty) & c < 0,\ k < c < l. \end{cases} \tag{1}$$

THEOREM B. *If $\varphi(t)$ is of bounded variation in some neighborhood of $t = t_0$ and if the integral*

$$f(s) = \int_0^{\infty} e^{-st}\varphi(t)\, dt$$

converges absolutely on the line $\sigma = c$, then

$$\lim_{T\to\infty} \frac{1}{2\pi i} \int_{c-iT}^{c+iT} f(s) e^{st_0}\, ds = \frac{\varphi(t_0^+) + \varphi(t_0^-)}{2}.$$

As an example we use the first illustrative distribution function of § 2.1. Equation (1) becomes

$$\lim_{T\to\infty} \frac{1}{2\pi} \int_{-T}^{T} \frac{e^{(1+iy)t}}{1 + iy}\, dy = \begin{cases} 0 & t < 0 \\ 1/2 & t = 0 \\ 1 & t > 0. \end{cases}$$

This is a familiar Fourier transform.

These two theorems are applicable only when the characteristic function is defined in a strip of positive width. This will be the case in the present chapter. However, if the strip reduces to a straight line, as it may for certain distribution functions, some modification of the above theorems is needed. We shall make this modification in § 5.2 of Chapter III, where it will be needed. It will develop, however, that all of our characteristic functions are defined by bilateral Laplace transforms which converge in strips of positive width, though this fact will not be known *a priori*.

5. CONVOLUTIONS

5.1. In the multiplication of characteristic functions the corresponding distribution functions are combined by *convolution*, a binary operation which is indeed the subject of the present study and which we now formally define.

DEFINITION 5.1. The Stieltjes convolution of two functions $\alpha(t)$ and $\beta(t)$, denoted by $\alpha(t) \# \beta(t)$, is the integral

$$\alpha \# \beta(t) = \alpha(t) \# \beta(t) = \int_{-\infty}^{\infty} \alpha(t-u)\, d\beta(u) \tag{1}$$

when that integral exists. The (Lebesgue) convolution of two functions $\varphi(t)$ and $\psi(t)$, denoted by $\varphi(t) * \psi(t)$, is the integral

$$\varphi * \psi(t) = \varphi(t) * \psi(t) = \int_{-\infty}^{\infty} \varphi(t-u)\psi(u)\, du, \tag{2}$$

when that integral exists.

In the integral (1) it will be sufficient to consider Riemann-Stieltjes integrals. If $\alpha(t)$ and $\beta(t)$ are distribution functions, $\alpha \# \beta(t)$ may fail to exist at a countable set of points, but it is non-decreasing in the remaining set of points and becomes a distribution function if suitably defined at the exceptional set, D. V. Widder [1946; 248]. The convolution $\varphi * \psi(t)$ of frequency functions is itself a frequency function. For, by Fubini's theorem

$$\int_{-\infty}^{\infty} dt \int_{-\infty}^{\infty} \varphi(t-u)\psi(u)\, du = \int_{-\infty}^{\infty} \psi(u)\, du \int_{-\infty}^{\infty} \varphi(t)\, dt = 1.$$

It is clear that both operations $\#$ and $*$ are commutative and associative.

Again using the functions of § 2.1 as examples, we have

$$\alpha \# \alpha(t) = \alpha(t) \qquad t \neq 0. \tag{3}$$

Here the exceptional set consists in the single point $t = 0$. Of course we define the convolution (3) so as to make equation (3) hold at $t = 0$ also. Similarly

$$\beta \# \beta(t) = \begin{cases} -te^t & -\infty < t \leqq 0 \\ 0 & 0 \leqq t < \infty. \end{cases}$$

If $\varphi(t)$ is the normal frequency function of § 3.1, then

$$\varphi * \varphi(t) = \frac{1}{2\pi}\int_{-\infty}^{\infty} e^{-(t-u)^2/2}\, e^{-u^2/2}\, du = \frac{1}{\sqrt{4\pi}}\, e^{-t^2/4}.$$

5.2. For our purposes it is important to know how to compute the mean and variance of $\gamma(t) = \alpha \# \beta(t)$ when those numbers are known for the separate functions $\alpha(t)$ and $\beta(t)$. We prove that it is only necessary to add in both cases.

THEOREM 5.2. *If $\alpha(t)$, $\beta(t)$ are distribution functions with mean and variance m_α, m_β and V_α, V_β, respectively, then $\alpha \# \beta(t)$ has mean and variance $m_\alpha + m_\beta$ and $V_\alpha + V_\beta$, respectively.*

This theorem is conveniently proved by use of characteristic functions, since the operation $\#$ for distribution functions corresponds to multiplication of their characteristic functions. For the reader's convenience we quote this result exactly; see D. V. Widder [1946; 257].

THEOREM C. *If the integrals*

$$f(s) = \int_{-\infty}^{\infty} e^{-st}\, d\alpha(t), \qquad g(s) = \int_{-\infty}^{\infty} e^{-st}\, d\beta(t)$$

converge absolutely for a common value of s, then for that value

$$f(s)g(s) = \int_{-\infty}^{\infty} e^{-st}\, d\gamma(t),$$

where $\gamma(t) = \alpha \# \beta(t)$.

From the definitions of characteristic function and mean it is clear that $\chi_\alpha(0) = 1$, $\chi_\alpha'(0) = -m_\alpha$ with corresponding equations for $\beta(t)$ and for $\gamma(t) = \alpha \# \beta(t)$. But by Theorem C,

$$-m_\gamma = \chi_\gamma'(0) = \chi_\alpha(0)\chi_\beta'(0) + \chi_\alpha'(0)\chi_\beta(0) = -m_\alpha - m_\beta.$$

Hence our result is proved in so far as it concerns the means.

Now set

$$A(s) = e^{m_\alpha s}\chi_\alpha(s) = \int_{-\infty}^{\infty} e^{-s(t-m_\alpha)}\, d\alpha(t)$$

$$B(s) = e^{m_\beta s}\chi_\beta(s)$$

$$C(s) = A(s)B(s).$$

Then $A(0) = 1$, $A'(0) = 0$, $A''(0) = V_\alpha$ with corresponding equations for $B(s)$. And by Theorem C,

$$C(s) = \int_{-\infty}^{\infty} e^{-s(t-m_\alpha-m_\beta)}\, d\gamma(t),$$

so that it is $C''(0) = m_\gamma$ that we wish to compute. But

$$C''(0) = A''(0)B(0) + 2A'(0)B'(0) + A(0)B''(0)$$
$$= V_\alpha + V_\beta,$$

and our theorem is proved.

The result of course extends to the convolution of any finite number of distribution functions. As a special case of Theorem 5.2 we have

COROLLARY 5.2. If $\varphi(t)$, $\psi(t)$ are frequency functions with mean and variance m_φ, m_ψ and V_φ, V_ψ, respectively, then $\varphi * \psi(t)$ has mean and variance $m_\varphi + m_\psi$ and $V_\varphi + V_\psi$, respectively.

6. THE FINITE KERNELS

6.1. All of the convolution transforms to be considered in this book, except those of Chapter VIII, will have kernels which can be synthesized from the following basic one

$$g(t) = \begin{cases} e^t & -\infty < t < 0 \\ 1/2 & t = 0 \\ 0 & 0 < t < \infty. \end{cases} \tag{1}$$

The finite kernels $G(t)$ will be made up by a finite number of convolutions of functions $|a_k| g(a_k t)$, $k = 1, 2, \cdots$. In the following chapters non-finite kernels will be synthesized from the same functions, but by use of infinitely many convolution operations.

As we have seen, $g(t)$ is a frequency function with mean -1 and variance 1. Its characteristic function is

$$\chi_g(s) = \int_{-\infty}^{\infty} e^{-st} g(t)\, dt = \frac{1}{1-s} \qquad -\infty < \sigma < 1. \tag{2}$$

Note that $|a| g(at)$ is again a frequency function if a is any real number not zero. Its characteristic function is $\left(1 - \frac{s}{a}\right)^{-1}$, the defining integral converging for $\sigma < a$ if $a > 0$ and for $\sigma > a$ if $a < 0$. Its mean and variance are $-1/a$ and $1/a^2$, respectively. These facts, together with Theorem C of § 5.2, show that the present description of the finite kernels is equivalent to that given in § 1.1.

6.2. Let $a_1, a_2, \cdots, a_n$ be any non-vanishing real constants, some or all of which may be coincident. We introduce the frequency functions

$$g_k(t) = |a_k| g(a_k t) \qquad k = 1, 2, \cdots, n, \tag{1}$$

and combine them by convolution to obtain new frequency functions, the finite kernels to be considered. Let us introduce the following definition.

DEFINITION 6.2. The function $\alpha_1 = \alpha_1(a_1, a_2, \cdots, a_n)$ is the largest negative a_k (or $-\infty$ if all a_k are positive); the function $\alpha_2 = \alpha_2(a_1, a_2, \cdots, a_n)$ is the smallest positive a_k (or $+\infty$ if all a_k are negative).

Thus

$$\alpha_2 = -\alpha_1(-a_1, -a_2, \cdots, -a_n).$$

For example,

$$\alpha_1(-3, 5, -2, 17) = -2$$

$$\alpha_2(-3, -1) = +\infty.$$

THEOREM 6.2. *If*

1. a_k *is real and* $\neq 0$ $\qquad k = 1, 2, \cdots, n$
2. $g_k(t)$ *is defined by* (1)
3. $G(t) = g_1 * g_2 * \cdots * g_n(t)$
4. $E(s) = \prod_1^n \left(1 - \frac{s}{a_k}\right)$
5. α_1, α_2 *are defined in Definition* 6.2,

then $G(t)$ *is a frequency function and*

$$\text{A.} \quad \chi_G(s) = \int_{-\infty}^{\infty} e^{-st} G(t)\, dt = \frac{1}{E(s)} \qquad \alpha_1 < \sigma < \alpha_2$$

$$\text{B.} \quad m_G = -\sum_1^n \frac{1}{a_k}$$

$$\text{C.} \quad V_G = \sum_1^n \frac{1}{a_k^2}.$$

This result is an immediate consequence of Corollary 5.2 and Theorem C. Note that the region of definition of the characteristic function for $g_k(t)$ is $\alpha_1(a_k) < \sigma < \alpha_2(a_k)$, namely a half-plane which includes the imaginary axis but not the point a_k. The intersection of all these half-planes, $k = 1, 2, \cdots, n$ defines the region of (absolute) convergence of the integral $\chi_G(s)$.

6.3. From the explicit formula expressing $G(t)$ in terms of $E(s)$ obtained from Theorem B, we can investigate the continuity properties of $G(t)$.

THEOREM 6.3. *If* $G(t)$ *is defined as in Theorem* 6.2, $n \geqq 2$, *then* $G(t) \in C^{n-2}$, $-\infty < t < \infty$.

By Theorem B,

$$G(t) = \frac{1}{2\pi i}\int_{-i\infty}^{i\infty} \frac{e^{st}}{E(s)}\, ds \qquad -\infty < t < \infty. \tag{1}$$

It is unnecessary to use the Cauchy value of this integral, as indicated in Theorem B, for $1/E(iy) = O(y^{-2})$ as $|y| \to \infty$ when $n \geqq 2$ and (1) converges absolutely. Differentiating, we obtain

$$G^{(n-2)}(t) = \frac{1}{2\pi i}\int_{-i\infty}^{i\infty} \frac{e^{st}s^{n-2}}{E(s)}\, ds. \tag{2}$$

This operation is valid since the integral (2) converges uniformly on any compact set of the t-axis. This follows since $y^{n-2}/E(iy) = O(y^{-2})$ as $|y| \to \infty$. Since the integral (2) converges uniformly, $G^{(n-2)}(t)$ is continuous, and the theorem is proved. To illustrate consider Example B, § 2 of Chapter I. There $n = 2$ and $G(t) = \frac{1}{2}e^{-|t|} \in C$, $-\infty < t < \infty$.

Note also that if $n \geqq 3$,

$$\left(1 - \frac{D}{b}\right)G(t) = \frac{1}{2\pi i}\int_{-i\infty}^{i\infty} \frac{e^{st}(1 - b^{-1}s)}{E(s)}\, ds$$

for any constant $b \neq 0$. If b coincides with a root of $E(s)$, a factor $\left(1 - \frac{s}{b}\right)$ will cancel in the integrand. We thus obtain

$$\prod_{k=3}^{n}\left(1 - \frac{D}{a_k}\right)G(t) = \frac{1}{2\pi i}\int_{-i\infty}^{i\infty} \frac{e^{st}}{\left(1 - \frac{s}{a_1}\right)\left(1 - \frac{s}{a_2}\right)}\, ds. \tag{3}$$

In fact, if $t \neq 0$ we may proceed one step further,

$$\prod_{k=2}^{n}\left(1 - \frac{D}{a_k}\right)G(t) = \frac{1}{2\pi i}\int_{-i\infty}^{i\infty} \frac{e^{st}}{\left(1 - \frac{s}{a_1}\right)}\, ds$$

$$= \frac{a_1}{2\pi}\int_{-\infty}^{\infty} \frac{e^{iyt}(a_1 + iy)}{a_1^2 + y^2}\, dy = g_1(t). \tag{4}$$

For, the integral (4) still converges uniformly* (though not absolutely) in any finite interval not including the origin due to the monotonic character of the function $y/(a_1^2 + y^2)$ near $y = \pm\infty$. The value of integral (3) is $g_1(t)$ by Theorem B.

*See, for example, S. Bochner [1932; 12]

Alternatively, we may compute the integral (3) directly by the calculus of residues or otherwise, obtaining $g_1 * g_2(t)$. Then applying the operator $\left(1 - \frac{D}{a_2}\right)$ to this explicitly we obtain $g_1(t)$ if $t \neq 0$. For example, if $a_1 = a_2 = 1$, we found in § 5.1 that

$$g(t) * g(t) = \begin{cases} -te^t & -\infty < t \leq 0 \\ 0 & 0 \leq t < \infty, \end{cases} \tag{5}$$

and $(1 - D)$ applied to this function gives $g(t)$ when $t \neq 0$.

6.4. Almost without exception in this book the convolution kernels used will be frequency rather than distribution functions. However, at the present juncture let us make a brief digression to show how frequency functions could arise quite naturally and could in fact be used basically. Recall that we have defined our inversion functions $E(s)$ not vanishing at the origin. If we had permitted a zero at $s = 0$, frequency functions would have been introduced, as indicated in the following theorem.

THEOREM 6.4. *If $G(t)$, $E(s)$ and α_2 are defined as in Theorem 6.2, then the function*

$$G^{(-1)}(t) = \int_{-\infty}^{t} G(u)\, du$$

is a distribution function whose bilateral Laplace transform is $[sE(s)]^{-1}$,

$$\frac{1}{sE(s)} = \int_{-\infty}^{\infty} e^{-st} G^{(-1)}(t)\, dt \qquad 0 < \sigma < \alpha_2.$$

For, by Fubini's theorem we may interchange the order of integration in

$$\int_{-\infty}^{\infty} e^{-st}\, dt \int_{-\infty}^{t} G(u)\, du$$

to obtain

$$\int_{-\infty}^{\infty} G(u)\, du \int_{u}^{\infty} e^{-st}\, dt = \frac{1}{s} \int_{-\infty}^{\infty} e^{-su} G(u)\, du, \tag{1}$$

provided that

$$\int_{-\infty}^{\infty} G(u)\, du \int_{u}^{\infty} e^{-\sigma t}\, dt < \infty.$$

But the inner integral converges for $\sigma > 0$ and the iterated integral is finite for $0 < \sigma < \alpha_2$ by Theorem 6.2. Also by that theorem, the right-hand side of (1) is $[sE(s)]^{-1}$, so that the desired result is proved.

7. INVERSION

7.1. We are now in a position to invert, by means of a linear differential operator, a convolution transform having a finite kernel.

THEOREM 7.1. If

1. *$G(t)$ and $E(s)$ are defined as in Theorem* 6.2

2.† $\varphi(t) \in B \cdot C$ $\qquad -\infty < t < \infty$

3. $f(x) = G * \varphi = \int_{-\infty}^{\infty} G(x-t)\varphi(t)\,dt,$

then

$$E(D)f(x) = \prod_1^n \left(1 - \frac{D}{a_k}\right) f(x) = \varphi(x) \qquad -\infty < x < \infty.$$

Any frequency function belongs to L (is absolutely integrable) on the whole real axis. Hence $G(x) * \varphi(x)$ is defined for all x. We saw in § 6.3 (3) that

$$\prod_3^n \left(1 - \frac{D}{a_k}\right) G(t) = g_1(t) * g_2(t) = h(t).$$

Hence

$$\prod_3^n \left(1 - \frac{D}{a_k}\right) f(x) = \int_{-\infty}^{\infty} h(x-t)\varphi(t)\,dt, \tag{1}$$

provided that this integral converges uniformly in the interval $-R \leqq x \leqq R$, every $R > 0$. It does so since

$$\left| \int_A^{\infty} h(x-t)\varphi(t)\,dt \right| \leqq M \int_{-\infty}^{R-A} h(t)\,dt \qquad |x| \leqq R$$

$$\left| \int_{-\infty}^{-B} h(x-t)\varphi(t)\,dt \right| \leqq M \int_{B-R}^{\infty} h(t)\,dt \qquad |x| \leqq R,$$

where M is an upper bound of $|\varphi(t)|$ for all t. Since the integrals on the right of these inequalities are independent of x and can be made arbitrarily small by choice of A and B, the desired uniform convergence is established.

A further direct application of the operator $\left(1 - \frac{D}{a_2}\right)$ under the

† This notation means that $\varphi(t)$ is bounded and continuous.

integral sign (1) is prevented by the fact that $h'(0)$ does not exist. However, if we write the integral (1) as the sum of two others corresponding to the intervals $(-\infty, x)$ and (x, ∞) we have

$$\frac{d}{dx}\int_{-\infty}^{\infty} h(x-t)\varphi(t)\,dt = h(0+)\varphi(x-) - h(0-)\varphi(x+)$$
$$+ \int_{-\infty}^{x} h'(x-t)\varphi(t)\,dt + \int_{x}^{\infty} h'(x-t)\varphi(t)\,dt.$$

Since $h(x)$ and $\varphi(x)$ are continuous and since

$$\left(1 - \frac{D}{a_2}\right) h(x) = g_1(x),$$

we have thus shown that

$$\prod_{2}^{n} \left(1 - \frac{D}{a_k}\right) f(x) = \int_{-\infty}^{\infty} g_1(x-t)\varphi(t)\,dt. \tag{2}$$

The differentiation under the integral signs of the two integrals above cannot be held in question since $h'(0+)$ and $h'(0-)$ both exist and the resulting integrals converge uniformly in any finite interval of the x-axis.

If desired, the above facts may be checked by use of the explicit formulas for $h(t)$. For example, if a_1 and a_2 are both positive, $a_1 \neq a_2$

$$h(t) = \frac{a_1 a_2}{a_2 - a_1}[e^{a_1 t} - e^{a_2 t}] \qquad -\infty < t \leqq 0$$
$$= 0 \qquad 0 \leqq t < \infty.$$

If $a_1 = a_2 > 0$, $h(t)$ is obtained from the function (5) § 6.3 by a change of variable:

$$h(t) = -a_1^2 t e^{a_1 t} \qquad -\infty < t \leqq 0$$
$$= 0 \qquad 0 \leqq t < \infty.$$

Now from equation (2) we may complete the proof at once by appeal to Theorem 3.1 of Chapter I. For,

$$\left(1 - \frac{D}{a_1}\right)\int_{-\infty}^{\infty} g_1(x-t)\varphi(t)\,dt = (1 - D)\int_{-\infty}^{\infty} g(x-t)\varphi(t)\,dt = \varphi(x)$$

for all x.

Corollary 7.1. The same conclusion holds if $\varphi(t) \in C \cdot L$.

The modifications in proof needed are slight and are omitted.

Hypothesis 2 of the theorem is stronger than needed, as already indicated in the corollary. Later we shall give best possible results in this direction, but for the present the principal features of the theory are put into evidence by use of simple assumptions on $\varphi(x)$. It should be observed, however, that the "local" conditions on $\varphi(x)$ are already weaker than in Theorem B, § 4.2, for example. This is brought about by the fact that the present kernel is positive in contrast with the Dirichlet kernel.

7.2. We prove next an analogous inversion theorem in which the frequency function $G(t)$ is replaced by the corresponding distribution function $G^{(-1)}(t)$.

THEOREM 7.2. *If*

1. $G^{-1}(t)$ *and* $E(s)$ *are defined as in Theorem* 6.4
2. $\varphi(t) \in L \cdot C \qquad -\infty < x < \infty$
3. $f(x) = \int_{-\infty}^{\infty} G^{(-1)}(x - t)\varphi(t)\, dt,$

then

$$DE(D)f(x) = D\prod_{1}^{n}\left(1 - \frac{D}{a_k}\right) f(x) = \varphi(x) \qquad -\infty < x < \infty.$$

Since $0 \leqq G^{(-1)}(t) \leqq 1$, the integral defining $f(x)$ converges absolutely for all x. Moreover, for $n > 1$

$$f'(x) = \int_{-\infty}^{\infty} G(x - t)\varphi(t)\, dt. \tag{1}$$

We have now only to apply Corollary 7.1 to obtain the desired result. If $n = 1$, the derivative of $G^{(-1)}(t)$ does not exist at $t = 0$, but equation (1) is still true trivially. Hence the theorem is established in all cases.

8. EXPONENTIAL POLYNOMIALS

8.1. We shall frequently need to consider exponential polynomials. A familiar fact about the number of their zeros will be recorded first.

DEFINITION 8.1. An exponential polynomial of degree d is a sum of the form

$$Q_d(t) = \sum_{k=1}^{p} e^{A_k t} P_k(t), \tag{1}$$

where $P_k(t)$ is a polynomial of degree $m_k - 1$, the A_k are distinct real numbers and

$$d = -1 + \sum_{k=1}^{p} m_k.$$

For example

$$Q_8(t) = e^{2t} + t^3e^t - t^2 + e^{-4t}.$$

Note that d is a unit less than the total number of terms if no coefficient is zero when each polynomial $P_k(t)$ is expanded in powers of t. If $Q_d(t)$ is identically zero we define its degree as -1.

THEOREM 8.1. *An exponential polynomial of non-negative degree d can have at most d zeros.*

It is to be understood as usual that the total number of zeros is the sum of the multiplicities of the distinct zeros. We prove the result by induction. If $d = 0$, $Q_0(t)$ consists of a single non-vanishing term of the form Be^{At}. Now assume the result true for exponential polynomials of degree $d - 1$, and suppose that $Q_d(t)$, defined by (1), had $d + 1$ zeros. The same would be true of $e^{-A_1t}Q_d(t)$. By Rolle's theorem, the derivative of this product would have at least d zeros. But this derivative is an exponential polynomial of degree at most $d - 1$, since $P_1'(t)$ is of degree less than $m_1 - 1$, and the induction assumption is contradicted. That d zeros are possible is evident from the example $(e^t - 1)^d$.

8.2. We shall show next that $G(t)$ is a combination of two exponential polynomials joined together at the origin.

THEOREM 8.2. *If $G(t)$ is the kernel of Theorem 6.2, where P of the roots $a_1, a_2, \cdots, a_n$ are positive and N are negative, then $G(t)$ is an exponential polynomial of degree $P - 1$ in $(-\infty, 0)$ and another of degree $N - 1$ in $(0, \infty)$.*

We use the explicit formula

$$G(t) = \frac{1}{2\pi i}\int_{-i\infty}^{i\infty} \frac{e^{st}}{E(s)}\, ds \qquad t \neq 0. \tag{1}$$

Suppose that $E(s)$, as defined in Theorem 6.2, has a root of order m_k at A_k, $k = 1, 2, \cdots, p$,

$$\sum_1^p m_k = n.$$

Each A_k is an a_k, but the A_k are distinct whereas the a_k need not be. To evaluate (1) let us expand $1/E(s)$ in partial fractions. Corresponding to the root A_k there will be m_k terms of the form

$$\frac{B_q}{(s - A_k)^q} \qquad q = 1, 2, \cdots, m_k.$$

Hence to compute (1) we need only evaluate integrals like

$$\frac{1}{2\pi i}\int_{-i\infty}^{i\infty} \frac{e^{st}}{(s - A)^q}\, ds \qquad t \neq 0. \tag{2}$$

But by Theorem B this formula inverts the familiar integrals

$$\frac{(q-1)!}{(s-A)^q} = \int_0^\infty e^{-(s-A)t}t^{q-1}\,dt \qquad \sigma > A, \quad A < 0$$

$$= -\int_{-\infty}^0 e^{-(s-A)t}t^{q-1}\,dt \qquad \sigma < A, \quad A > 0,$$

both of which converge on the imaginary axis. Consequently (2) is equal to

$$\begin{cases} \dfrac{t^{q-1}e^{At}}{(q-1)!} & 0 < t < \infty \\ 0 & -\infty < t < 0 \end{cases}$$

when $A < 0$ and to

$$\begin{cases} \dfrac{-t^{q-1}e^{A}}{(q-1)!} & -\infty < t < 0 \\ 0 & 0 < t < \infty \end{cases}$$

when $A > 0$. Hence in the computation of (1) a positive root A_k of multiplicity m_k contributes nothing on the positive t-axis and an exponential polynomial of degree $m_k - 1$ on the negative real axis. When $A_k < 0$ the contribution is nothing on $(-\infty, 0)$ and is an exponential polynomial of degree $m_k - 1$ on $(0, \infty)$. Since none of the roots of $E(s)$ are at the origin $A_k \neq 0$. Adding all contributions, $k = 1, 2, \cdots, p$, and recalling that the A_k are distinct, we obtain the desired result.

COROLLARY 8.2. If all a_k are positive, $G(t)$ is identically zero on $(0, \infty)$; if all are negative, $G(t)$ is identically zero on $(-\infty, 0)$.

These results are the special cases $N = 0$ and $P = 0$, respectively, of the theorem.

9. GREEN'S FUNCTIONS

9.1. Let us define a Green's function for the linear differential system

$$E(D)f(x) = \varphi(x) \tag{1}$$

$$f(x) = o(e^{\alpha_2 x}) \qquad x \to +\infty \tag{2}$$

$$f(x) = o(e^{\alpha_1 x}) \qquad x \to -\infty. \tag{3}$$

Here $E(s)$ is the polynomial of § 6.2 with roots $a_1, a_2, \cdots, a_n$ and α_1, α_2 are the numbers defined there. If $\alpha_2 = +\infty$, for example, (2) is to be understood as $f(x) = o(e^{kx})$, $x \to +\infty$, for *every* positive k.

DEFINITION 9.1.* The Green's function of the system (1) (2) (3) is a function $G(t)$ with the following properties:

a. $E(D)G(t) = 0$ $\qquad t \neq 0$

b. $G(t)$ satisfies (2) and (3)

c. $G(t) \in C^{n-2}$ $\qquad -\infty < t < \infty$

d. $G^{(n-1)}(0+) - G^{(n-1)}(0-) = (-1)^n a_1 a_2 \cdots a_n$.

For example, if $E(s)$ has only the two roots ± 1, then $\alpha_1 = -1$, $\alpha_2 = 1$ and $G(t) = e^{-|t|}/2$, the kernel of Example B, § 2, Chapter I.

Let us first establish the uniqueness of the Green's function.

THEOREM 9.1. *There is at most one Green's function for the system* (1) (2) (3).

Use the notation of § 8.2, indicating the distinct roots of $E(s)$ by $A_1, A_2, \cdots, A_p$ with multiplicities $m_1, m_2, \cdots, m_p$, respectively. Then the general solution of the homogeneous equation $E(D)f(x) = 0$ is the function $Q_d(x)$ of equation (1) § 8.1 with $d = n - 1$. If such a function is to satisfy (2) the terms involving positive A_k must drop (all such A_k are $\geqq \alpha_2$); to satisfy (3) the terms involving negative A_k must disappear. Define

$$G_1(t) = \sum_{k=1}^{N} e^{A_k t} P_k(t) \tag{4}$$

$$G_2(t) = \sum_{k=1}^{P} e^{A_k t} P_k(t), \tag{5}$$

where the first sum is over negative A_k only, the second is over positive A_k only. Hence $G_1(t)$ is a solution of $E(D)f(t) = 0$ on $(0, \infty)$ which satisfies (2); $G_2(t)$ is a solution on $(-\infty, 0)$ which satisfies (3). We show that the coefficients of the $P_k(t)$ can be determined in at most one way so that when $G(t)$ is defined as $G_1(t)$ in $(0, \infty)$ and $G_2(t)$ in $(-\infty, 0)$ then conditions c and d of Definition 9.1 will be satisfied. We must have

$$G_1^{(k)}(0) = G_2^{(k)}(0) \qquad k = 0, 1, \cdots, n-2 \tag{6}$$

$$G_1^{(n-1)}(0) - G_2^{(n-1)}(0) = (-1)^n a_1 a_2 \cdots a_n. \tag{7}$$

That is, the exponential polynomial $G_1(t) - G_2(t)$ of degree $n - 1$ must have $n - 1$ zeros at the origin. If it were possible to determine the coefficients in $G_1(t)$ and $G_2(t)$ in a second way to satisfy (6) and (7), we would have a second exponential polynomial of degree $n - 1$ with $n - 1$ zeros at the origin. Since by (7) both must have the same value for the

* Compare M. Bocher [1917; 98].

derivative of order $n - 1$ at the origin, their difference would be an exponential polynomial of degree $n - 1$ at most with n zeros, contradicting Theorem 8.1.

9.2. The previous theorem does not show the existence of a Green's function for the given system. We now show its existence by exhibiting it.

THEOREM 9.2. *The kernel $G(t)$ of Theorem 6.2 is the Green's function of the system* (1) (2) (3), § 9.1.

In the proof of Theorem 8.2 we showed that the kernel $G(t)$ of Theorem 6.2 has the forms (4) and (5), § 9.1, in the intervals $(0, \infty)$ and $(-\infty, 0)$, respectively. Hence property a of Definition 9.1 is satisfied. Property b is satisfied trivially since $G(t)$, being a frequency function, vanishes at $\pm\infty$. Property c follows from Theorem 6.3 when $n \geqq 2$ and is to be ignored when $n = 1$. Finally, to establish d we use equation (4), § 6.3,

$$\prod_{2}^{n} \left(1 - \frac{D}{a_k}\right) G(t) = |a_1| g(a_1 t) \qquad t \neq 0.$$

Expand the left-hand member and compare the jump of the highest order derivative at the origin (the derivatives of lower order are continuous) with the jump of the right-hand side:

$$\frac{(-1)^{n-1}}{a_2 a_2 \cdots a_n} [G^{(n-1)}(0+) - G^{(n-1)}(0-)] = -a_1.$$

This completes the proof of the theorem.

9.3. In conditions a and d of Definition 9.1 the origin plays a special role. It could be replaced by an arbitrary point t. The resulting Green's function would be $G(x - t)$. This is a consequence of the fact that the coefficients of the differential operator $E(D)$ are constants. Thus

$$E(D)G(x - t) = 0 \qquad x \neq t,$$

the indicated differentiation being with respect to x.

A characteristic property of a Green's function is that it enables one to solve a non-homogeneous system explicitly by an integral. It serves this purpose in the present case.

THEOREM 9.3 *If $\varphi(x) \in B \cdot C$, $-\infty < x < \infty$, then the unique solution of the system* (1) (2) (3) § 9.1 *is*

$$f(x) = \int_{-\infty}^{\infty} G(x - t)\varphi(t)\, dt. \tag{1}$$

The solution is unique since the exponential polynomial (1) § 8.1, $d = n - 1$, the general solution of the corresponding homogeneous equation, can satisfy (2) and (3) of § 9.1 only if it is identically zero.

This follows since no positive A_k is less than α_2 and no negative A_k is greater than α_1.

That the function $f(x)$ defined by (1) satisfies (1) § 9.1 was proved in Theorem 7.1. The boundary conditions (2) (3) § 9.1 are satisfied trivially since $f(x)$ is bounded:

$$|f(x)| \leq \sup_{-\infty<t<\infty} |\varphi(t)| \qquad -\infty < x < \infty.$$

This completes the proof.

10. EXAMPLES

10.1. Let us give here two examples that will be of special interest to us later.

Example A. Choose $a_k = k$, $k = 1, 2, \cdots, n$. Then

$$E(s) = \left(1 - \frac{s}{1}\right)\left(1 - \frac{s}{2}\right)\cdots\left(1 - \frac{s}{n}\right)$$

$$E'(k) = \frac{(-1)^k}{k}\frac{1}{\binom{n}{k}} = (-1)^k\frac{(n-k)!(k-1)!}{n!}.$$

Hence

$$\frac{1}{E(s)} = \sum_{k=1}^{n}(-1)^{k-1}\binom{n}{k}\frac{1}{1 - \frac{s}{k}}.$$

But

$$\frac{1}{1 - \frac{s}{k}} = k\int_{-\infty}^{0} e^{-st}e^{kt}\,dt \qquad \sigma < k,$$

so that

$$G(t) = \sum_{k=1}^{n}(-1)^{k-1}\binom{n}{k} ke^{kt} \qquad -\infty < t < 0$$

$$= -\frac{d}{dt}(1 - e^t)^n = ne^t(1 - e^t)^{n-1} \qquad -\infty < t < 0$$

$$= 0 \qquad 0 < t < \infty.$$

Note that when $n = 1$ this reduces to the unit kernel $g(t)$ of § 6.1.

Example B. Choose $a_k = k$, $k = \pm 1, \pm 2, \cdots, \pm n$. Then

$$E(s) = (1 - s^2)\left(1 - \frac{s^2}{2^2}\right)\cdots\left(1 - \frac{s^2}{n^2}\right)$$

$$= E_0(s)E_0(-s),$$

where $E_0(s)$ is the polynomial $E(s)$ of Example A. If $G_0(t)$ is the kernel of that example, we must clearly compute

$$G(t) = G_0(t) * G_0(-t)$$

$$G(x) = \int_0^\infty G_0(x-t)G_0(-t)\,dt \qquad x < 0$$

$$G(x) = \int_x^\infty G_0(x-t)G_0(-t)\,dt \qquad x > 0. \tag{1}$$

That is, after replacing $t - x$ by a new variable in (1)

$$G(x) = n^2 \int_0^\infty e^{x-t}(1 - e^{x-t})^{n-1}(1 - e^{-t})^{n-1}e^{-t}\,dt \qquad x < 0$$

$$= n^2 \int_0^\infty e^{-t}(1 - e^{-t})^{n-1}e^{-x-t}(1 - e^{-x-t})^{n-1}\,dt \qquad x > 0,$$

so that $G(-x) = G(x)$, as would be expected from the symmetric distribution of the zeros of $E(s)$. If $n = 1$, $G(t) = e^{-|t|}/2$, the illustrative Green's function of § 9.1, or the kernel of Example B, § 2.1 of Chapter I.

These examples will be of interest to us because of the relation of the polynomials $E(s)$ to the infinite product expansions of $1/\Gamma(s)$ and $\sin \pi s/\pi s$.

11. SUMMARY

11.1. The chief result of the present chapter is that if $E(s)$ is a polynomial with real roots only, $E(0) = 1$, then its reciprocal is the bilateral Laplace transform of a frequency function $G(t)$ and that $E(D)\{G(x) * \varphi(x)\} = \varphi(x)$. The kernel $G(t)$ of the convolution transform was identified with the Green's function of a certain differential system and was used to solve the system by an explicit convolution.

CHAPTER III

The Non-Finite Kernels

1. INTRODUCTION

1.1. In Chapter II we confined our attention to convolution transforms having "finite" kernels, those whose bilateral Laplace transforms are reciprocals of polynomials with real roots. In the present chapter we enlarge the class of kernels greatly, including a class whose Laplace transforms are reciprocals of entire functions of genus one and having real roots. We shall show that the inversion theorems of Chapter II generalize completely here, the Weierstrass infinite product expansion of the above mentioned entire functions leading in the expected way to the inversion operators for the more general transforms. Thus a convolution transform with one of these new kernels is still inverted by a linear differential operator with constant coefficients. Whereas the order for the finite kernels was equal to the number of roots of the inversion polynomial, the order of the operator is now infinite.

In order to see clearly how the enlarged class of kernels should be chosen to produce the maximum degree of generality within the framework of our methods, we begin with a preliminary study involving limits of sequences of polynomials. In the previous chapter the fact that the roots of the inversion polynomials had real roots only evidently played a fundamental role. Since we will naturally introduce the differential operators of infinite order as limits of others of finite order, it becomes imperative to investigate the class of functions, the prospective inversion functions, which can be the uniform limits of polynomials with real roots only. E. Laguerre had shown that all such functions $E(s)$ for which $E(0) = 1$ can be put in the form

$$E(s) = e^{-cs^2+bs} \prod_{k=1}^{\infty} \left(1 - \frac{s}{a_k}\right) e^{s/a_k}$$

$$\sum_{k=1}^{\infty} \frac{1}{a_k^2} < \infty,$$

where the constants c, b, a_k are real and $c \geq 0$. In particular, the infinite product may have only a finite number of factors, so that the inversion polynomials of Chapter II are included. We shall show that it is these functions $E(s)$ of Laguerre which we may use as our inversion operators, for each one except e^{bs} is the reciprocal of the characteristic function of some frequency function $G(t)$. The latter we take as the kernel of a convolution transform

$$f = G * \varphi$$

and show that under suitable restrictions on φ

$$E(D)f = \varphi.$$

In the present chapter we restrict attention to the case $c = 0$. In the concluding section we specialize the results to obtain inversion theorems for the Laplace, Stieltjes, and Meijer transforms together with certain iterations thereof.

2. LIMITS OF DISTRIBUTION FUNCTIONS

2.1. It will be convenient to have a notation for the class of normalized distribution functions defined in § 2.1 of Chapter II.

DEFINITION 2.1a. The function $\alpha(t)$ belongs to the class D, $\alpha(t) \in D$, if and only if it is a normalized distribution function.

The limit process which we shall find appropriate in the class D is a pointwise limit at all points of continuity of the limit function. Rather than introduce a separate notation for this operation, we use instead a symbol for the set of points of continuity of the limit function.

DEFINITION 2.1b. A point t is in the set C_α, $t \in C_\alpha$, if and only if it is a point of continuity for the function $\alpha(t)$.

The desired limit operation can now be written as follows:

$$\lim_{n\to\infty} \alpha_n(t) = \alpha(t) \qquad t \in C_\alpha.$$

The equality is to hold for *each* point t of C_α. For example, the functions

$$\alpha_n(t) = \begin{cases} \dfrac{e^{t/n}}{2} & -\infty < t \leqq 0 \\[2ex] 1 - \dfrac{e^{-nt}}{2} & 0 \leqq t < \infty, \end{cases}$$

$n = 1, 2, \cdots$, all belong to D. If $\alpha(t)$ is the normalized function

$$\alpha(t) = \begin{cases} 1/2 & -\infty < t < 0 \\ 3/4 & t = 0 \\ 1 & 0 < t < \infty, \end{cases}$$

then C_α is the set $t \neq 0$, and

$$\lim_{n\to\infty} \alpha_n(t) = \alpha(t) \qquad t \in C_\alpha.$$

This example shows that the limit of a sequence of functions of D, though non-decreasing, need not belong to D because (a) it may not be normalized and (b) the difference $\alpha(\infty) - \alpha(-\infty)$ may fail to be 1.

2.2. We investigate the relation of the above limit process for distribution functions to the process for the corresponding characteristic functions. The result is a known theorem of P. Lévy [1925; 195] often referred to as the "continuity theorem" in statistical studies. We reproduce it here in the form needed.

THEOREM 2.2. *If*

1. $\alpha_n(t) \in D \qquad n = 0, 1, 2, \cdots$

2. $\chi_{\alpha_n}(s) = \int_{-\infty}^{\infty} e^{-st}\, d\alpha_n(t)$

3. $\lim_{n\to\infty} \alpha_n(t) = \alpha_0(t) \qquad t \in C_{\alpha_0}$,

then

$$\lim_{n\to\infty} \chi_{\alpha_n}(iy) = \chi_{\alpha_0}(iy)$$

uniformly in $-A \leq y \leq A$ *for every* $A > 0$.

We must show that when $n \to \infty$ the integral

$$I_n(y) = \int_{-\infty}^{\infty} e^{-iyt}\, d[\alpha_n(t) - \alpha_0(t)]$$

approaches zero uniformly in $|y| \leq A$. Given an arbitrary $\epsilon > 0$, we choose R so that R and $-R$ belong to C_α and so that

$$\alpha_0(-R) < \epsilon, \qquad 1 - \alpha_0(R) < \epsilon.$$

This is possible by the definition of a distribution function. With this R we may now choose, by hypothesis 3, an integer n_1 such that for $n > n_1$

$$\alpha_n(-R) < \alpha_0(-R) + \epsilon < 2\epsilon$$

$$1 - \alpha_n(R) < 1 - \alpha_0(R) + \epsilon < 2\epsilon.$$

Now write $I_n(y)$ as the sum of three integrals, $I_n'(y)$, $I_n''(y)$, $I_n'''(y)$, corresponding respectively to the intervals of integration $(-\infty, -R)$, $(-R, R)$, (R, ∞). An integration by parts shows that

$$|I_n''(y)| \leq |\alpha_n(-R) - \alpha_0(-R)| + |\alpha_n(R) - \alpha_0(R)|$$
$$+ A \int_{-R}^{R} |\alpha_n(t) - \alpha_0(t)|\, dt$$

for $|y| \leqq A$. By Lebesgue's limit theorem, we may choose $n_2 \geqq n_1$ such that

$$|I''_n(y)| < \epsilon \qquad n > n_2, \ |y| \leqq A.$$

But clearly for the same values of n and y

$$|I'_n(y)| \leqq \alpha_n(-R) + \alpha_0(-R) < 3\epsilon$$

$$|I'''_n(y)| \leqq [1 - \alpha_n(R)] + [1 - \alpha_0(R)] < 3\epsilon.$$

Hence for $n > n_2$

$$|I_n(y)| < 7\epsilon \qquad |y| \leqq A,$$

and the proof is complete.

2.3. We turn next to the converse of Theorem 2.2.

THEOREM 2.3. *If hypotheses* 1 *and* 2 *of Theorem* 2.2 *hold, and if*

3. $\lim_{n \to \infty} \chi_{\alpha_n}(iy) = \chi_{\alpha_0}(iy) \qquad -\infty < y < \infty,$

uniformly in $-A \leqq y \leqq A$ *for some* $A > 0$,

then

$$\lim_{n \to \infty} \alpha_n(t) = \alpha_0(t) \qquad t \in C_{\alpha_0}.$$

Since the functions $\alpha_n(t)$ are distribution functions they are uniformly bounded and Helly's theorem, Widder [1946; 27], is applicable. Hence there exists a subsequence $\{\beta_k(t)\}_1^\infty$ of the set $\{\alpha_n(t)\}_1^\infty$ and a function $\alpha(t)$ such that

$$\lim_{k \to \infty} \beta_k(t) = \alpha(t) \qquad -\infty < t < \infty. \tag{1}$$

If $\alpha(t)$ is now normalized, equation (1) still holds for $t \in C_\alpha$. We show first that $\alpha(t) \in D$.

Form the function

$$I_k(y) = \int_{-\infty}^{\infty} \frac{1 - \cos ry}{ry^2} \chi_{\beta_k}(iy)\, dy \tag{2}$$

$$= \int_{-\infty}^{\infty} \frac{1 - \cos ry}{ry^2}\, dy \int_{-\infty}^{\infty} e^{-iyt}\, d\beta_k(t),$$

where r is any positive integer. The Fubini theorem is clearly applicable, so that

$$I_k(y) = \int_{-\infty}^{\infty} d\beta_k(t) \int_{-\infty}^{\infty} e^{-iyt} \frac{1 - \cos ry}{ry^2}\, dy.$$

Since the inner integral is a familiar Fourier transform, we have

$$I_k(y) = \pi \int_{-r}^{r} \left(1 - \frac{|t|}{r}\right) d\beta_k(t) = \frac{\pi}{r} \int_0^r \beta_k(t)\, dt - \frac{\pi}{r} \int_{-r}^0 \beta_k(t)\, dt. \tag{3}$$

Applying Lebesgue's limit theorem to the integrals (2) and (3) and making use of hypothesis 3 and equation (1), we obtain

$$\frac{1}{r}\int_0^r \alpha(t)\,dt - \frac{1}{r}\int_{-r}^0 \alpha(t)\,dt = \frac{1}{\pi}\int_{-\infty}^{\infty} \frac{1-\cos ry}{ry^2}\chi_{\alpha_0}(iy)\,dy$$

$$= \frac{1}{\pi}\int_{-\infty}^{\infty} \frac{1-\cos y}{y^2}\chi_{\alpha_0}\left(\frac{iy}{r}\right)dy. \tag{4}$$

Finally, we let $r \to \infty$, again applying Lebesgue's theorem to the integral (4). Since $\chi_{\alpha_0}(iy)$ is continuous at $y = 0$ and has the value 1 there, we have

$$\alpha(\infty) - \alpha(-\infty) = \frac{1}{\pi}\int_{-\infty}^{\infty} \frac{1-\cos y}{y^2}\,dy = 1.$$

That is, $\alpha(t) \in D$ as stated.

But by hypothesis 3 and Theorem 2.2, $\chi_\alpha(iy) = \chi_{\alpha_0}(iy)$. By the uniqueness theorem for Fourier-Stieltjes transforms, see Corollary 5.2 below, $\alpha(t) = \alpha_0(t)$. Since $\{\beta_k(t)\}_0^\infty$ in the above argument may be chosen from an arbitrary infinite subsequence of $\{\alpha_n(t)\}_0^\infty$, and since the limit $\alpha_0(t)$ is independent of the choice of subsequence, it is clear that

$$\lim_{n\to\infty} \alpha_n(t) = \alpha_0(t) \qquad t \in C_{\alpha_0},$$

and the proof is complete.

Corollary 2.3. If hypotheses 1 and 2 of Theorem 2.3 hold for $n = 1, 2, 3, \cdots$, and if

3. $\lim_{n\to\infty} \chi_{\alpha_n}(iy) = f(y) \qquad -\infty < y < \infty,$

uniformly in $-A \leq y \leq A$ for some $A > 0$,

then there exists a function $\alpha(x)$ of D such that

$$\lim_{n\to\infty} \alpha_n(t) = \alpha(t) \qquad t \in C_\alpha, \tag{5}$$

and $f(y)$ is the characteristic function of $\alpha(t)$,

$$f(y) = \int_{-\infty}^{\infty} e^{-iyt}\,d\alpha(t) \qquad -\infty < y < \infty. \tag{6}$$

In fact we define $\alpha(t)$ by equation (1). Then as before $\alpha(t) \in D$. Hence by Theorem 2.2 $\chi_{\beta_k}(iy) \to \chi_\alpha(iy)$ and $\chi_\alpha(iy) = f(y)$ by hypothesis 3. This proves (6), and (5) now follows by Theorem 2.3 itself.

The theorems of this section show that in dealing with limits of distribution functions, one may equally well consider uniform limits of their characteristic functions.

3. PÓLYA'S CLASS OF ENTIRE FUNCTIONS

3.1. The functions which we shall use as our most general inversion functions belong to a class originally considered by E. Laguerre [1882; 174]. They are the uniform limits of polynomials with real roots. G. Pólya [1913; 224] has designated the class by the number II, thus contrasting it with the class I of functions which are the uniform limits of polynomials with real *positive* roots. We shall be concerned here only with the former class and shall redesignate it as E to avoid confusion with the class II already introduced in Chapter I.

DEFINITION 3.1. An entire function $E(s)$ belongs to class E, $E(s) \in E$, if and only if it has the form

$$E(s) = e^{-cs^2+bs} \prod_{k=1}^{\infty} \left(1 - \frac{s}{a_k}\right) e^{s/a_k}, \tag{1}$$

where $c \geqq 0$, b, $a_k (k = 1, 2, \cdots)$ are real, and

$$\sum_{k=1}^{\infty} \frac{1}{a_k^2} < \infty. \tag{2}$$

We wish to include the case in which the product (1) has a finite number of factors or indeed reduces to 1. To include these cases without additional notation, we agree that from a certain point on, all a_k may $= \infty$. Examples of functions belonging to class E are:

$$1, \quad (1-s), \quad e^s, \quad e^{-s^2}, \quad (\sin s)/s \quad 1/\Gamma(1-s).$$

Observe that the product of two functions of the class again belongs to it.

3.2. We show now that any function of class E is the uniform limit of polynomials with real roots. We need a preliminary result.

LEMMA 3.2. If $|z| \leqq A$ and $n \geqq 2A$, then

$$\left| z + \log\left(1 - \frac{z}{n}\right)^n \right| \leqq \frac{A^2}{n}.$$

That branch of the logarithm which reduces to 0 when $z = 0$ is intended. Using the Maclaurin expansion we have

$$\left| z + \log\left(1 - \frac{z}{n}\right)^n \right| \leqq n \left[\frac{1}{2}\left|\frac{z}{n}\right|^2 + \frac{1}{3}\left|\frac{z}{n}\right|^3 + \cdots\right] \tag{1}$$

$$\leqq \frac{n}{2} \frac{|z|^2/n^2}{1 - (|z|/n)} \leqq \frac{A^2}{n},$$

as stated.

THEOREM 3.2. *If $E(s) \in E$, there exists a sequence of polynomials $E_1(s)$, $E_2(s)$, $\cdots$, each with real roots only, such that*

$$\lim_{n\to\infty} E_n(s) = E(s)$$

uniformly in the circle $|s| \leqq A$ for every $A > 0$.

It is a consequence of the familiar Weierstrass factor theorem that when 3.1 (2) holds, the infinite product 3.1 (1) is uniformly approximated in $|s| \leqq A$ by its partial products. Since a partial product is a polynomial with real roots multiplied by an exponential e^{cs} (which can be combined with e^{bs}) it is clearly sufficient to show that e^{-cs^2} and e^{bs}, $cb \neq 0$, can also be uniformly approximated by polynomials with real roots only. But if we set $z = -bs$ in Lemma 3.2, we have for $|s| \leqq A/|b|$

$$\left| -bs + \log\left(1 + \frac{bs}{n}\right)^n \right| \leqq \frac{A^2}{n}.$$

Using the inequality

$$|e^z - 1| \leqq e^{|z|} - 1,$$

this gives

$$\left| \left(1 + \frac{bs}{n}\right)^n e^{-bs} - 1 \right| \leqq e^{A^2/n} - 1,$$

from which

$$\lim_{n \to \infty} \left(1 + \frac{bs}{n}\right)^n = e^{bs}$$

uniformly in $|s| \leqq A/|b|$. Replacing bs by $-cs^2$, we obtain

$$\lim_{n \to \infty} \left(1 - \frac{cs^2}{n}\right)^n = e^{-cs^2}$$

uniformly in $|s| \leqq (A/c)^{1/2}$. In each case the approximating polynomials have real roots only (since $c > 0$).

3.3. It was the converse of Theorem 3.2 which was proved by Laguerre. We prove the more general theorem of Pólya in which the uniform convergence in every circle is replaced by uniform convergence in a single circle. We follow the method of proof given by N. Obrechkoff [1941; 11].

THEOREM 3.3. *If*

1. $a_k = a_k(n)$, $k = 1, 2, \cdots, n$, *are real*

2. $E_n(s) = \prod_{k=1}^{n} \left(1 - \frac{s}{a_k}\right)$

3. $\lim_{n \to \infty} E_n(s) = E(s)$ *uniformly in* $|s| \leqq A$ *for some* $A > 0$,

then $E(s) \in E$.

As in the convention introduced for Definition 3.1, we may have $a_k = \infty$ for all k sufficiently large. Hence $E_n(s)$ may be of degree $<n$ or indeed may reduce to 1. Thus we are considering an arbitrary uniformly convergent sequence of polynomials $\{E_n(s)\}_1^\infty$ with real roots and such that $E_n(0) = 1$.

By Weierstrass's theorem $E(s)$ is analytic at $s = 0$ and

$$\lim_{n\to\infty} E_n'(0) = E'(0), \qquad \lim_{n\to\infty} E_n''(0) = E''(0). \tag{1}$$

If we set

$$p_n = \sum_{k=1}^{n} \frac{1}{a_k(n)}, \qquad q_n = \sum_{k=1}^{n} \frac{1}{a_k^2(n)},$$

equations (1) are equivalent to

$$\lim_{n\to\infty} p_n = -E'(0) = p, \qquad \lim_{n\to\infty} q_n = [E'(0)]^2 - E''(0) = q. \tag{2}$$

As a consequence, the sequences $\{p_n\}_1^\infty$ and $\{q_n\}_1^\infty$ are certainly bounded, say by the constant M:

$$|p_n| < M, \qquad 0 \leqq q_n < M \qquad n = 1, 2, \cdots \tag{3}$$

From inequality (1) § 3.2 with $n = 1$, we have

$$|(1 - z)e^z| \leqq e^{|z|^2} \qquad |z| \leqq 1/2; \tag{4}$$

and from elementary inequalities

$$|(1 - z)e^z| \leqq (1 + |z|)e^{|z|} \leqq e^{2|z|} \leqq e^{4|z|^2} \qquad |z| \geqq 1/2.$$

Hence for all s

$$\left| E_n(s)e^{p_n s} \right| = \left| \prod_1^n \left(1 - \frac{s}{a_k}\right) e^{s/a_k} \right| \leqq e^{4q_n|s|^2}$$

$$|E_n(s)| \leqq e^{|p_n s| + 4q_n|s|^2} \leqq e^{M|s| + 4M|s|^2}.$$

That is, the sequence of polynomials $\{E_n(s)\}_1^\infty$ is uniformly bounded in any circle. By Vitali's convergence theorem, E. C. Titchmarsh [1939; 168], it converges uniformly in any circle, so that $E(s)$ is entire.

By a theorem of A. Hurwitz, E. C. Titchmarsh [1939; 119], $E(s)$ will have only real zeros, if any. Let us denote them by $\alpha_1, \alpha_2, \alpha_3, \cdots$, in the order of increasing absolute value, with the usual convention about multiple roots and admitting $\alpha_k = \infty$ from a certain point on. For example, if $E(s) = e^s$, all $\alpha_k = \infty$. By another appeal to Hurwitz's theorem we may assume that the roots $a_k(n)$ are arranged in such a way that

$$\lim_{n\to\infty} a_k(n) = \alpha_k \qquad k = 1, 2, 3, \cdots \tag{5}$$

This is possible since in a small circle about a root α of order r, $E_n(s)$ will have exactly r roots for all n sufficiently large. By (2), (3), (5) we have for any integer p, $n \geqq p$, that

$$\sum_{k=1}^{p} \frac{1}{a_k^2(n)} \leqq q_n < M, \qquad \sum_{k=1}^{p} \frac{1}{\alpha_k^2} \leqq q \leqq M$$

$$a = \sum_{k=1}^{\infty} \frac{1}{\alpha_k^2} \leqq q. \tag{6}$$

Now form the functions

$$g(s) = as + \sum_{k=1}^{\infty} \left[\frac{1}{s-\alpha_k} + \frac{1}{\alpha_k}\right] = \sum_{k=1}^{\infty} \frac{s^2}{\alpha_k^2(s-\alpha_k)}$$

$$g_n(s) = q_n s + \sum_{k=1}^{n} \left[\frac{1}{s-a_k} + \frac{1}{a_k}\right] = \sum_{k=1}^{n} \frac{s^2}{a_k^2(s-a_k)}.$$

[Both functions are identically zero if $E(s)$ has no roots.] Let D be an arbitrary bounded region, say inside the circle $|s| < R$, containing no α_k. We shall show that $g_n(s)$ tends uniformly to $g(s)$ in D as $n \to \infty$.

Let T be an arbitrary number $> R$, and denote by N the number of the α_k inside the circle $|s| < T$. By Hurwitz's theorem there will also be just N of the $a_k(n)$ in this circle for all n sufficiently large:

$$|\alpha_k| < T, \qquad |a_k(n)| < T, \qquad k = 1, 2, \cdots, N$$

$$|\alpha_k| \geqq T, \qquad |a_k(n)| \geqq T, \qquad k = N+1, \quad N+2, \cdots$$

That is,

$$\left|\sum_{k=N+1}^{\infty} \frac{s^2}{\alpha_k^2(s-\alpha_k)}\right| \leqq \frac{R^2}{T-R} \sum_{k=N+1}^{\infty} \frac{1}{\alpha_k^2} \leqq \frac{MR^2}{T-R} \qquad |s| \leqq R$$

$$\left|\sum_{k=N+1}^{n} \frac{s^2}{a_k^2(s-a_k)}\right| \leqq \frac{R^2}{T-R} \sum_{k=N+1}^{n} \frac{1}{a_k^2(n)} \leqq \frac{MR^2}{T-R} \qquad n > N.$$

Hence

$$|g(s) - g_n(s)| \leqq \left|\sum_{k=1}^{N} \frac{s^2}{\alpha_k^2(s-\alpha_k)} - \frac{s^2}{a_k^2(s-a_k)}\right| + \frac{2MR^2}{T-R}. \tag{7}$$

The first term on the right clearly approaches zero uniformly in D as $n \to \infty$. Hence the right-hand side of (7) can be made less than an arbitrary positive ϵ for all s in D first by choice of T and then by choice of n.

But

$$g_n(s) = \frac{E_n'(s)}{E_n(s)} + p_n + q_n s.$$

Hence, by Weierstrass's theorem

$$\frac{E'(s)}{E(s)} + p + qs = as + \sum_{k=1}^{\infty} \left[\frac{1}{s - \alpha_k} + \frac{1}{\alpha_k} \right].$$

The series converges uniformly in D, so that we may integrate term by term from 0 to s,

$$E(s)e^{ps+qs^2/2} = e^{as^2/2} \prod_{k=1}^{\infty} \left(1 - \frac{s}{\alpha_k}\right) e^{s/\alpha_k}.$$

Since $a \leqq q$ by (6), $E(s) \in E$, and the proof is complete.

3.4. For our purposes we need to generalize the result of Pólya, replacing the hypothesis of uniform convergence in a single circle by that of uniform convergence on a segment of the imaginary-axis. As a preliminary to this result we prove a lemma, somewhat stronger than needed for that result, but needed later in its full generality.

LEMMA 3.4. If

1. $E_n(s) \in E$ $\qquad n = 1, 2, \cdots$
2. $\lim_{n \to \infty} E_n(iy)$ exists uniformly in $|y| \leqq A$ for some A,

then $\lim_{n \to \infty} E_n(s)$ exists uniformly in $|s| \leqq B$ for some B.

Set

$$E_n(s) = e^{-c(n)s^2 - b(n)s} \prod_{k=1}^{\infty} \left(1 - \frac{s}{a_k(n)}\right) e^{s/a_k(n)}$$

$$Q_n(s) = e^{b(n)s} E_n(s).$$

By the definition of class E, $c(n) \geqq 0$, so that

$$(1) \qquad |E_n(iA)|^2 = e^{2c(n)A^2} \prod_{k=1}^{\infty} \left(1 + \frac{A^2}{a_k^2(n)}\right)$$

$$\geqq 2c(n)A^2 + A^2 \sum_{k=1}^{\infty} a_k^{-2}(n).$$

Since the left-hand side of (1) approaches a limit as $n \to \infty$ there exists a constant M such that

$$(2) \qquad h(n) = c(n) + \sum_{k=1}^{\infty} a_k^{-2}(n) \leqq M^2, \qquad n = 1, 2, \cdots.$$

Each term of the series (2) is bounded by M so that

$$\frac{|s|^2}{a_k^2(n)} \leqq \frac{1}{4}$$

inside the circle $|s| \leqq 1/(2M)$. Hence we may apply 3.3 (4) there to obtain

$$\left|\left(1 - \frac{s}{a_k(n)}\right) e^{s/a_k(n)}\right| \leqq e^{|s|^2/a_k^2(n)}$$

$$|Q_n(s)| \leqq e^{h(n)|s|^2} \qquad |s| \leqq (2M)^{-1}.$$

Again using (2) we see that

$$|Q_n(s)| \leqq e^{1/4} \qquad |s| \leqq (2M)^{-1}.$$

Out of each infinite subset of positive integers, by E. C. Titchmarsh [1939; 169], we can pick a sequence $\{m\}$ such that $Q_m(s)$ approaches a limit uniformly in any smaller circle $|s| \leqq B < (2M)^{-1}$ Since one factor, $E_m(iy)$, of $Q_m(iy)$ approaches uniformly a non-zero limit [$E_m(0) = 1$ for all m], the other factor $e^{ib(m)y}$ must approach a limit uniformly in some interval $|y| \leqq \delta$. That is, to an arbitrary $\epsilon > 0$ there corresponds an integer m_0 such that when m and μ are numbers of the sequence $\{m\}$ greater than m_0

$$|\sin [b(m) - b(\mu)]y| \leqq |e^{iy[b(m)-b(\mu)]} - 1| < \epsilon$$

$$|y[b(m) - b(\mu)] - k\pi| < \sin^{-1}\epsilon \qquad -\delta \leqq y \leqq \delta,$$

for $k = 0, \pm 1, \cdots$. But this latter inequality is untenable for $y = 0$ unless $k = 0$, so that when $y = \delta$

$$|b(m) - b(\mu)| < \delta^{-1} \sin^{-1} \epsilon.$$

Consequently $b(m)$ approaches a limit as $m \to \infty$ and $Q_m(s)e^{-b(m)s}$ must approach a limit uniformly in the circle $|s| \leqq B$. This limit, an analytic function, must coincide with $\lim E_n(s)$ on $s = iy$, no matter what subset of integers $\{m\}$ was selected from. This proves the theorem.

THEOREM 3.4. *If*

1. $a_k(n)$ *is real,* $\qquad k = 1, 2, \cdots$

2. $E_n(s) = \prod_{k=1}^{n} \left(1 - \frac{s}{a_k(n)}\right) \qquad n = 1, 2, \cdots$

3. $\lim_{n\to\infty} E_n(iy) = E(iy)$ *uniformly in* $-A \leqq y \leqq A$, *for some* $A > 0$,

then $E(s) \in E$.

In the first instance $E(s)$ is defined only on a segment of the imaginary axis, but the intended meaning is that this function can be extended analytically throughout the plane. The functions $E_n(s)$ belong to class E so that Lemma 3.4 is applicable. That is, $E_n(s)$ approaches a limit $E(s)$ uniformly in some circle $|s| \leqq B$, which of course coincides with the limit of hypothesis 3 on $s = iy, -A \leqq y \leqq A$. By Theorem 3.3, $E(s) \in E$, and the proof is complete.

4. THE CLOSURE OF A CLASS OF DISTRIBUTION FUNCTIONS

4.1. We showed in Chapter II that the reciprocal of a polynomial $P(s)$ with real roots only, $P(0) = 1$, is always the characteristic function of a distribution function. Let us say that all such distribution functions belong to the subclass D_f of D. The derivatives of functions of D_f are precisely the finite kernels treated in Chapter II.

DEFINITION 4.1a. A function $\alpha(t)$ belongs to D_f if and only if its characteristic function $\chi_\alpha(s)$ is the reciprocal of a polynomial $P(s)$ with real roots only, $P(0) = 1$,

$$\chi_\alpha(s) = \int_{-\infty}^{\infty} e^{-st}\, d\alpha(t) = \frac{1}{P(s)} \cdot$$

Let us next consider the closure of D_f in D, under the limit operation defined in § 2.1.

DEFINITION 4.1b. A function $\alpha(t)$ belongs to $\bar{D}_f$, the closure of D_f, if and only if it belongs to D and if there exist functions $\alpha_1(t)$, $\alpha_2(t)$, $\cdots$, of D_f such that

$$\lim_{n\to\infty} \alpha_n(t) = \alpha(t) \qquad t \in C_\alpha. \tag{1}$$

We shall show that $\bar{D}_f$ is the set of those distribution functions whose characteristic functions are the reciprocals of Pólya's functions of class E (§ 3).

THEOREM 4.1. *A function* $\alpha(t) \in \bar{D}_f$ *if and only if* $\alpha(t) \in D$ *and* $\chi_\alpha(s)$ *is the reciprocal of a function of class* E.

We prove first the necessity of the conditions. If $\alpha(t) \in \bar{D}_f$, then $\alpha(t) \in D$ by Definition 4.1b and its characteristic function

$$\chi_\alpha(s) = \int_{-\infty}^{\infty} e^{-st}\, d\alpha(t) \tag{2}$$

is defined at least on the imaginary axis and has the value 1 at $s = 0$. Let $\{\alpha_n(t)\}_1^\infty$ be a sequence satisfying equation (1) and such that

$$\frac{1}{E_n(s)} = \int_{-\infty}^{\infty} e^{-st}\, d\alpha_n(t) \qquad n = 1, 2, \cdots, \tag{3}$$

where $E_n(s)$ is a polynomial with real roots only, $E_n(0) = 1$. By Theorem 2.2

$$\lim_{n\to\infty} \frac{1}{E_n(s)} = \chi_\alpha(s)$$

uniformly on any closed interval of the imaginary axis. Since $\chi_\alpha(s)$ is not zero on $s = iy$, $|y| \leq \delta$ for some $\delta > 0$,

$$\lim_{n\to\infty} E_n(s) = \frac{1}{\chi_\alpha(s)}$$

uniformly on $s = iy$, $|y| \leqq \delta$. By Theorem 3.4, $1/\chi_\alpha(s)$ is a function of class E, as we wished to prove.

Conversely, let $\alpha(t) \in D$ and let its characteristic function (2) have a reciprocal, $E(s)$ belonging to E. Then by Theorem 3.2 there exists a sequence $\{E_n(s)\}_1^\infty$ of polynomials with real roots only, $E_n(0) = 1$, such that

$$\lim_{n\to\infty} E_n(s) = E(s)$$

uniformly in every circle. By Theorem 6.2 of Chapter II, equation (3) holds, where $\alpha_1(t)$, $\alpha_2(t)$, $\cdots$ are distribution functions. Since $E(0) = 1$, there is an interval $s = iy$, $|y| \leqq \delta$, such that

$$\lim_{n\to\infty} \frac{1}{E_n(s)} = \frac{1}{E(s)}$$

uniformly thereon. By Theorem 2.3, equation (1) holds and $\alpha(t) \in \bar{D}_f$. This completes the proof.

For the validity of this theorem it is important that the closure of D_f should be taken in D. For, a sequence of functions belonging to D_f may well approach a function not in D. For example, if

$$\begin{aligned} \alpha_n(t) &= e^{t/n} & -\infty < t \leqq 0 \\ &= 1 & 0 \leqq t < \infty, \end{aligned}$$

the limit function is constantly equal to 1. The bilateral Laplace transform of this function is 0, and this is certainly not the reciprocal of a function of class E.

5. THE NON-FINITE KERNELS

5.1. We have shown that any distribution function which has a characteristic function whose reciprocal belongs to class E is a member of $\bar{D}_f$. But we have not shown that every function $1/E(s)$, $E(s) \in E$, is the characteristic function of some distribution function. We shall now prove this if $E(s) \neq e^{bs}$. Indeed, we shall show that $1/E(s)$ is the characteristic function of a frequency function $G(t)$,

$$\frac{1}{E(s)} = \int_{-\infty}^{\infty} e^{-st} G(t)\, dt.$$

The resulting functions $G(t)$ are the ones which we shall use as kernels for the convolution transform.

5.2. Let us first establish an inversion formula which will enable us to obtain distribution functions from their corresponding characteristic functions. It is due to Lévy [1925; 167]. We present it here in the form it takes when applied to Laplace transforms.

THEOREM 5.2. *If $\alpha(t)$ is a normalized function of D, and if*

$$f(s) = \chi_\alpha(s) = \int_{-\infty}^{\infty} e^{-st}\, d\alpha(t), \tag{1}$$

then

$$\alpha(t) - \alpha(0) = \lim_{R\to\infty} \frac{1}{2\pi i} \int_{-iR}^{iR} \frac{e^{st}-1}{s} f(s)\, ds.$$

Note that this result comes formally from the classical inversion formula for the Laplace transform, D. V. Widder [1946; 241]. However, the usual proof does not apply since the integral (1) may converge on a line only, $s = iy$, rather than in a strip.

Set

$$I(R) = \frac{1}{2\pi i}\int_{-iR}^{iR} \frac{e^{st}-1}{s} f(s)\, ds = \frac{1}{2\pi}\int_{-R}^{R} \frac{e^{iyt}-1}{iy}\, dy \int_{-\infty}^{\infty} e^{-iyu}\, d\alpha(u).$$

By the uniform convergence of the integral (1) on the imaginary axis we have

$$I(R) = \frac{1}{\pi}\int_{-\infty}^{\infty} d\alpha(u) \int_0^R \left[\frac{\sin y(t-u)}{y} + \frac{\sin yu}{y}\right] dy$$

$$= \frac{1}{\pi}\int_{-\infty}^{\infty} d\alpha(u) \int_0^{R(t-u)} \frac{\sin y}{y}\, dy + \frac{1}{\pi}\int_{-\infty}^{\infty} d\alpha(u)\int_0^{Ru} \frac{\sin y}{y}\, dy.$$

Integration by parts gives

$$I(R) = -\frac{1}{2} + \frac{1}{2} + \frac{1}{\pi}\int_{-\infty}^{\infty} \alpha(u)\frac{\sin R(t-u)}{t-u}\, du - \frac{1}{\pi}\int_{-\infty}^{\infty} \alpha(u)\frac{\sin Ru}{u}\, du. \tag{2}$$

Allowing R to become infinite in these familiar Dirichlet integrals we obtain

$$I(+\infty) = \alpha(t) - \alpha(0).$$

However, we give the details of proof since the present hypotheses on $\alpha(t)$ are not the most usual ones; see D. V. Widder [1946; 65]. It will be enough to show that the integral

$$J(R) = \frac{1}{\pi}\int_{-\infty}^{\infty} \alpha(t)\frac{\sin Rt}{t}\, dt \tag{3}$$

tends to $\alpha(0)$ as $R \to \infty$ since the first integral on the right of (2) reduces to one of type (3) after a translation in the variable of integration. Corresponding to an arbitrary positive ϵ we can determine A_0 so that

$$\left|\frac{1}{\pi}\int_A^B \frac{\sin t}{t}\, dt\right| < \epsilon, \tag{4}$$

when $A > A_0$, $B > A_0$. By the second law of the mean

$$\frac{1}{\pi}\int_A^B \alpha(t)\frac{\sin Rt}{t}\,dt = \frac{\alpha(B)}{\pi}\int_{\xi R}^{BR}\frac{\sin t}{t}\,dt \qquad A \leqq \xi \leqq B$$

$$\frac{1}{\pi}\int_{-B}^{-A} \alpha(t)\frac{\sin Rt}{t}\,dt = \frac{\alpha(-A)}{\pi}\int_{-\eta R}^{-BR}\frac{\sin t}{t}\,dt \qquad A \leqq \eta \leqq B.$$

Hence for $R > 1$, $A > A_0$, $B > B_0$ we have by (4)

$$\left|\frac{1}{\pi}\int_A^B \alpha(t)\frac{\sin Rt}{t}\,dt\right| < \epsilon$$

$$\left|\frac{1}{\pi}\int_{-B}^{-A} \alpha(t)\frac{\sin Rt}{t}\,dt\right| < \epsilon;$$

or allowing B to become infinite,

$$\left|J(R) - \frac{1}{\pi}\int_{-A}^{A} \alpha(t)\frac{\sin Rt}{t}\,dt\right| \leqq 2\epsilon.$$

But this integral, over a finite range, tends to $\alpha(0)$ as $R \to \infty$ under the present hypotheses; see Theorem 7.2, D. V. Widder [1946; 65]. Hence $J(+\infty) = \alpha(0)$ as desired.

Corollary 5.2. Two distinct normalized distribution functions cannot have the same characteristic function.

For if $f(s)$ in Theorem 5.2 is identically zero, then $\alpha(t) - \alpha(0)$ and hence $\alpha(t)$ (since $\alpha(t)$ is normalized) is identically zero.

5.3. Another preliminary result concerns the behaviour of $1/E(s)$ on vertical lines.

Lemma 5.3. If $E(s) \in E$, then

$$|E(\sigma + i\tau)| \geqq E(\sigma).$$

For, if

$$E(s) = e^{-cs^2+bs}\prod_{k=1}^{\infty}\left(1 - \frac{s}{a_k}\right)e^{s/a_k} \qquad c \geqq 0, \tag{1}$$

then

$$|E(\sigma + i\tau)| = e^{-c(\sigma^2-\tau^2)+b\sigma}\prod_{k=1}^{\infty}\left[\left(1 - \frac{\sigma}{a_k}\right)^2 + \frac{\tau^2}{a_k^2}\right]^{1/2} e^{\sigma/a_k}$$

$$\geqq e^{-c\sigma^2+b\sigma}\prod_{k=1}^{\infty}\left|1 - \frac{\sigma}{a_k}\right| e^{\sigma/a_k} = |E(\sigma)|.$$

Note that the result holds equally well if the infinite product is replaced by a finite product.

THEOREM 5.3. *If* $E(s) \in E$ *and* $E(s) \neq e^{bs}P(s)$, *where* $P(s)$ *is a polynomial, then for any positive numbers* p *and* R

$$\frac{1}{|E(\sigma + i\tau)|} = O\left(\frac{1}{|\tau|^p}\right) \qquad |\tau| \to \infty \tag{2}$$

uniformly in the strip $|\sigma| \leqq R$.

If $E(s)$ has at most a finite number n of zeros then $c > 0$ and we have

$$|E(\sigma + i\tau)| \geqq e^{-c(\sigma^2 - \tau^2) + b\sigma} \prod_1^n \left|\frac{\tau}{a_k}\right| e^{\sigma/a_k}$$

$$\geqq e^{-cR^2 + c\tau^2 - |b|R} \prod_1^n \left|\frac{\tau}{a_k}\right| e^{-R/|a_k|}$$

throughout the strip $|\sigma| \leqq R$. From this (2) follows immediately by virtue of the factor $e^{c\tau^2}$

If $E(s)$ has infinitely many zeros ($c \geqq 0$), choose $N > p$ and so large that $|a_k| \geqq R$ when $k > N$. Set

$$E_N(s) = \prod_{k=N+1}^{\infty} \left(1 - \frac{s}{a_k}\right) e^{s/a_k}.$$

Then by Lemma 5.3

$$E|(\sigma + i\tau)| \geqq e^{-cR^2 - |b|R} \prod_1^N \left|\frac{\tau}{a_k}\right| e^{-R/|a_k|} |E_N(\sigma)|.$$

Since $E_N(\sigma)$ is not zero in $-R \leqq \sigma \leqq R$ its reciprocal has a maximum in that interval. Hence

$$\frac{1}{|E(\sigma + i\tau)|} = O\left(\frac{1}{|\tau|^N}\right) \qquad |\tau| \to \infty$$

uniformly in $|\sigma| \leqq R$. This proves the theorem.

5.4. We are now in a position to show that every function $E(s)$ of class $E \neq e^{bs}$ is the reciprocal of the characteristic function of a function $\alpha(t)$ of D. In fact $\alpha(t)$ will possess a derivative $G(t)$ which is then a frequency function. If $E(s)$ is a polynomial $P(s)$, this result was already proved in Chapter II,

$$\frac{1}{P(s)} = \int_{-\infty}^{\infty} e^{-st} G(t)\, dt.$$

If $E(s) = P(s)e^{bs}$, then

$$\frac{1}{E(s)} = \frac{e^{-bs}}{P(s)} = \int_{-\infty}^{\infty} e^{-s(t+b)} G(t)\, dt$$

$$= \int_{-\infty}^{\infty} e^{-st} G(t - b)\, dt,$$

and $G(t - b)$ is clearly a frequency function. Hence we may confine our attention to the case in which $E(s)$ is not the product of a polynomial by an exponential e^{bs}.

THEOREM 5.4. *If $E(s)$ is the function of Theorem 5.3, then there exists a function $\alpha(t) \in D$ such that*

$$\frac{1}{E(s)} = \int_{-\infty}^{\infty} e^{-st}\, d\alpha(t) \qquad s = i\tau,\ -\infty < \tau < \infty \tag{1}$$

$$\alpha(t) - \alpha(0) = \frac{1}{2\pi i} \int_{-i\infty}^{i\infty} \frac{e^{st} - 1}{sE(s)}\, ds. \tag{2}$$

For, by Theorem 3.2, there exists a sequence of polynomials $E_n(s)$, $E_n(0) = 1$, such that

$$\lim_{n \to \infty} E_n(s) = E(s)$$

uniformly in every circle. By Lemma 5.3,

$$|E(iy)| \geqq E(0) = 1,$$

so that

$$\lim_{n \to \infty} \frac{1}{E_n(iy)} = \frac{1}{E(iy)}$$

uniformly in $-A \leqq y \leqq A$ for every $A > 0$. By Theorem 6.2 of Chapter II

$$\frac{1}{E_n(iy)} = \int_{-\infty}^{\infty} e^{-iyt}\, d\alpha_n(t) \qquad n = 1, 2, \cdots,$$

for some $\alpha_n(t) \in D$. Hence by Corollary 2.3 a function $\alpha(t) \in D$ must exist so that (1) holds for $s = iy$. Finally equation (2) holds as a result of Theorem 5.2. Observe that no Cauchy value is needed for the improper integral (2) since it converges absolutely by virtue of Theorem 5.3.

COROLLARY 5.4. Under the conditions of Theorem 5.4,

$$\frac{1}{E(s)} = \int_{-\infty}^{\infty} e^{-st} G(t)\, dt,$$

where $G(t)$ is the frequency function

$$G(t) = \alpha'(t) = \frac{1}{2\pi i} \int_{-i\infty}^{i\infty} \frac{e^{st}}{E(s)}\, ds \qquad -\infty < t < \infty. \tag{3}$$

This follows at once by differentiation under the integral sign in equation (2). The process is justified since the resulting integral (3) is uniformly convergent in $-\infty < t < \infty$ by Theorem 5.3.

It is helpful to consider the foregoing theory from the point of view of abstract topology. The class D of distribution functions introduced by Definition 2.1a becomes a topological Abelian semigroup if the group operation is defined as $\#$ (convolution) and the topology is introduced by the limit operation

$$\alpha_n(t) \rightrightarrows \alpha(t),$$

where the double arrow is an abbreviation for the limit operation defined in § 2.1 [$\alpha_n(t)$ approaches $\alpha(t)$ at all points of continuity of $\alpha(t)$]. To verify that D is an Abelian semigroup we have only to recall (E. Hille [1948; 147]) that for elements α, β, γ of D

$$\alpha \# \beta = \beta \# \alpha \in D$$

$$\alpha \# (\beta \# \gamma) = (\alpha \# \beta) \# \gamma.$$

The unit function $u(t)$ which is 0 and 1 in $(-\infty, 0)$, $(0, \infty)$, respectively, evidently satisfies the relation

$$\alpha \# u = \alpha$$

for all α of D, but D is not a group since there is generally no inverse α^{-1} of a given element α:

$$\alpha \# \alpha^{-1} = u.$$

For example, the element $\alpha(t)$ which is e^t and 1 in $(-\infty, 0)$, $(0, \infty)$, respectively, has no inverse. For, if $\alpha^{-1}(t)$ existed its characteristic function would be $1 - s$ by Theorem C, § 5.2, Chapter II, an obvious absurdity.

The transformation which replaces an element α of D by its characteristic function $\chi_\alpha(s)$ sets up a homomorphic mapping of D onto D'. In D' the group operation is ordinary multiplication of complex numbers [Theorem C quoted above], and the topology is set up by uniform limits:

$$\chi_{\alpha_n}(iy) \rightarrow f(y)$$

uniformly in $|y| \leq A$ for some $A > 0$ [Theorems 2.2 and 2.3]. That the homomorphism is also an isomorphism follows from Corollary 5.2.

The subclass D_f of Definition 4.1a is a sub-semigroup of D. This is most easily seen in the isomorphic semigroup D'_f of reciprocals of polynomials with real roots only and equal to 1 at the origin. For, the product of two such polynomials is another of the same type.

Finally the class $\bar{D}_f$ of Definition 4.1b is also a sub-semigroup of D. The corresponding isomorphic class $\bar{D}'_f$ is here the class of functions $1/E(s)$, $E(s) \in E$. It was precisely the purpose of Theorem 4.1 to prove this. The semigroup property for $\bar{D}'_f$ is evident from the fact, already observed in § 3.1, that the product of two functions of E is again in E.

6. PROPERTIES OF THE NON-FINITE KERNELS

6.1. Let us develop now some further properties of the frequency function $G(t)$.

THEOREM 6.1. *If $E(s)$ is the function of Theorem 5.3, then*

$$\frac{1}{E(s)} = \int_{-\infty}^{\infty} e^{-st}G(t)\,dt, \tag{1}$$

where $G(t) \in C^\infty$, $-\infty < t < \infty$, and the integral converges in the largest vertical strip which contains the origin and is free of zeros of $E(s)$.

As in § 6.2 of Chapter 2, we define α_1 as the largest of the negative roots ($\alpha_1 = -\infty$ if there are none) of $E(s)$ and α_2 as the smallest of the positive roots ($\alpha_2 = +\infty$ if there are none). Then we must show that the integral (1) converges for $\alpha_1 < \sigma < \alpha_2$. This is an immediate consequence of Theorem 5.3. For, by use of Cauchy's integral theorem we may shift the path of integration in the integral (3), § 5.4, to the line $\sigma = \alpha_2 - \epsilon$, $0 < \epsilon < \alpha_2$,

$$G(t) = \frac{e^{(\alpha_2-\epsilon)t}}{2\pi}\int_{-\infty}^{\infty} \frac{e^{iyt}\,dy}{E(\alpha_2 - \epsilon + iy)},$$

so that

$$\alpha(t) = O(e^{(\alpha_2-\epsilon)t}) \qquad t \to -\infty.$$

Similarly, by shifting the path of integration to the line $\sigma = \alpha_1 + \epsilon$, we see that

$$\alpha(t) = o(e^{(\alpha_1+\epsilon)t}) \qquad t \to +\infty.$$

Hence (1) converges for $\alpha_1 < \sigma < \alpha_2$. That it cannot converge in any larger vertical strip is evident since $1/E(s)$ has poles at α_1 and α_2 (if these are finite numbers).

To show that $G(t)$ has derivatives of all orders we may differentiate the integral defining $G(t)$,

$$G^{(p)}(t) = \frac{1}{2\pi i}\int_{-\infty}^{\infty} \frac{s^p e^{st}}{E(s)}\,ds, \qquad -\infty < t < \infty, \qquad p = 1, 2, \cdots \tag{2}$$

The integrals (2) converge uniformly in $-\infty < t < \infty$ by Theorem 5.3. This validates the differentiation and also shows that $G^{(p)}(t) \in C$.

6.2. We next compute the mean and variance of $G(t)$.

THEOREM 6.2. *If $E(s) \in E$,*

$$E(s) = e^{-cs^2+bs}\prod_{k=1}^{\infty}\left(1 - \frac{s}{a_k}\right)e^{s/a_k},$$

then the frequency function $G(t)$ for which

$$\frac{1}{E(s)} = \int_{-\infty}^{\infty} e^{-st} G(t)\, dt \tag{1}$$

has mean b and variance $2c + \sum_{k=1}^{\infty} 1/a_k^2$.

Here the product and sum are extended over all the zeros of $E(s)$, if any. The cases in which $E(s)$ has no zeros or a finite number of them may be considered as included by permitting all or some of the a_k to be ∞. If $E(s)$ is a polynomial, the result is already included in Theorem 6.2, Chapter II. In any case we have, after defining $F(s)$ as

$$F(s) = \frac{e^{bs}}{E(s)}, \tag{2}$$

that

$$\frac{F'(s)}{F(s)} = 2cs - \sum_{k=1}^{\infty} \left(\frac{1}{s - a_k} + \frac{1}{a_k} \right), \tag{3}$$

$$\frac{F(s)F''(s) - [F'(s)]^2}{[F(s)]^2} = 2c + \sum_{k=1}^{\infty} \frac{1}{(s - a_k)^2}. \tag{4}$$

On the other hand

$$F(s) = \int_{-\infty}^{\infty} e^{-s(t-b)} G(t)\, dt$$

$$-F'(0) = \int_{-\infty}^{\infty} (t - b) G(t)\, dt$$

$$V_G = F''(0) = \int_{-\infty}^{\infty} (t - b)^2 G(t)\, dt.$$

But from equations (2), (3), (4) it is clear that $F(0) = 1$, $F'(0) = 0$,

$$F''(0) = 2c + \sum_{k=1}^{\infty} \frac{1}{a_k^2},$$

so that the theorem is proved.

7. INVERSION

7.1. We are now in a position to develop an inversion theory for the convolution transforms whose kernels are the functions of class E introduced in § 3. For the present we limit attention to the case in which no factor e^{-cs^2} is present in the inversion function $E(s)$. That is, we assume that $c = 0$ in formula (1), § 3.1. We will treat this factor separately in a later chapter since it requires somewhat different methods.

THEOREM 7.1. *If*

1. $E(s) \in E$ *and* $E(s) = e^{bs} \prod_{k=1}^{\infty} \left(1 - \frac{s}{a_k}\right) e^{s/a_k}$

2. $G(t) = \frac{1}{2\pi i} \int_{-i\infty}^{i\infty} \frac{e^{st}}{E(s)}\, ds \qquad -\infty < t < \infty$

3. $\varphi(t) \in B \cdot C \qquad -\infty < t < \infty$

4. $f(x) = \int_{-\infty}^{\infty} G(x - t)\varphi(t)\, dt$

5. $P_n(s) = e^{(b-\epsilon_n)s} \prod_{k=1}^{n} \left(1 - \frac{s}{a_k}\right) e^{s/a_k}, \qquad \epsilon_n = o(1),\ n \to \infty,$

then

$$E(D)f(x) = \lim_{n\to\infty} P_n(D)f(x) = \varphi(x) \qquad -\infty < x < \infty.$$

We are assuming that there are infinitely many roots a_k of $E(s)$ since the case of a finite number of roots was treated in Chapter II. For each positive integer n set

$$E_n(s) = e^{\epsilon_n s} \prod_{k=n+1}^{\infty} \left(1 - \frac{s}{a_k}\right) e^{s/a_k} = \frac{E(s)}{P_n(s)}$$

$$G_n(t) = \frac{1}{2\pi i} \int_{-i\infty}^{i\infty} \frac{e^{st}}{E_n(s)}\, ds \qquad -\infty < t < \infty. \tag{1}$$

By Theorem 5.3 this integral converges absolutely and uniformly on $(-\infty, \infty)$. If D stands for differentiation with respect to x it is clear that

$$P_n(D)e^{sx} = e^{sx}P_n(s),$$

so that

$$P_n(D)G(x) = \frac{1}{2\pi i} \int_{-i\infty}^{i\infty} \frac{P_n(s)e^{sx}}{E(s)}\, ds$$

$$= \frac{1}{2\pi i} \int_{-i\infty}^{i\infty} \frac{e^{sx}}{E_n(s)}\, ds = G_n(x).$$

Differentiation under the integral sign is justified by the uniform convergence of the integral (1). Hence

$$P_n(D)f(x) = \int_{-\infty}^{\infty} G_n(x - t)\varphi(t)\, dt. \tag{2}$$

This step is valid if the integral (2) converges uniformly. It does so for x in any finite interval since $\varphi(x) \in B$ and since $G_n(t)$ is a frequency function and hence absolutely integrable.

Since

$$\varphi(x) = \int_{-\infty}^{\infty} G_n(t)\varphi(x)\,dt,$$

we have

$$P_n(D)f(x) - \varphi(x) = \int_{-\infty}^{\infty} G_n(t)\,[\varphi(x-t) - \varphi(x)]\,dt. \tag{3}$$

For an arbitrary $\epsilon > 0$ choose δ so that when $|t| \leqq \delta$, x fixed,

$$|\varphi(x-t) - \varphi(x)| \leqq \epsilon. \tag{4}$$

Now write the integral (3) as the sum of two others I_1 and I_2 corresponding respectively to the ranges of integration $|t| \leqq \delta$ and $|t| \geqq \delta$. By (4) it is clear that

$$|I_1| \leqq \epsilon \int_{-\delta}^{\delta} G_n(t)\,dt \leqq \epsilon \int_{-\infty}^{\infty} G_n(t)\,dt = \epsilon.$$

Suppose that M is an upper bound for $|\varphi(t)|$ and that n is large enough to make $|\epsilon_n| < \delta/2$. Then since $(t - \epsilon_n)^2 \geqq \delta^2/4$ for t on the range of I_2 we have

$$|I_2| \leqq \frac{8M}{\delta^2} \int_{|t| \geqq \delta} G_n(t)\,(t-\epsilon_n)^2\,dt \leqq \frac{8M}{\delta^2} \int_{-\infty}^{\infty} G_n(t)\,(t-\epsilon_n)^2\,dt\,.$$

By Theorem 6.2 the mean of $G_n(t)$ is ϵ_n and

$$|I_2| \leqq \frac{8M}{\delta^2} \sum_{k=n+1}^{\infty} \frac{1}{a_k^2}\,.$$

Hence

$$\overline{\lim_{n\to\infty}}\, |P_n(D)f(x) - \varphi(x)| \leqq \epsilon,$$

and our result is established. As in Theorem 7.1 of Chapter II, our assumptions about $\varphi(t)$ are stronger than needed, and will be weakened later.

COROLLARY 7.1. If

1. $E(s) = \prod_{k=1}^{\infty} \left(1 - \frac{s}{a_k}\right)$, $\qquad \sum_{k=1}^{\infty} 1/|a_k| < \infty$
2. $G(t) = \frac{1}{2\pi i} \int_{-i\infty}^{i\infty} \frac{e^{st}}{E(s)}\,ds$ $\qquad -\infty < t < \infty$
3. $\varphi(t) \in B \cdot C$ $\qquad -\infty < t < \infty$
4. $f(x) = \int_{-\infty}^{\infty} G(x-t)\varphi(t)\,dt$
5. $P_n(s) = \prod_{k=1}^{n} \left(1 - \frac{s}{a_k}\right)$,

then

$$\lim_{n\to\infty} P_n(D)f(x) = \varphi(x) \qquad -\infty < x < \infty.$$

The infinite product of hypothesis 1 is known to converge when the series $\sum 1/|a_k|$ converges. Since $|a_k| > 1$ for large k the series $\sum 1/a_k^2$ will also converge and we may apply Theorem 7.1. In that theorem take

$$b = -\sum_1^\infty 1/a_k$$

$$\epsilon_n = \sum_1^n 1/a_k + b = -\sum_{n+1}^\infty 1/a_k.$$

Then

$$E(s) = \prod_{k=1}^\infty \left(1 - \frac{s}{a_k}\right) = e^{bs} \prod_{k=1}^\infty \left(1 - \frac{s}{a_k}\right) e^{s/a_k}$$

$$P_n(s) = \prod_{k=1}^n \left(1 - \frac{s}{a_k}\right) = e^{(b-\epsilon_n)s} \prod_{k=1}^n \left(1 - \frac{s}{a_k}\right) e^{s/a_k}$$

Hence the conclusions of corollary and theorem are equivalent, and the proof is complete.

8. GREEN'S FUNCTIONS

8.1. In § 9.1 of Chapter II we defined the Green's function of a certain linear differential system of finite order. Here we extend the definition to include a corresponding system of infinite order. Let $E(s) \in E$ and have the specific definition

$$E(s) = \prod_{k=1}^\infty \left(1 - \frac{s}{a_k}\right) e^{s/a_k}.$$

As in § 6.1 and earlier we denote by α_1 the largest negative root (or $-\infty$) of $E(s)$, by α_2 the smallest positive root (or $+\infty$). The differential system of infinite order under consideration will be

$$E(D)f(x) = \varphi(x) \tag{1}$$

$$f(x) = o(e^{\alpha_2 x}) \qquad x \to +\infty \tag{2}$$

$$f(x) = o(e^{\alpha_1 x}) \qquad x \to -\infty. \tag{3}$$

To define the Green's function of this system we first replace it by the "truncated" system in which the differential equation (1) is replaced by

$$P_n(D)f(x) = \varphi(x), \tag{4}$$

where, as in Theorem 7.1,

$$P_n(s) = \prod_{k=1}^n \left(1 - \frac{s}{a_k}\right) e^{s/a_k}. \tag{5}$$

Following § 9 of Chapter II, it is clear that the Green's function of the truncated system (4) (2) (3) should be

$$G_n(t) = \frac{1}{2\pi i} \int_{-i\infty}^{i\infty} \frac{e^{st}}{P_n(s)}\, ds. \tag{6}$$

If the exponential factors in (5) were removed, this would be equation (1) of § 6.3, Chapter II. Since the exponential factors in (5) produce a translation in $f(x)$ in (4), it is clear that the Green's function (6) should be obtainable from that of § 9, Chapter II, by an equivalent translation. But this is precisely what has been accomplished in (6) by the introduction of the exponential factor into $P_n(s)$.

We are now able to give our definition as follows.

DEFINITION 8.1. The Green's function of the system (1) (2) (3) is defined as

$$\lim_{n\to\infty} G_n(t),$$

where $G_n(t)$ is given by equations (5), (6).

8.2. It is now easy to establish the existence of the Green's function. It is, in fact, the kernel of the convolution transform treated in Theorem 7.1.

THEOREM 8.2. *The Green's function of the system* (1) (2) (3) § 8.1 *exists and is equal to*

$$G(t) = \frac{1}{2\pi i} \int_{-i\infty}^{i\infty} \frac{e^{st}}{E(s)}\, ds\,. \tag{1}$$

From the definition of $P_n(s)$ it is evident that

$$\frac{|e^{iyt}|}{|P_n(iy)|} \leqq \frac{|a_1 a_2|}{(a_1^2 + y^2)^{1/2}(a_2^2 + y_2^2)^{1/2}} \qquad n \geqq 2.$$

The dominant function is integrable on $(-\infty, \infty)$ and is independent of n. Hence we may apply Lebesgue's limit theorem to equation 8.1 (6) to obtain (1), since it is known from elementary theory that $P_n(s)$ tends to $E(s)$ for every s. Since $E(iy) \neq 0$, the theorem is established.

Note that we already knew from §§ 4,5 that

$$\lim_{n\to\infty} \int_{-\infty}^{x} G_n(t)\, dt = \int_{-\infty}^{x} G(t)\, dt \qquad -\infty < x < \infty.$$

We have now shown in addition that

$$\lim_{n\to\infty} G_n(t) = G(t) \tag{2}$$

in the pointwise sense.

8.3. As in § 9.3 of Chapter II, the Green's function provides an explicit solution for the differential system under consideration.

THEOREM 8.3. *If* $\varphi(t) \in B \cdot C$, $-\infty < t < \infty$, *then a solution of the system* (1) (2) (3) *is*

$$f(x) = \int_{-\infty}^{\infty} G(x - t)\varphi(t)\, dt. \tag{1}$$

That equation 8.1 (1) is satisfied by this function $f(x)$ was proved in Theorem 7.1. Since $f(x) \in B$, the boundary conditions are satisfied trivially.

8.4. Observe that in the statement of Theorem 8.3 it is not stated that $G * \varphi$ is the *unique* solution of the system 8.1 (1) (2) (3), as in Theorem 9.3, Chapter II, the corresponding result for a system of finite order. If we had adopted as our definition of $E(D)$

$$E(D) f(x) = \lim_{n \to \infty} P_n(D) f(x) \tag{1}$$

$$P_n(D) f(x) = \prod_{k=1}^{n} \left(1 - \frac{D}{a_k}\right) e^{D/a_k} f(x),$$

where the limit exists boundedly or "in the mean" on $(-\infty, \infty)$, for example, then the solution 8.2 (1) could be shown to be unique. But our definition requires only that the limit (1) should exist for every real x, and in this case no uniqueness proof has been given. In fact, we can give an example of a function $f(x)$ satisfying 8.1 (2) (3) and for which the limit (1) is zero as n tends to ∞ through the *even* integers.

We choose the roots of $E(s)$ as $\pm 1, \pm 2, \cdots$. By grouping the factors in the infinite product expansion we obtain

$$E(s) = \frac{\sin \pi s}{\pi s} = \lim_{n \to \infty} P_{2n}(s)$$

$$E_n(s) = P_{2n}(s) = \prod_{k=1}^{n} \left(1 - \frac{s^2}{k^2}\right).$$

Define a function

$$h(x) = \frac{e^{-x}}{(1 + e^{-x})^2}.$$

Then $f(x) = h'(x)$ will be the required function. It is evident that $h'(\pm\infty) = 0$, so that 8.1 (2) (3) are satisfied with $\alpha_1 = -1$, $\alpha_2 = 1$. It remains to show that

$$\lim_{n \to \infty} E_n(D) h'(x) = 0 \qquad -\infty < x < \infty.$$

Preliminary to this result we prove by induction that

(2) $$E_n(D)h(x) = c_n[h(x)]^{n+1}$$

$$c_n = \frac{(2n+1)!}{n!n!}.$$

Simple computation gives

$$(1 - D^2)h(x) = \frac{6e^{-2x}}{(1+e^{-x})^4} = c_1[h(x)]^2,$$

so that (2) is valid for $n = 1$. Assume it true when n is replaced by $n - 1$ and differentiate the equation twice:

$$D^2E_{n-1}(D)h(x) = -c_{n-1}\left\{\frac{2n(2n+1)e^{-(n+1)x}}{(1+e^{-x})^{2n+2}} - \frac{n^2e^{-nx}}{(1+e^{-x})^{2n}}\right\}$$

$$\left(1 - \frac{D^2}{n^2}\right)E_{n-1}(D)h(x) = \frac{c_{n-1}}{n^2}\frac{2n(2n+1)e^{-(n+1)x}}{(1+e^{-x})^{2n+2}},$$

$$= c_n[h(x)]^{n+1}.$$

Thus equation (2) is established.

By differentiating equation (2), we have

(3) $$E_n(D)h'(x) = c_n(n+1)\,[h(x)]^nh'(x).$$

Since $h(x)$ is an even function $E_n(D)h'(x)$ must vanish at $x = 0$ for all n. For each $x \neq 0$ the right-hand side of equation (8) is the general term of a convergent series and hence tends to zero with n. The test ratio of the series is

$$\lim_{n\to\infty} \frac{c_n}{c_{n-1}}\left(\frac{n+1}{n}\right)h(x) = 4h(x),$$

and this is less than unity. Hence for all x

$$\lim_{n\to\infty} E_n(D)h'(x) = 0.$$

8.5. Let us now use the two examples of § 10 in Chapter II to illustrate Theorem 8.2. In Example A of that section we showed that the kernel

$$G_n(t) = ne^t(1-e^t)^{n-1} \qquad -\infty < t \leq 0$$
$$= 0 \qquad 0 < t < \infty$$

corresponded to the inversion function

$$E_n(s) = \prod_1^n \left(1 - \frac{s}{k}\right).$$

But

$$\lim_{n\to\infty} n^sE_n(s) = \frac{1}{\Gamma(1-s)},$$

and

$$\frac{1}{n^sE_n(s)} = \int_{-\infty}^{\infty} e^{-st}G_n(t - \log n)\,dt.$$

The kernel corresponding to $E(s) = 1/\Gamma(1 - s)$ is

$$G(t) = e^{-e^t}e^t$$

[see (A) of the table in § 9.10 of the present chapter]. Hence we should expect that

$$\lim_{n\to\infty} G_n(t - \log n) = G(t) \qquad -\infty < t < \infty. \tag{1}$$

Since

$$G_n(t - \log n) = e^t \left(1 - \frac{e^t}{n}\right)^{n-1} \qquad -\infty < t \leqq \log n,$$

equation (1) is equivalent to the obvious relation

$$\lim_{n\to\infty} \left(1 - \frac{x}{n}\right)^{n-1} = e^{-x} \qquad 0 < x < \infty;$$

so that 8.2 (2) is verified directly in this special case.

In Example B of § 10 Chapter II

$$E_n(s) = \prod_1^n \left(1 - \frac{s^2}{k^2}\right)$$

$$G_n(\log x) = n^2 \int_0^1 xt(1 - xt)^{n-1}(1 - t)^{n-1}\, dt \qquad 0 < x \leqq 1$$

$$= n^2 \int_0^1 x^{-1}t(1 - x^{-1}t)^{n-1}(1 - t)^{n-1}\, dt \qquad 1 \leqq x < \infty.$$

Since $E_n(s) \to \sin \pi s/(\pi s)$ as $n \to \infty$ and since the latter inversion function corresponds to the kernel $e^t/(e^t + 1)^2$ [see (B) of the table], 8.2 (2) will be verified if

$$\lim_{n\to\infty} n^2 \int_0^1 xt(1 - xt)^{n-1}(1 - t)^{n-1}\, dt = x/(1 + x)^2 \qquad 0 < x \leqq 1.$$

But by the Laplace asymptotic method, G. Pólya and G. Szegö [1925; 81], we have as $n \to \infty$

$$\int_0^1 t(1 - xt)^n(1 - t)^n\, dt \sim \int_0^\infty te^{-nxt}e^{-nt}\, dt = \frac{1}{[n(x + 1)]^2}.$$

Compare also D. V. Widder [1940; 213].

8.6. In § 8.2 we showed that every kernel $G(t)$ of the type considered in Theorem 7.1 is the pointwise limit of finite kernels. Later we shall need a corresponding result involving limits of more general kernels (perhaps non-finite). In order to treat finite and non-finite kernels simultaneously, we introduce the *degree* of a kernel by the following definition.

DEFINITION 8.6. If $E(s) \in E$ and $G(t)$ is the corresponding kernel

$$G(t) = \frac{1}{2\pi i}\int_{-i\infty}^{i\infty} \frac{e^{st}}{E(s)}\,ds, \tag{1}$$

then the degree of $G(t)$, denoted by $d(G)$, is N if

$$E(s) = e^{bs}\prod_{k=1}^{N}\left(1 - \frac{s}{a_k}\right); \tag{2}$$

otherwise $d(G) = \infty$.

For example, if $g(t)$ is the unit kernel of II § 6.1, then $d(g) = 1$; if $E(s) = e^{-s^2}$, $G(t) = (4\pi)^{-1/2}e^{-t^2/4}$, then $d(G) = \infty$. Theorem 6.3, Chapter II, shows that for finite kernels $G(t) \in C^{N-2}$, $N = d(G)$; Theorem 6.1 of the present chapter shows that for non-finite kernels $G(t) \in C^{\infty}$.

THEOREM 8.6. *If $E_n(s) \in E$ and $G_n(t)$ is the corresponding kernel, $n = 0, 1, 2, \cdots$, if*

$$\lim_{n\to\infty}\int_{-\infty}^{x} G_n(t)\,dt = \int_{-\infty}^{x} G_0(t)\,dt \qquad -\infty < x < \infty; \tag{3}$$

then

$$\lim_{n\to\infty} \| G_n^{(p)}(t) - G_0^{(p)}(t) \|_\infty = 0 \qquad p = 0, 1, \cdots, d(G_0) - 2. \tag{4}$$

Here, the superscript p indicates a pth derivative, and

$$\|f(x)\|_\infty = \underset{-\infty<x<\infty}{\text{l.u.b.}} |f(x)|.$$

By Theorem 2.2 equation (3) implies that $1/E_n(iy) \to 1/E_0(iy)$ uniformly for every interval $|y| \leqq A$. Since $|E_n(iy)| \geqq 1$ it follows that $E_n(iy) \to E_0(iy)$ uniformly in the same intervals and by Lemma 3.4, $E_n(s) \to E_0(s)$ uniformly in every circle. If

$$E_n(s) = \sum_{k=0}^{\infty} d_k(n)s^k$$

we have by a classical theorem of Weierstrass that $d_k(\infty) = d_k(0)$, $k = 0, 1, \cdots$. From the definition of class E it is clear that the coefficients in the expansion

$$|E_n(iy)|^2 = \sum_{k=0}^{\infty} D_k(n)y^{2k} \tag{5}$$

are all non-negative. On the other hand

$$D_k(n) = \sum_{j=0}^{k} (-1)^{j+(k/2)} d_j(n)d_{k-j}(n),$$

so that $D_k(\infty) = D_k(0)$, $k = 0, 1, \cdots$

Now if $d(G_0) = N < \infty$, we see from the explicit formula (2) that

$$|E_0(iy)|^2 \geqq 1 + a_1^{-2}a_2^{-2}\cdots a_N^{-2}y^{2N} = 1 + D_N(0)y^{2N},$$

$D_N(0) \neq 0$. Since $D_N(\infty) = D_N(0)$, we have for a positive constant $B < D_N(0)$ that $D_N(n) \geqq B$ when n is sufficiently large, $n > n_0$. Using the fact that the coefficients in (5) are $\geqq 0$ we obtain

$$|E_n(iy)|^2 \geqq 1 + By^{2N} \qquad n > n_0, \qquad n = 0. \tag{6}$$

By (1)

$$\| G_n^{(p)}(t) - G_0^{(p)}(t) \|_\infty \leqq \frac{1}{2\pi} \int_{-\infty}^{\infty} \left| \frac{y^p}{E_n(iy)} - \frac{y^p}{E_0(iy)} \right| dy. \tag{7}$$

Inequalities (6) show that the integral (7) is dominated by

$$\frac{1}{\pi} \int_{-\infty}^{\infty} \frac{|y^p|}{[1 + By^{2N}]^{1/2}} dy,$$

and this converges if $p \leq N - 2$. Hence for these values of p we may apply Lebesgue's limit theorem to the integral (7) to obtain (4).

If $d(G_0) = \infty$, then $E_0(s)$ must have the form

$$E_0(s) = e^{-cs^2+bs} \prod \left(1 - \frac{s}{a_k}\right)$$

where either $c > 0$ or the product must have infinitely many factors (or both). But

$$e^{2cy^2} = \sum_{k=0}^{\infty} \frac{(2c)^k}{k!} y^{2k}$$

$$\prod_{k=1}^{\infty} \left(1 + \frac{y^2}{a_k^2}\right) = \sum_{k=0}^{\infty} p_k y^{2k} \qquad p_k > 0.$$

Hence in every case the series (5), $n = 0$, has *all* of its coefficients greater than 0. As before we may conclude (6), but now with arbitrary N. Hence there need now be no restriction on p in (7) and the conclusion (4) holds for all p. This completes the proof.

9. EXAMPLES

9.1. In § 5 of Chapter I we have already given several illustrations of our inversion theory. There we took the point of view that those examples, known from other sources, corroborated conjectures made by use of the operational calculus. Now with Theorem 7.1 at our disposal we may specialize the kernel $G(t)$ in various ways, either obtaining new results or reestablishing old ones by the new method. In particular the inversion theories for the Laplace and Stieltjes transforms now appear as very special cases of the developments of the present chapter.

9.2. We record first the following simple results concerning change of variable.

THEOREM 9.2. *If* $F(x) \in C^n$, $0 < x < \infty$, *then*

$$(1)\qquad \prod_{k=1}^{n}\left(1-\frac{D}{k}\right)[e^xF(e^x)] = \frac{(-1)^n}{n!}\, e^{(n+1)x}F^{(n)}(e^x)$$

$$(2)\qquad \prod_{k=1}^{n}\left(1+\frac{D}{k}\right)[e^{-nx}F(e^x)] = \frac{1}{n!}\, F^{(n)}(e^x)$$

$$(3)\qquad \prod_{k=1}^{n}\left(1-\frac{2D}{2k-1}\right)[e^{x/2}F(e^x)] = (-1)^n c_n e^{(2n+1)x/2}F^{(n)}(e^x)$$

$$(4)\qquad \prod_{k=1}^{n}\left(1+\frac{2D}{2k-1}\right)[e^{-(2n-1)x/2}F(e^x)] = c_n e^{x/2}F^{(n)}(e^x)$$

$$(5)\qquad \prod_{k=1}^{n}\left(1-\frac{D}{2k-1}\right)[e^xF(e^{2x})] = (-1)^n c_n e^{(2n+1)x}F^{(n)}(e^{2x})$$

$$c_n = \frac{4^n n!}{(2n)!}$$

for $-\infty < x < \infty$.

The first of these identities was essentially proved in § 5, Chapter I. For the second, we have

$$\left(1+\frac{D}{n}\right)e^{-nx}F(e^x) = \frac{e^{-(n-1)x}}{n}\,F'(e^x).$$

Hence if the operations (2) are performed in the order $k = n$, $n-1, \cdots, 2, 1$, the result is immediate. It is a familiar fact that the order of application of the factor operators is immaterial when the coefficients are constants.

The other identities may be established in a similar way.

9.3. The Laplace transform

$$(1)\qquad F(x) = \int_0^\infty e^{-xt}\Phi(t)\,dt$$

becomes after exponential change of the variables x and t

$$(2)\qquad e^xF(e^x) = \int_{-\infty}^{\infty} G(x-t)\Phi(e^{-t})\,dt$$

$$(3)\qquad G(x) = e^{-e^x}e^x.$$

Since for $\sigma < 1$

$$(4)\qquad \int_{-\infty}^{\infty} e^{-st}G(t)\,dt = \int_0^\infty e^{-t}t^{-s}\,dt = \Gamma(1-s),$$

the inversion formula of Theorem 7.1 becomes

$$(5)\qquad \frac{1}{\Gamma(1-D)}\,e^xF(e^x) = \Phi(e^{-x}), \qquad -\infty < x < \infty.$$

Using the familiar infinite product expansion

$$\frac{1}{\Gamma(s)} = e^{\gamma s} s \prod_{k=1}^{\infty} \left(1 + \frac{s}{k}\right) e^{-s/k}$$

$$\gamma = \lim_{n\to\infty} \left[1 + \frac{1}{2} + \cdots + \frac{1}{n} - \log n\right]$$

and the identity

$$\frac{1}{\Gamma(s)} \frac{1}{\Gamma(1-s)} = \frac{\sin \pi s}{\pi} = s \prod_{k=1}^{\infty} \left(1 - \frac{s^2}{k^2}\right),$$

we obtain

$$\frac{1}{\Gamma(1-s)} = e^{-\gamma s} \prod_{k=1}^{\infty} \left(1 - \frac{s}{k}\right) e^{s/k}.$$

In Corollary 7.1, choose

$$\epsilon_n = \log n + \gamma - 1 - \frac{1}{2} - \cdots - \frac{1}{n},$$

which evidently tends to zero with $1/n$ as required. Hence (5) may be replaced by

$$\lim_{n\to\infty} e^{D \log n} \prod_{k=1}^{n} \left(1 - \frac{D}{k}\right) [e^x F(e^x)] = \Phi(e^{-x}) \qquad -\infty < x < \infty.$$

By equation 9.2 (1) this becomes

$$\lim_{n\to\infty} \frac{(-1)^n}{n!} e^{(n+1)(x+\log n)} F^{(n)}(e^{x+\log n}) = \Phi(e^{-x}).$$

We have thus proved the following result.

THEOREM 9.3. *If $\Phi(x) \in C \cdot B$ on $0 < x < \infty$ and if*

$$F(x) = \int_0^\infty e^{-xt} \Phi(t)\, dt,$$

then

$$\lim_{n\to\infty} \frac{(-1)^n}{n!} \left(\frac{n}{x}\right)^{n+1} F^{(n)}\left(\frac{n}{x}\right) = \Phi(x) \qquad 0 < x < \infty.$$

As previously pointed out, this is the familiar real inversion of the Laplace transform; see D. V. Widder [1946; 288].

9.4. As our next example, let us choose

$$E(s) = \frac{\sin \pi s}{\pi s} = \prod_{k=1}^{\infty} \left(1 - \frac{s^2}{k^2}\right)$$

$$= \frac{1}{\Gamma(1+s)\Gamma(1-s)}.$$

If we replace s by $-s$ and t by $-t$ in 9.3 (4) we obtain

$$\Gamma(1-s)\,\Gamma(1+s) = \int_{-\infty}^{\infty} e^{-st}[G(t) * G(-t)]\,dt.$$

But

$$G(x) * G(-x) = \int_{-\infty}^{\infty} e^{-e^{x-t}} e^{x-t} e^{-e^{-t}} e^{-t}\,dt$$

$$(1) \qquad = \frac{e^x}{(e^x+1)^2} \qquad -\infty < x < \infty.$$

That is,

$$(2) \qquad \frac{\pi s}{\sin \pi s} = \int_{-\infty}^{\infty} e^{-st} \frac{e^t}{(e^t+1)^2}\,dt \qquad -1 < \sigma < 1.$$

On the other hand, this kernel arises after a change of variable in the Stieltjes transform

$$(3) \qquad F(x) = \int_0^{\infty} \frac{\Phi(t)}{x+t}\,dt.$$

First differentiate with respect to x and then make exponential changes of variable as follows

$$-e^x F'(e^x) = \int_{-\infty}^{\infty} \frac{e^{x-t}\Phi(e^t)}{(e^{x-t}+1)^2}\,dt.$$

This is a convolution transform with the function (1) as kernel. Hence by Theorem 7.1

$$-\frac{\sin \pi D}{\pi D}\,[e^x F'(e^x)] = \Phi(e^x) \qquad -\infty < x < \infty$$

$$-\lim_{n\to\infty} \prod_{k=-n+1}^{n+1}{}' \left(1 + \frac{D}{k}\right) [e^x F'(e^x)] = \Phi(e^x),$$

where the prime indicates that there is no factor corresponding to $k = 0$.

To interpret this in terms of the original variables, we again make use of Theorem 9.2. By equation (1), § 9.2,

$$(4) \qquad -\prod_{k=1}^{n-1}\left(1 - \frac{D}{k}\right)[e^x F'(e^x)] = \frac{(-1)^n}{(n-1)!}\, e^{nx} F^{(n)}(e^x).$$

Now define a function $H(x)$ by the equation

$$e^{nx} F^{(n)}(e^x) = e^{-(n+1)x} H(e^x)$$

so as to make formula 9.2 (2) applicable to the right-hand side of (4). We obtain

$$\prod_{k=1}^{n+1}\left(1 + \frac{D}{k}\right) e^{-(n+1)x} H(e^x) = \frac{1}{(n+1)!}\, H^{(n+1)}(e^x).$$

Now eliminating the function $H(x)$ and returning to the original variables, we find that

$$\lim_{n\to\infty} \frac{(-1)^n}{(n+1)!(n-1)!} [x^{2n+1}F^{(n)}(x)]^{(n+1)} = \Phi(x) \qquad 0 < x < \infty. \tag{5}$$

This is a known inversion of the Stieltjes transform; see D. V. Widder [1946; 350]. We state the result as follows.

THEOREM 9.4. *If $\Phi(x) \in C \cdot B$ on $0 < x < \infty$, and if the integral*

$$F(x) = \int_0^\infty \frac{\Phi(t)}{x+t}\, dt \tag{6}$$

converges for $x > 0$, then equation (5) *holds.*

Note that we have proved a little more than this. For, our assumption on $\Phi(x)$ is only that it should be bounded and continuous. This is sufficient to guarantee the convergence of the integral

$$-F'(x) = \int_0^\infty \frac{\Phi(t)}{(x+t)^2}\, dt \tag{7}$$

and the validity of the inversion (5). But of course it does not imply the convergence of the integral (6). As an example consider the function $F(x) = \log(1/x)$. It has no representation in the form (6) although its derivative satisfies equation (7) with $\Phi(t) = 1$. Equation (5) is trivially satisfied for this function.

9.5. The Stieltjes transform arises as a special case of our general theory in another way. From equation 9.4 (6) we have

$$F(e^x)e^{x/2} = \int_{-\infty}^{\infty} \frac{e^{t/2}\Phi(e^t)\, dt}{e^{(x-t)/2} + e^{(t-x)/2}}$$

$$f(x) = G * \varphi(x),$$

where

$$f(x) = F(e^x)e^{x/2}, \qquad \varphi(x) = e^{x/2}\Phi(e^x), \qquad G(x) = \frac{1}{2}\operatorname{sech}\frac{x}{2}.$$

Integrating by parts in 9.4 (2), the following Laplace transforms result

$$\frac{\pi s}{\sin \pi s} = -s \int_{-\infty}^{\infty} \frac{e^{-st}}{e^t + 1}\, dt \qquad -1 < \sigma < 0$$

$$\frac{\pi}{\cos \pi s} = \int_{-\infty}^{\infty} \frac{e^{-st}}{e^{t/2} + e^{-t/2}}\, dt \qquad -\tfrac{1}{2} < \sigma < \tfrac{1}{2}. \tag{1}$$

Thus in the present case the inversion function is $E(s) = \pi^{-1}\cos \pi s$, so that

$$\pi^{-1}[\cos \pi D] f(x) = \varphi(x) \qquad -\infty < x < \infty.$$

Using the identities (3) and (4) of § 9.2 as (1) and (2) of § 9.2 were used in the previous section, we can prove the following result.

THEOREM 9.5. *If* $\Phi(t)\sqrt{t} \in C \cdot B$ *in* $(0, \infty)$, *and if*

$$F(x) = \int_0^\infty \frac{\Phi(t)}{x+t}\,dt,$$

then

$$\lim_{n\to\infty} \frac{(-1)^n}{2\pi}\left(\frac{e}{n}\right)^{2n} [x^{2n}F^{(n)}(x)]^{(n)} = \Phi(x) \tag{2}$$

for $0 < x < \infty$.

This is essentially a known result; compare D. V. Widder [1946; 345]. As an example consider the familiar formula

$$\frac{\pi}{\sqrt{x}} = \int_0^\infty \frac{dt}{\sqrt{t}(x+t)}.$$

When $F(x) = \pi/\sqrt{x}$, the left-hand side of (2) becomes

$$\lim_{n\to\infty} \frac{1}{2}\left[\frac{(2n)!}{4^n n!}\right]^2 \left[\frac{e}{n}\right]^{2n} \frac{1}{\sqrt{x}},$$

and by Stirling's formula this is seen to equal $x^{-1/2}$, as predicted.

This example is the special case $\nu = 1$ of the following:

$$E(s) = \frac{2^\nu}{B\left(\dfrac{\nu}{2}+s, \dfrac{\nu}{2}-s\right)}, \qquad G(t) = \operatorname{sech}^\nu \frac{t}{2} \qquad \nu > 0.$$

This pair results from the equation

$$\int_{-\infty}^\infty \frac{e^{-st}}{(e^{t/2}+e^{-t/2})^\nu}\,dt = \int_0^\infty \frac{u^{(\nu-2-2s)/2}}{(1+u)^\nu}\,du \qquad -\frac{\nu}{2} < \sigma < \frac{\nu}{2}, \tag{3}$$

where the integral on the right is a familiar form of the Beta-function. When $\nu = 1$ equation (3) reduces to (1).

9.6. Theorem 7.1 may equally well be applied to iterated transforms. Each integration of a given convolution transform, with inversion function $E(s)$, gives rise to an additional repetition of the operator $E(D)$ in the inversion. Thus, for example,

$$E^3(D)\,[G * G * G * \varphi] = \varphi.$$

We shall illustrate by the first iterated Laplace transform.

We consider the transform

$$f = G * G * \varphi, \tag{1}$$

where G is defined by equation 9.3 (3). By the product theorem the inversion function for this transform is

$$E(s) = \Gamma^2(1-s) = \int_{-\infty}^{\infty} e^{-st} G * G(t)\, dt.$$

As in § 9.3 it will be useful to intercept our result after exponential changes of variable to the range $(0, \infty)$. Consider the transform

$$F(x) = \int_0^\infty \frac{e^{-xt}}{t}\, dt \int_0^\infty e^{-u/t}\Phi(u)\, du \tag{2}$$

$$= \int_0^\infty \Phi(u)\, du \int_0^\infty e^{-xt-(u/t)} \frac{dt}{t}$$

$$= 2\int_0^\infty K_0(2\sqrt{xu})\Phi(u)\, du,$$

where $K_0(x)$ is the modified Bessel function, W. Magnus and F. Oberhettinger [1948; 39],

$$K_0(2x) = \frac{1}{2}\int_0^\infty e^{-t-(x^2/t)} \frac{dt}{t}.$$

With appropriate changes of variable equation (2) reduces to the form (1), with $f(x) = e^x F(e^x)$, $\varphi(x) = \Phi(e^{-x})$:

$$e^x F(e^x) = \int_{-\infty}^{\infty} \Phi(e^{-u})\, du \int_{-\infty}^{\infty} G(x-u-t)G(t)\, dt.$$

Since

$$\frac{1}{\Gamma^2(1-s)} = \lim_{n\to\infty} e^{2s\log n} \prod_{k=1}^{n}\left(1-\frac{s}{k}\right)^2,$$

the inversion formula of § 7 becomes

$$\lim_{n\to\infty} \prod_{k=1}^{n}\left(1-\frac{D}{k}\right)^2 e^x F(e^x)\Big|_{x=t+2\log n} = \Phi(e^{-t}). \tag{3}$$

By identity 9.2 (1),

$$\prod_{k=1}^{n}\left(1-\frac{D}{k}\right) e^x F(e^x) = \frac{(-1)^n}{n!} e^{(n+1)x}F^{(n)}(e^x).$$

Set

$$e^{(n+1)x}F^{(n)}(e^x) = e^x R(e^x)$$

so as to make the same identity applicable a second time. Then

$$\prod_{k=1}^{n}\left(1-\frac{D}{k}\right)^2 e^x F(e^x) = \frac{1}{n!n!} e^{(n+1)x}R^{(n)}(e^x).$$

From the definition of $R(x)$ it is clear that

$$R^{(n)}(x) = D^n x^n D^n F(x),$$

so that equation (3) becomes

$$\lim_{n\to\infty} \frac{x^{n+1}}{n!\,n!} D^n x^n D^n F(x) \bigg|_{x=n^2/t} = \Phi(t). \tag{4}$$

We state the result.

THEOREM 9.6. *If $\Phi(x) \in C \cdot B$ in $(0, \infty)$, and if*

$$F(x) = 2\int_0^\infty K_0(2\sqrt{xt})\Phi(t)\,dt, \tag{5}$$

$$K_0(2x) = \frac{1}{2}\int_0^\infty e^{-t-(x^2/t)}\frac{dt}{t},$$

then equation (4) holds for $0 < t < \infty$.

Note that the integral (5) converges for $0 < x < \infty$, as one sees by applying Fubini's theorem to the iterated integral (2), using the boundedness of Φ. As an illustration of formula (4), take $\Phi(t) = t$. From (2) we find $F(x) = 1/x^2$ and equation (4) becomes

$$\lim_{n\to\infty} \frac{(n+1)^2}{x}\bigg|_{x=n^2/t} = t.$$

This example shows, as we have remarked several times, that the conditions imposed on $\Phi(t)$ at present are stronger than needed. For, the inversion formula is valid even though $\Phi(t) = t$ is not bounded.

It should be pointed out that the iteration of a convolution transform is not equivalent to the iteration of the transform produced from it by the usual exponential change of variables.

Thus, the iteration of 9.3 (2) produces (5), but the iteration of 9.3 (1) yields the Stieltjes transform 9.4 (6).

9.7. Next let us consider the first iterate of the Stieltjes transform,

$$f = G * G * \varphi, \qquad G = \frac{1}{2}\operatorname{sech}\frac{x}{2}. \tag{1}$$

By the product theorem, the inversion function is

$$E(s) = \frac{\cos^2 \pi s}{\pi^2}.$$

But

$$\frac{1}{4}\operatorname{sech}\frac{x}{2} * \operatorname{sech}\frac{x}{2} = \int_{-\infty}^{\infty} \frac{dt}{[e^{(x-t)/2} + e^{(t-x)/2}]\,[e^{t/2} + e^{-t/2}]}$$

$$= \frac{xe^{x/2}}{e^x - 1}.$$

Hence equation (1) becomes explicitly, using the definitions of § 9.5 for f and φ,

$$F(e^x)e^{x/2} = \int_{-\infty}^{\infty} \frac{(x-t)e^{x/2}\Phi(e^t)}{e^{x-t}-1}\,dt$$

$$F(x) = \int_0^{\infty} \frac{\log x - \log t}{x-t}\Phi(t)\,dt\,.$$

This is the familiar form of the iterated Stieltjes transform; compare R. P. Boas and D. V. Widder [1931; 1].

By use of the product expansion of the inversion function we obtain

$$\lim_{n\to\infty} \frac{1}{\pi^2}\prod_{k=1}^{n}\left(1-\frac{2D}{2k-1}\right)^2\left(1+\frac{2D}{2k-1}\right)^2 f(x) = \varphi(x)\,.$$

Replacing f and φ by their values in terms of the capital letters and using the identities 9.2 (3) and 9.2 (4) as in § 9.5, we interpret Theorem 7.1 as follows.

THEOREM 9.7. *If* $\Phi(t)\sqrt{t} \in C \cdot B$ *on* $(0, \infty)$ *and if*

$$F(x) = \int_0^{\infty} \frac{\log (x/t)\Phi(t)}{x-t}\,dt,$$

then

(2) $$\lim_{n\to\infty} \frac{1}{4\pi^2}\left(\frac{e}{n}\right)^{4n} D^n x^{2n} D^{2n} x^{2n} D^n F(x) = \Phi(x) \qquad 0 < x < \infty.$$

As an example we may consider the equation

$$\frac{\pi^2}{\sqrt{x}} = \int_0^{\infty} \frac{\log x/t}{(x-t)\sqrt{t}}\,dt \qquad 0 < x < \infty.$$

Since

(3) $$D^n x^{-1/2} = \frac{(-1)^n}{c_n} x^{-(2n+1)/2},$$

we have at once, using the number c_n of equation 9.2 (5),

$$D^n x^{2n} D^{2n} x^{2n} D^n x^{-1/2} = c_n^{-4} x^{-1/2}.$$

By Stirling's formula

$$c_n \sim \frac{1}{\sqrt{2}}\left(\frac{e}{n}\right)^n \qquad n \to \infty,$$

so that equation (2) is verified for this special function.

9.8. As our next example, let us use the Meijer transform

$$F(x) = \int_0^{\infty} \sqrt{xt}\,K_0(xt)\Phi(t)\,dt,$$

where $K_0(t)$ is the modified Bessel function

$$K_0(x) = -\gamma - \log\frac{x}{2} + \sum_{k=1}^{\infty}\frac{1}{k!k!}\left(1 + \frac{1}{2} + \cdots + \frac{1}{k} - \gamma\right)\left(\frac{x}{2}\right)^{2k}$$

$$(1) \qquad K_0(x) = \frac{1}{2}\int_{-\infty}^{\infty} e^{-x\cosh t}\,dt \qquad 0 < x < \infty.$$

For the latter integral formula, see, for example, W. Magnus and F. Oberhettinger [1948; 39]. After exponential change of variables this becomes

$$f(x) = e^{x/2}F(e^x) = \int_{-\infty}^{\infty} e^{x-t}K_0(e^{x-t})e^{-t/2}\Phi(e^{-t})\,dt$$

$$= G * \varphi,$$

where

$$G(x) = e^x K_0(e^x), \qquad \varphi(x) = e^{-x/2}\Phi(e^{-x}).$$

The bilateral Laplace transform of this kernel is known, but for the reader's convenience, we derive it here with a minimum of preliminary material. Use the product theorem to obtain the inverse Laplace transform of $\Gamma^2[(1 - s)/2]$:

$$\Gamma\left(\frac{1-s}{2}\right) = 2\int_{-\infty}^{\infty} e^{-st}e^{t}e^{-e^{2t}}\,dt = 2\int_{-\infty}^{\infty} e^{-st}g(t)\,dt$$

$$\Gamma^2\left(\frac{1-s}{2}\right) = 4\int_{-\infty}^{\infty} e^{-st}g * g(t)\,dt \qquad -\infty < \sigma < 1$$

$$g * g(x) = e^x\int_{-\infty}^{\infty} e^{-e^{2x-2t}}e^{-e^{2t}}\,dt$$

$$= e^x\int_{-\infty}^{\infty} e^{-2e^x\cosh(x-2t)}\,dt$$

$$= e^x K_0(2e^x).$$

For the last equation we have used (1). Hence the desired transform is

$$\int_{-\infty}^{\infty} e^{-st}G(t)\,dt = 2^{-s-1}\Gamma^2\left(\frac{1-s}{2}\right).$$

From the usual product expansion for $\Gamma(s)$ we see that

$$\frac{1}{\Gamma\left(\dfrac{1-s}{2}\right)} = \lim_{n\to\infty} \frac{n^{s/2}}{\sqrt{\pi}} \prod_{k=1}^{n} \left(1 - \frac{s}{2k-1}\right),$$

so that the inversion formula becomes

$$\frac{2^{D+1}}{\Gamma^2\left(\dfrac{1-D}{2}\right)} f(x) = \varphi(x)$$

$$\lim_{n\to\infty} \frac{2}{\pi} (2n)^D \prod_{k=1}^{n} \left(1 - \frac{D}{2k-1}\right)^2 f(x) = \varphi(x). \tag{2}$$

We now interpret this in terms of the original functions F and Φ. To make formula 9.2 (5) applicable, set

$$e^{x/2}F(e^x) = e^x R(e^{2x}).$$

Then

$$\prod_{k=1}^{n} \left(1 - \frac{D}{2k-1}\right) e^{x/2}F(e^x) = (-1)^n c_n e^{(2n+1)x} R^{(n)}(e^{2x}), \tag{3}$$

where c_n was defined in 9.2 (5). Set

$$e^{(2n+1)x} R^{(n)}(e^{2x}) = e^x H(e^{2x})$$

and apply the above operator again:

$$\prod_{k=1}^{n} \left(1 - \frac{D}{2k-1}\right)^2 e^{x/2}F(e^x) = c_n^2 e^{(2n+1)x} H^{(n)}(e^{2x}).$$

The inversion (2) now becomes

$$\lim_{n\to\infty} \frac{2}{\pi} c_n^2 (2ne^x)^{2n+1} H^{(n)}(4n^2e^{2x}) = e^{-x/2}\Phi(e^{-x}).$$

If we set $4n^2e^{2x} = z$ and eliminate the functions H and R we obtain the following special case of Theorem 7.1.

THEOREM 9.8. *If $\Phi(t)\sqrt{t} \in C \cdot B$ on $(0, \infty)$, and if*

$$F(x) = \int_0^\infty \sqrt{xt} K_0(xt)\Phi(t)\, dt,$$

then

$$\lim_{n\to\infty} \frac{1}{\pi\sqrt{x}} \left(\frac{e}{n}\right)^{2n} z^{n+(1/2)} D^n z^n D^n F(\sqrt{z}) z^{-1/4} \Big|_{z=(2n/x)^2} = \Phi(x). \tag{4}$$

As an example we may apply (4) to invert

$$\frac{\pi}{2\sqrt{x}} = \int_0^\infty \sqrt{xt}K_0(xt)\frac{dt}{\sqrt{t}}.$$

This identity follows from

$$\int_0^\infty K_0(t)\,dt = \frac{\pi}{2},$$

which in turn is derived from (1). By use of 9.7 (3) we find that

$$z^{n+(1/2)}D^n z^n D^n z^{-1/2} = c_n^{-2},$$

so that (4) is verified trivially.

The inversion (4) is in a somewhat different form from that originally obtained by R. P. Boas [1942; 21]. That his differential operator is equivalent to (4) may be verified by noting that both have the fundamental solutions

$$x^{2k+(1/2)}, \qquad x^{2k+(1/2)}\log x, \qquad k = 0, 1, 2, \cdots .$$

We may treat the general Meijer transform

$$f(x) = \int_0^\infty \sqrt{xt}K_\nu(xt)\Phi(t)\,dt$$

$$K_\nu(x) = \frac{1}{2}\int_{-\infty}^\infty e^{-x\cosh t}e^{-\nu t}\,dt \qquad 0 < x, -1 < \nu < 1,$$

in exactly the same way. The kernel $G(x)$ is now $e^x K_\nu(e^x)$. Set

$$g_\nu(t) = 2e^{(\nu+1)t}e^{-e^{2t}}$$

so that

$$\Gamma\left(\frac{1+\nu-s}{2}\right) = \int_{-\infty}^\infty e^{-st}g_\nu(t)\,dt \qquad \sigma < 1+\nu$$

and

$$g_\nu * g_{-\nu}(t) = 4e^t K_\nu(2e^t).$$

Hence the inversion function is

$$E(s) = \frac{2^{s+1}}{\Gamma\left(\frac{1+\nu-s}{2}\right)\Gamma\left(\frac{1-\nu-s}{2}\right)} \qquad \sigma < 1+\nu, \quad \sigma < 1-\nu. \tag{5}$$

The inversion analogous to (2) is now interpreted by the infinite product expansion of (5). We leave it to the reader to obtain the analogue of (4) if desired. The similarity of Theorems 9.6 and 9.8 was to be expected in view of the similar disposition of the roots of $E(s)$ in the two cases.

9.9. As our final example we consider a kernel of class III. Of course any of the finite kernels of Chapter II are of this class if all of the zeros of $E(s)$ are taken of the same sign. However, the present example will be our first non-finite kernel of this class.

We take for the zeros of the inversion function the numbers $a_k = -k^2\pi^2$, $k = 1, 2, \cdots$. They are all of one sign and such that the series $\sum_1^\infty 1/a_k$ converges (see definition in § 8 of Chapter I). From the infinite product expansion of $s^{-1} \sin s$ and the partial fraction expansion of $\csc s$ (see E. C. Titchmarsh [1939; 113]),

$$\csc s = \frac{1}{s} + \sum_{k=1}^{\infty} \frac{(-1)^k 2s}{s^2 - k^2\pi^2},$$

we obtain by change of variable

$$E(s) = \frac{\sinh \sqrt{s}}{\sqrt{s}} = \prod_{k=1}^{\infty} \left(1 + \frac{s}{k^2\pi^2}\right)$$

$$\frac{1}{\sqrt{s} \sinh \sqrt{s}} = \frac{1}{s} + \sum_{k=1}^{\infty} \frac{2(-1)^k}{s + k^2\pi^2}.$$

Now recall the definition of the Jacobi Theta-function

$$\vartheta_3(x, t) = 1 + 2 \sum_{k=1}^{\infty} e^{-k^2\pi^2 t} \cos 2k\pi x$$

(see, for example, G. Doetsch [1937; 26]). Term by term integration, easily justified, gives

$$\int_0^\infty e^{-st} \vartheta_3(\tfrac{1}{2}, t)\, dt = \frac{1}{s} + \sum_{k=1}^{\infty} \frac{2(-1)^k}{s + k^2\pi^2} \qquad \sigma > 0,$$

$$= \frac{1}{\sqrt{s} \sinh \sqrt{s}}.$$

Using the relations $\vartheta_3(\frac{1}{2}, 0+) = 0$, $\vartheta_3(\frac{1}{2}, \infty) = 1$, we see after integration by parts that

$$\frac{1}{E(s)} = \frac{\sqrt{s}}{\sinh \sqrt{s}} = \int_0^\infty e^{-st} \vartheta_3'(\tfrac{1}{2}, t)\, dt, \tag{1}$$

where the prime indicates differentiation with respect to t. Thus the kernel $G(t)$ in the present case is

$$\begin{aligned} G(t) &= \vartheta_3'(\tfrac{1}{2}, t) && 0 < t < \infty \\ &= 0 && -\infty < t \leq 0. \end{aligned}$$

Consequently Corollary 7.1 becomes in this special case

THEOREM 9.9. *If* $\varphi(t) \in B \cdot C$, $-\infty < t < \infty$, *and if*

$$f(x) = \int_0^\infty \varphi(x-t)\vartheta_3'(\tfrac{1}{2}, t)\,dt, \tag{2}$$

then

$$\lim_{n\to\infty} \prod_{k=1}^{n} \left(1 + \frac{D}{k^2\pi^2}\right) f(x) = \varphi(x) \qquad -\infty < x < \infty. \tag{3}$$

For lack of an established name let us refer to (2) as a theta-transform. As an example take $\varphi(t) = t$, $f(x) = x - (1/6)$, to which the inversion (3) will apply even though $\varphi(t)$ is not bounded. In this case the transform $f(x)$ is easily computed from its definition (2) and from (1), since $E(0) = 1$ and $E'(0) = 1/6$. On the other hand

$$\prod_{k=1}^{\infty} \left(1 + \frac{D}{k^2\pi^2}\right)\left(x - \frac{1}{6}\right) = x - \frac{1}{6} + \sum_{k=1}^{\infty} \frac{1}{k^2\pi^2} = x,$$

so that equation (3) is verified.

9.10. Let us summarize our applications in tabular form, listing the definitions of $E(s)$ and $G(t)$ in the cases treated. They were:

(A)	Laplace	$\int_0^\infty e^{-xt}\Phi(t)\,dt$	
(B)	Derived Stieltjes	$\int_0^\infty \frac{\Phi(t)}{(x+t)^2}\,dt$	
(C)	Stieltjes	$\int_0^\infty \frac{\Phi(t)}{x+t}\,dt$	
(D)	Generalized Stieltjes	$\int_0^\infty \frac{\Phi(t)}{(x+t)^\nu}\,dt$	$\nu > 0$
(E)	Iterated Laplace	$\int_0^\infty \frac{e^{-xt}}{t}\,dt \int_0^\infty e^{-u/t}\Phi(u)\,du$	
(F)	Iterated Stieltjes	$\int_0^\infty \frac{\log(x/t)}{x-t}\Phi(t)\,dt$	
(G)	Meijer ($\nu = 0$)	$\int_0^\infty \sqrt{xt}\,K_0(xt)\Phi(t)\,dt$	
(H)	Meijer ($-1 < \nu < 1$)	$\int_0^\infty \sqrt{xt}\,K_\nu(xt)\Phi(t)\,dt$	
(I)	Theta	$\int_0^\infty \varphi(x-t)\vartheta_3'(\tfrac{1}{2}, t)\,dt.$	

Of course (B) and (C) are special cases of (D), and (G) of (H). They are listed separately since they are likely to be of more use than the general cases. The table is in no sense exhaustive.

	$G(t)$	$E(s)$
(A)	$e^{-e^t}e^t$	$\dfrac{1}{\Gamma(1-s)}$
(B)	$e^t(e^t+1)^{-2}$	$\dfrac{\sin \pi s}{\pi s}$
(C)	$\dfrac{1}{2}\operatorname{sech}\dfrac{t}{2}$	$\pi^{-1}\cos \pi s$
(D)	$2^{-\nu}\operatorname{sech}^{\nu}\dfrac{t}{2}$	$\dfrac{1}{B\left(\dfrac{\nu}{2}+s, \dfrac{\nu}{2}-s\right)}$
(E)	$e^{-e^t} * e^{-e^t}$	$\dfrac{1}{\Gamma^2(1-s)}$
(F)	$\dfrac{t}{2}\operatorname{csch}\dfrac{t}{2}$	$\pi^{-2}\cos^2 \pi s$
(G)	$K_0(e^t)e^t$	$\dfrac{2^{s+1}}{\Gamma^2\left(\dfrac{1-s}{2}\right)}$
(H)	$K_\nu(e^t)e^t$	$\dfrac{2^{s+1}}{\Gamma\left(\dfrac{1+\nu-s}{2}\right)\Gamma\left(\dfrac{1-\nu-s}{2}\right)}$
(I)	$\begin{cases}\vartheta_3'(\frac{1}{2}, t), & 0<t\\ 0, & t \leqq 0\end{cases}$	$\dfrac{\sinh\sqrt{s}}{\sqrt{s}}$

10. ASSOCIATED KERNELS

10.1. We conclude this chapter by generalizing in two respects the inversion theory of § 7. We have hitherto demanded that the inversion function $E(s)$ should belong to class E and, as one consequence, that $E(0) = 1$. Moreover the kernel $G(t)$ was defined by the integral

$$G(t) = \frac{1}{2\pi i}\int_{-i\infty}^{i\infty} \frac{e^{st}}{E(s)}\, ds \qquad -\infty < t < \infty, \tag{1}$$

the path of integration being the imaginary axis. We now allow the inversion function to vanish at the origin and permit the path of integration in (1) to be any vertical line not passing through a zero of $E(s)$. In this way (1) gives rise to a whole series of associated kernels corresponding to the intervals of the σ-axis which are free of zeros. Only one of these will be a frequency function but all of the corresponding convolution transforms will be inverted by the same inversion operator $E(D)$.

10.2. Let us begin by proving two preliminary results.

LEMMA 10.2a. *If*

1. r is a non-negative integer
2. $a_k^* = a_k - c \neq 0, \qquad k = 1, 2, \cdots; \qquad \sum_{k=1}^{\infty} 1/a_k^2 < \infty$
3. $b^* = b + \frac{r}{c} - \sum_{k=1}^{\infty} \frac{c}{a_k a_k^*}$
4. $E(s) = s^r e^{bs} \prod_{k=1}^{\infty} \left(1 - \frac{s}{a_k}\right) e^{s/a_k}$
5. $E^*(s) = e^{b^*s} \left(1 + \frac{s}{c}\right)^r e^{-sr/c} \prod_{k=1}^{\infty} \left(1 - \frac{s}{a_k^*}\right) e^{s/a_k^*}$,

then

$$E(s + c) = E(c)E^*(s).$$

It is clear that $E(s + c)$ and $E^*(s)$ have the same zeros. That their non-vanishing quotient is constant, and hence equal to $E(c)$, is easily verified algebraically by use of the identities

$$\left(1 - \frac{s}{a_k^*}\right)\left(1 - \frac{c}{a_k}\right) = 1 - \frac{s + c}{a_k}$$

$$\frac{1}{a_k^*} - \frac{1}{a_k} = \frac{c}{a_k a_k^*}.$$

LEMMA 10.2b. If $E(s)$ and $E*(s)$ are defined as in the previous lemma, then

$$E(D)f(x) = E(c)e^{cx}E*(D)[e^{-cx}f(x)].$$

As always the differential operators are to be interpreted as limits corresponding to the limits defining the infinite products involved. Hence the result is true if established for each of the corresponding factors of these products, and follows from the differential identities

$$(c + D)[e^{-cx}f(x)] = e^{-cx}Df(x)$$

$$(a_k^* - D)[e^{-cx}f(x)] = e^{-cx}(a_k - D)f(x).$$

10.3. We turn now to the generalized inversion theorem.

THEOREM 10.3. *If*

1. $E(s)s^{-r} \in E$ *for some integer* $r \geqq 0$

2. $G(t) = \dfrac{1}{2\pi i} \displaystyle\int_{c-i\infty}^{c+i\infty} \frac{e^{st}}{E(s)}\, ds \qquad E(c) \neq 0, \qquad c \neq 0$

3. $e^{-cx}\varphi(x) \in C \cdot B \qquad -\infty < x < \infty$

4. $f(x) = \displaystyle\int_{-\infty}^{\infty} G(x-t)\varphi(t)\, dt,$

then

$$E(D)f(x) = \varphi(x) \qquad -\infty < x < \infty.$$

Let $E(s)$ be the function of Lemma 10.2a. Since $E(c) \neq 0$ and $c \neq 0$ hypothesis 2 of that lemma is satisfied. By the conclusion and by change of variable

$$G(t) = \frac{1}{2\pi i} \int_{-i\infty}^{i\infty} \frac{e^{(s+c)t}}{E(c)E^*(s)}\, ds$$

$$= e^{ct}G^*(t)/E(c), \tag{1}$$

where

$$G^*(t) = \frac{1}{2\pi i} \int_{-i\infty}^{i\infty} \frac{e^{st}ds}{E^*(s)}.$$

Since $E^*(s) \in E$, $G^*(t)$ is a frequency function to which Theorem 7.1 is applicable. In terms of this kernel hypothesis 4 becomes, using (1),

$$E(c)e^{-cx}f(x) = \int_{-\infty}^{\infty} G^*(x-t)e^{-ct}\varphi(t)\, dt.$$

Hence by Theorem 7.1

$$E(c)E^*(D)\,[e^{-cx}f(x)] = e^{-cx}f(x) \qquad -\infty < x < \infty. \tag{2}$$

But by Lemma 10.2b the left-hand side of (2) multiplied by e^{cx} is $E(D)f(x)$, and the theorem is proved.

For example, choose $E(s)$ the function $1/\Gamma(1-s)$ of Example A, § 9.10, with $1 < c < 2$. Since the residue of $\Gamma(1-s)$ at $s = 1$ is -1 we have

$$G(t) = \frac{1}{2\pi i} \int_{c-i\infty}^{c+i\infty} e^{st}\Gamma(1-s)\, ds = e^{-e^t}e^t - e^t.$$

After the usual changes of variable we find that the transform

$$F(x) = \int_0^{\infty} [e^{-xt} - 1]\Phi(t)\, dt \tag{3}$$

is inverted by the operator which inverts the Laplace transform:

$$\lim_{k\to\infty} \frac{(-1)^k}{k!} F^{(k)}\left(\frac{k}{t}\right)\left(\frac{k}{t}\right)^{k+1} = \Phi(t), \qquad 0 < t < \infty, \tag{4}$$

provided that $\Phi(t)t^c \in B \cdot C$ on $(0, \infty)$. More generally, if $n < c < n + 1$ the kernel of (3) may be replaced by

$$\left[e^{-xt} - \sum_{k=0}^{n-1} \frac{(-xt)^k}{k!}\right],$$

and the same inversion (4) applies.

11. SUMMARY

11.1. The principal result of the present chapter is the inversion formula

$$E(D) \int_{-\infty}^{\infty} G(x - t)\varphi(t)\, dt = \varphi(x),$$

where

$$G(t) = \frac{1}{2\pi i} \int_{-i\infty}^{i\infty} \frac{e^{st}}{E(s)}\, ds$$

$$E(s) = e^{bs} \prod_{1}^{\infty} \left(1 - \frac{s}{a_k}\right) e^{s/a_k} \qquad \sum_{1}^{\infty} 1/a_k^2 < \infty. \tag{1}$$

As justification for this choice of kernels the chapter began with a proof of Laguerre's important result (somewhat generalized) to the effect that the uniform limit of a sequence of polynomials with real roots only and equal to 1 at $s = 0$ is always a function (1) multiplied by a factor e^{-cs^2}, $c \geqq 0$. [A discussion of the operator e^{cD^2} is deferred to Chapter VIII.] By use of this theorem of Laguerre we were able to show that the kernels $G(t)$ considered in this book form a semi-group and hence from a certain point of view constitute *all* of the "natural" kernels. Finally, in an effort to make the theory as concrete as possible a large number of special kernels was treated.

CHAPTER IV

Variation Diminishing Transforms

1. INTRODUCTION

1.1. It is natural to ask whether there exists an intrinsic description of our class of kernels $G(t)$,

$$G(t) = (2\pi i)^{-1} \int_{-i\infty}^{i\infty} [E(s)]^{-1} e^{st}\, ds\,,$$

$$E(s) = e^{-cs^2+bs} \prod_k \left(1 - \frac{s}{a_k}\right) e^{s/a_k}\,, \tag{1}$$

$$c \geqq 0,\ \sum_k a_k^{-2} < \infty\,,$$

or whether these kernels have been selected only for reasons of convenience. Theorem 4.1 of Chapter III may be thought of as giving an affirmative response to this question. In the present chapter we shall show in an even more striking fashion that this is true.

Let us denote by $\mathscr{V}[a(1), \cdots, a(n)]$ the number of changes of sign of the real sequence $a(1), \cdots, a(n)$. Thus, for example,

$$\mathscr{V}[1, 0, 1, 0, -1] = 1,$$

$$\mathscr{V}[1, 0, 1, 0, 1] = 0,$$

$$\mathscr{V}[0, 0, 0, 0, 0] = -1.$$

If $f(x)$ is a real function defined for $-\infty < x < \infty$ then $\mathscr{V}[f(x)] = \text{l.u.b.}\,\mathscr{V}[f(x_1), \cdots, f(x_n)]$, taken over all sets $-\infty < x_1 < x_2 < \cdots < x_n < +\infty$. It is of course possible that $\mathscr{V}[f(x)] = \infty$.

DEFINITION 1.1a. A frequency function $\varphi(t)$ is said to be variation diminishing if $\mathscr{V}[h * \varphi] \leqq \mathscr{V}[h]$ for every $h \in B \cdot C(-\infty, \infty)$.

DEFINITION 1.1b. A frequency function $\varphi(t)$ is said to be totally positive if and only if $x_1 < x_2 < \cdots < x_n$, $t_1 < t_2 < \cdots < t_n$ implies that

$$D_n = \det\,[\varphi(x_i - t_j)]_n \geqq 0.$$

This inequality should hold for $n = 1, 2, \cdots$

We shall prove that every kernel (1) is both variation diminishing and totally positive, and we shall show that every variation diminishing frequency function, and every totally positive frequency function is of the form (1). These remarkable results are due to I. J. Schoenberg [1947, et seq.].

From these fundamental properties other important facts concerning $G(t)$ can be derived. For example, we will prove that for each value of $n = 0, 1, \cdots, G^{(n)}(t)$ has exactly n changes of sign, and that $-\log G(t)$ is convex. Such properties play an important role in later chapters.

2. GENERATION OF VARIATION DIMINISHING FREQUENCY FUNCTIONS

2.1. In this section we shall prove that $G(t)$ defined by 1.1 (1) is variation diminishing. Let $\varphi(t)$, $\varphi_n(t)$ $(n = 1, 2, \cdots)$ be frequency functions. We shall write

$$\varphi_n(t) \rightrightarrows \varphi(t) \qquad n \to \infty$$

if, for every t_1 and t_2,

$$\lim_{n\to\infty} \int_{t_1}^{t_2} \varphi_n(t)\, dt = \int_{t_1}^{t_2} \varphi(t)\, dt,$$

thus extending the symbol of p. 54 to include frequency functions.

Let sgn x be 1, 0, or -1 as $x > 0$, $x = 0$, or $x < 0$. We say that x and y are of opposite sign if $\operatorname{sgn}(xy) = -1$ and of the same sign if $\operatorname{sgn}(xy) = 0$ or 1.

LEMMA 2.1a. If $\{\varphi_n(t)\}_{n=1}^{N}$ are variation diminishing frequency functions and if $\varphi = \varphi_1 * \varphi_2 * \cdots * \varphi_N$ then φ is a variation diminishing frequency function.

We proceed by induction on N. Our lemma is true for $N = 1$. Suppose that it has been established for $n = 1, \cdots, N - 1$; we shall show that it is true for N. Let $\varphi_1 * \varphi_2 * \cdots * \varphi_{n-1} = \varphi'$. We have

$$h * \varphi = (h * \varphi') * \varphi_N.$$

Since φ_N is variation diminishing $\mathscr{V}[h * \varphi] \leqq \mathscr{V}[h * \varphi']$. By our induction assumption $\mathscr{V}[h * \varphi'] \leqq \mathscr{V}[h]$. Our desired conclusion follows.

LEMMA 2.1b. Let $\psi(t)$, $\{\psi_k(t)\}_1^\infty$ be real functions defined for $-\infty < t < \infty$. If

1. $\mathscr{V}[\psi_k(t)] \leqq n \qquad k = 1, 2, \cdots,$
2. $\lim_{k\to\infty} \psi_k(t) = \psi(t) \qquad -\infty < t < \infty,$

then $\mathscr{V}[\psi(t)] \leqq n$.

If $\mathscr{V}[\psi(t)] = N$ then there exist points $-\infty < t_0 < t_1 \cdots < t_N < \infty$, such that $\psi(t_{j-1})$ and $\psi(t_j)$ are of opposite sign, $j = 1, \cdots, N$. If k is sufficiently large then $\operatorname{sgn}[\psi(t_j) \cdot \psi_k(t_j)] = 1$ for $j = 0, \cdots, N$, and thus for k sufficiently large $\mathscr{V}[\psi_k(t)] \geqq \mathscr{V}[\psi(t)]$, which proves the lemma.

LEMMA 2.1c. Let $\varphi(t)$, $\{\varphi_n(t)\}_1^\infty$ be frequency functions. If

1. $\varphi_n(t)$ is variation diminishing, $n = 1, 2, \cdots,$
2. $\varphi_n(t) \rightrightarrows \varphi(t) \qquad n \to \infty,$

then $\varphi(t)$ is variation diminishing.

Since φ_n is variation diminishing $\mathscr{V}[h * \varphi_n] \leqq \mathscr{V}[h], n = 1, 2, \cdots$. We have $h * \varphi = \lim_{n\to\infty} h * \varphi_n$. It follows from Lemma 2.1b that $\mathscr{V}[h * \varphi] \leqq \mathscr{V}[h]$.

As in Chapter II we set

$$g(t) = \begin{cases} e^t & -\infty < t < 0 \\ \dfrac{1}{2} & t = 0 \\ 0 & 0 < t < \infty. \end{cases} \tag{1}$$

LEMMA 2.1d. If a is any real number not zero then $|a| g(at)$ is a variation diminishing frequency function.

Since if $\varphi(t)$ is a variation diminishing frequency function and if $a \neq 0$, then $|a|\varphi(at)$ is also a variation diminishing frequency function, it is sufficient to prove our lemma for $a = 1$. Let $h(t) \in B \cdot C\ (-\infty, \infty)$ and consider

$$f(x) = \int_{-\infty}^{\infty} g(x - t)h(t)\, dt.$$

We must show that $\mathscr{V}[f(x)] \leqq \mathscr{V}[h(t)]$. We have

$$h(x) = (1 - D)f(x) = -e^x De^{-x}f(x).$$

Appealing to Rolle's theorem and remembering that $e^{-x}f(x)$ vanishes at $+\infty$ we obtain our desired result.

Combining our lemmas and Theorem 4.1 of Chapter III we obtain

THEOREM 2.1. *$G(t)$ defined by* 1.1 (1) *is a variation diminishing frequency function.*

3. LOGARITHMIC CONVEXITY

3.1. In this section we shall show that if $\varphi(t)$ is a variation diminishing frequency function and if $\varphi(t) = \frac{1}{2}[\varphi(t+) + \varphi(t-)]$ for all t then $-\log \varphi(t)$ is convex. Let us recall the formal definition of convexity.

DEFINITION 3.1. Let $\varphi(t)$ be defined for $T \leqq t \leqq T'$ ($\varphi(t)$ may assume the value $+\infty$). $\varphi(t)$ is said to be convex if for $T \leqq t_1, t_2 \leqq T'$ and $0 \leqq \theta \leqq 1$ we have

$$\varphi[(1 - \theta)t_1 + \theta t_2] \leqq (1 - \theta)\varphi(t_1) + \theta\varphi(t_2). \tag{1}$$

LEMMA 3.1a. A necessary and sufficient condition that $\varphi(t)$ be convex is that for every $T \leqq t_1 \leqq t_2 \leqq T'$ and every a, $-\infty < a < \infty$,

$$\text{(2)} \qquad \underset{t_1 \leqq t \leqq t_2}{\text{l.u.b.}} [\varphi(t) - at] \leqq \max [\varphi(t_1) - at_1;\ \varphi(t_2) - at_2].$$

Necessity. Suppose that (2) is not satisfied for some t_1, t_2, and a. There would then exist t, $t_1 < t < t_2$ such that

$$\text{(3)} \qquad \varphi(t) - at > \varphi(t_1) - at_1,$$

$$\text{(3')} \qquad \varphi(t) - at > \varphi(t_2) - at_2.$$

We may write t in the form $(1 - \theta)t_1 + \theta t_2$ where $0 < \theta < 1$. Multiply (3) by $(1 - \theta)$ and (3′) by θ and add. Simplifying we obtain

$$\varphi(t) > (1 - \theta)\varphi(t_1) + \theta\varphi(t_2)$$

so that $\varphi(t)$ cannot be convex.

Sufficiency. Let $t_1 < t_2$ and $0 \leqq \theta \leqq 1$ be given. If $t = (1 - \theta)t_1 + \theta t_2$ then $t_1 \leqq t \leqq t_2$, and hence by (2) we have for every a that

$$\varphi(t) - at \leqq \max [\varphi(t_1) - at_1;\ \varphi(t_2) - at_2].$$

Choosing $a = [\varphi(t_2) - \varphi(t_1)]/(t_2 - t_1)$ and simplifying we find that

$$\varphi(t) \leqq (1 - \theta)\varphi(t_1) + \theta\varphi(t_2)$$

as desired.

LEMMA 3.1b. If

1. $\psi(x) \in B(-\infty, \infty)$,
2. $\psi(x) = \frac{1}{2}[\psi(x+) + \psi(x-)] \qquad -\infty < x < \infty,$

then

$$\psi(x) = \lim_{\epsilon \to 0+} \int_{-\infty}^{\infty} \psi(x - t)\, [\pi^{-1/2}\epsilon^{-1}e^{-(t/\epsilon)^2}]\, dt.$$

We have

$$\int_{-\infty}^{\infty} \psi(x - t)\, [\pi^{-1/2}\epsilon^{-1}e^{-(t/\epsilon)^2}]\, dt = I_1 + I_2$$

where

$$I_1 = \pi^{-1/2} \int_0^{\infty} \psi(x - \epsilon u)e^{-u^2}\, du,$$

$$I_2 = \pi^{-1/2} \int_0^{\infty} \psi(x + \epsilon u)e^{-u^2}\, du.$$

Now

$$\lim_{\epsilon \to 0+} \psi(x - \epsilon u)e^{-u^2} = e^{-u^2}\psi(x-) \qquad (0 < u < \infty),$$

$$|\psi(x - \epsilon u)e^{-u^2}| \leqq e^{-u^2}\, ||\psi||_{\infty} \qquad (0 < u < \infty, \quad 0 < \epsilon).$$

Applying Lebesgue's limit theorem we find that

$$\lim_{\epsilon\to 0^+} I_1 = \tfrac{1}{2}\psi(x^-), \qquad \lim_{\epsilon\to 0^+} I_2 = \tfrac{1}{2}\psi(x^+),$$

as desired.

Let us define

$$\Delta_h\psi(x) = \psi(x+h) - \psi(x-h) \qquad (h > 0).$$

LEMMA 3.1c. If

1. $\varphi(x)$ is a variation diminishing frequency function,

2. $\varphi(x) \in B(-\infty, \infty)$, $\varphi(x) = \frac{1}{2}[\varphi(x+) + \varphi(x-)]$ $(-\infty < x < \infty)$,

then for any real a and $h > 0$, $\Delta_h e^{ax}\varphi(x)$ has at most one change of sign.

We apply Lemma 3.1b with $\psi(x) = e^{-ax}\Delta_h e^{ax}\varphi(x)$ to obtain

$$e^{-ax}\Delta_h e^{ax}\varphi(x) = \lim_{\epsilon\to 0^+}\int_{-\infty}^{\infty}[e^{ah}\varphi(x+h-t)$$
$$- e^{-ah}\varphi(x-h-t)]\,[\pi^{-1/2}\epsilon^{-1}e^{-(t/\epsilon)^2}]\,dt.$$

From this we deduce that

$$e^{-ax}\Delta_h e^{ax}\varphi(x) = \lim_{\epsilon\to 0^+}\int_{-\infty}^{\infty}\varphi(x-t)\pi^{-1/2}\epsilon^{-1}[e^{ah-(t+h)^2/\epsilon^2} - e^{-ah-(t-h)^2/\epsilon^2}]\,dt.$$

Since

$$e^{ah-(t+h)^2/\epsilon^2} - e^{-ah-(t-h)^2/\epsilon^2}$$

has one change of sign and since φ is variation diminishing it follows that for each $\epsilon > 0$

$$\int_{-\infty}^{\infty}\varphi(x-t)\pi^{-1/2}\epsilon^{-1}[e^{ah-(t+h)^2/\epsilon^2} - e^{-ah-(t-h)^2/\epsilon^2}]\,dt$$

has at most one change of sign. Applying Lemma 2.1b we obtain our desired result.

LEMMA 3.1d. If

1. $\varphi(t)$ is a variation diminishing frequency function,

2. $\varphi(t) \in B(-\infty, \infty)$, $\varphi(t) = \frac{1}{2}[\varphi(t+) + \varphi(t-)]$ $(-\infty < t < \infty)$,

then $-\log\varphi(t)$ is convex.

In view of Lemma 3.1a it is enough to show that

$$\underset{t_1\leq t\leq t_2}{\text{l.u.b.}}\,[-\log\varphi(t) - at] \leqq \max\,[-\log\varphi(t_1) - at_1;\quad -\log\varphi(t_2) - at_2]$$

or equivalently that

$$\underset{t_1\leq t\leq t_2}{\text{g.l.b.}}\,[\varphi(t)e^{at}] \geqq \min\,[\varphi(t_1)e^{at_1};\quad \varphi(t_2)e^{at_2}]$$

for every t_1, t_2 and a. Suppose that for some t_1, t_2 and a this were not true. Noting that $\varphi(t)e^{at}$ vanishes at $+\infty$ or $-\infty$ (or both) and drawing pictures it is easily seen that this would imply that for h sufficiently small $\Delta_h e^{ax}\varphi(x)$ has two or more changes of sign, which we know to be impossible.

THEOREM 3.1. *If* $G(t)$ *is defined by* 1.1 (1) *then* $-\log G(t)$ *is convex.*

This is an immediate corollary of Theorem 2.1 and Lemma 3.1d.

4. CHARACTERIZATION OF VARIATION DIMINISHING FUNCTIONS

4.1. In this section we shall establish the converse of Theorem 2.1.

LEMMA 4.1. If

1. $\varphi(t)$ is a variation diminishing frequency function,

2. $h(t) \in C(-\infty, \infty)$,

3. $h(t)\varphi(x-t) \in L^1(-\infty, \infty) \qquad (-\infty < x < \infty)$,

then

$$\mathscr{V}[h * \varphi] \leqq \mathscr{V}[h].$$

Let us set $\chi(T, t) = 1$ for $|t| \leqq T$, $2 - |t|/T$ for $T < |t| \leqq 2T$, and 0 for $|t| > 2T$. By Lebesgue's limit theorem

$$h(t) * \varphi(t) = \lim_{T\to\infty} \int_{-\infty}^{\infty} \chi(T, u)h(u)\varphi(t-u)\,du.$$

Let $m \leqq \mathscr{V}[h * \varphi]$. It follows from Lemma 2.1b that for T sufficiently large, $T = T'$,

$$\mathscr{V}\left[\int_{-\infty}^{\infty} \chi(T', u)h(u)\varphi(t-u)\,du\right] \geqq m.$$

The function $\chi(T', u)h(u) \in B \cdot C \cdot (\infty, \infty)$ and thus since φ is variation diminishing

$$\mathscr{V}[\chi(T', u)h(u)] \geqq m.$$

Since

$$\mathscr{V}[h(t)] \geqq \mathscr{V}[\chi(T', t)h(t)]$$

we have that

$$\mathscr{V}[h(t)] \geqq m.$$

Our desired result follows.

THEOREM 4.1. *If* $\varphi(t)$ *is a variation diminishing frequency function then there exists a kernel* $G(t)$ *defined by* 1.1 (1) *such that* $\varphi(t) = G(t)$ *almost everywhere.*

The function $\pi^{-\frac{1}{2}}e^{-t^2}$ is variation diminishing and thus $\Phi(t) = \varphi(t) * \pi^{-\frac{1}{2}} e^{-t^2}$ is variation diminishing. Further $\Phi(t)$ is bounded, infinitely differentiable, and positive. By Lemma 3.1d $-\log \Phi(t)$ is convex which implies that $\Phi'(t)/\Phi(t) \in \downarrow$. We define

$$\beta_1 = \lim_{t \to +\infty} \Phi'(t)/\Phi(t),$$

$$\beta_2 = \lim_{t \to -\infty} \Phi'(t)/\Phi(t).$$

(It is possible for β_1 to be $-\infty$ and for β_2 to be $+\infty$). It is easily verified that because $\Phi(t)$ is a frequency function

$$\beta_1 < 0, \qquad \beta_2 > 0,$$

and that if $\beta_1' > \beta_1$, $\beta_2' < \beta_2$ then

$$\Phi(t) = 0(e^{\beta_1' t}) \qquad t \to +\infty,$$

$$\Phi(t) = 0(e^{\beta_2' t}) \qquad t \to -\infty.$$

It follows that the bilateral Laplace transform

$$\int_{-\infty}^{\infty} e^{-st}\Phi(t)\,dt = \frac{1}{\Omega(s)}$$

converges absolutely in the strip $\beta_1 < \text{Rl}\, s < \beta_2$. Since

$$\int_{-\infty}^{\infty} \Phi(t)\,dt = 1$$

the function $\Omega(s)$ is analytic in some circle (of radius R) about $s = 0$. We set

$$\Omega(s) = \sum_{k=0}^{\infty} \omega_k s^k.$$

The relation defining $\Omega(s)$ can be put in the form

$$\int_{-\infty}^{\infty} e^{st}\Phi(x - t)\Omega(s)\,dt = e^{sx}.$$

Let $p_{\epsilon,n}(x) = x[x + \epsilon] \cdots [x + (n - 1)\epsilon]$. We have

$$\int_{-\infty}^{\infty} \Phi(x - t) \left[p_{\epsilon,n}\left(\frac{\partial}{\partial s}\right) e^{st}\Omega(s) \right]_{s=0} dt = \left[p_{\epsilon,n}\left(\frac{\partial}{\partial s}\right) e^{sx} \right]_{s=0} = p_{\epsilon,n}(x).$$

If

$$P_{\epsilon,n}(t) = \left[p_{\epsilon,n}\left(\frac{\partial}{\partial s}\right) e^{st}\Omega(s) \right]_{s=0}$$

then $P_{\epsilon,n}(t)$ is a real polynomial in t of degree n. The formula above may be rewritten in the form

$$\int_{-\infty}^{\infty} P_{\epsilon,n}(t)\Phi(x-t)\,dt = p_{\epsilon,n}(x).$$

By Lemma 4.1

$$\mathscr{V}[P_{\epsilon,n}(t)] \geqq n.$$

This implies that $P_{\epsilon,n}(t)$ has only real zeros. Let

$$P_n(t) = \left[\left(\frac{\partial}{\partial s}\right)^n e^{st}\Omega(s)\right]_{s=0}.$$

We have

$$\lim_{\epsilon\to 0+} P_{\epsilon,n}(t) = P_n(t).$$

By a theorem of A. Hurwitz, E. C. Titchmarsh [1939; 119], $P_n(t)$ has only real zeros. If we set

$$Q_n(s) = \left(\frac{s}{n}\right)^n P_n\left(\frac{n}{s}\right)$$

then $Q_n(s)$ has only real zeros. An easy computation gives

$$Q_n(s) = \sum_{k=0}^{n} s^k \omega_k \frac{n!}{(n-k)!n^k}$$

from which it follows that

$$\lim_{n\to\infty} Q_n(s) = \Omega(s)$$

uniformly in every circle about $s = 0$ of radius less than R. By Theorem 3.3 of Chapter III $\Omega(s)$ is of the form

$$\Omega(s) = e^{-c's^2+bs}\prod_k \left(1-\frac{s}{a_k}\right) e^{s/a_k}$$

where a_k, b, c' are real constants and

$$c' \geqq 0, \qquad \sum_k a_k^{-2} < \infty.$$

Now by the convolution theorem

$$\int_{-\infty}^{\infty} \Phi(t)e^{-st}\,dt = e^{s^2/4}\int_{-\infty}^{\infty} \varphi(t)e^{-st}\,dt$$

at least for Rl $s = 0$. Thus

$$\int_{-\infty}^{\infty} \varphi(t)e^{-st}\,dt = e^{-s^2/4}/\Omega(s) \qquad \text{Rl } s = 0. \tag{1}$$

Since $\varphi(t) \in L^1(-\infty, \infty)$ the function $e^{-s^2/4}/\Omega(s)$ must be bounded on the line $\mathrm{Rl}\, s = 0$. This is possible only if $c = c' - \frac{1}{4} \geqq 0$. If

$$E(s) = e^{-cs^2+bs} \prod_k \left(1 - \frac{s}{a_k}\right) e^{s/a_k},$$

$$G(t) = (2\pi i)^{-1} \int_{-i\infty}^{i\infty} [E(s)]^{-1} e^{st}\, ds,$$

then

$$\int_{-\infty}^{\infty} G(t) e^{-st}\, dt = e^{-s^2/4}/\Omega(s) \qquad \alpha_1 < \mathrm{Rl}\, s < \alpha_2. \tag{2}$$

It follows from (1) and (2) that $\varphi(t) = G(t)$ almost everywhere.

5. THE CHANGES OF SIGN OF $G^{(n)}(t)$

5.1. If $E(s)$ is of the form

$$E(s) = e^{bs} \prod_{k=1}^{N} \left(1 - \frac{s}{a_k}\right) e^{s/a_k}$$

then we say that the corresponding kernel $G(t)$ is of degree N. If $E(s)$ is not of the above form the degree of $G(t)$ is said to be infinite. Let $G(t)$ be of degree N then $G(t)$ has $N - 2$ continuous derivatives (see Chapter II); the $N - 1$ *st* derivative is continuous except at $t_0 = b + \sum_1^N a_k^{-1}$, where $G^{(N-1)}(t_0+)$ and $G^{(N-1)}(t_0-)$ both exist. Let us set $G^{(N-1)}(t_0) = \frac{1}{2}[G^{(N-1)}(t_0+) + G^{(N-1)}(t_0-)]$. With this agreement $G(t)$ of degree N has (for N finite or infinite) $N - 1$ derivatives.

THEOREM 5.1. *If $G(t)$ defined by* 1.1 (1) *is of degree N then $(d/dt)^n G(t)$ has exactly n changes of sign for $n = 0, 1, \cdots, N - 1$.*

Applying Lemma 3.1b to $G^{(n)}(x)$ we have

$$G^{(n)}(x) = \lim_{\epsilon \to 0+} \int_{-\infty}^{\infty} G^{(n)}(x - t)\, [\pi^{-1/2} \epsilon^{-1} e^{-(t/\epsilon)^2}]\, dt.$$

After n integrations by parts this becomes

$$G^{(n)}(x) = \lim_{\epsilon \to 0+} \int_{-\infty}^{\infty} G(x - t)\, \{\epsilon^{-1} \pi^{-1/2} (d/dt)^n e^{-(t/\epsilon)^2}\}\, dt.$$

The function $(d/dt)^n e^{-(t/\epsilon)^2}$ has exactly n changes of sign. Since $G(t)$ is variation diminishing

$$\int_{-\infty}^{\infty} G(x - t)\, \{\epsilon^{-1} \pi^{-1/2} (d/dt)^n e^{-(t/\epsilon)^2}\}\, dt$$

has, for each $\epsilon > 0$, at most n changes of sign. Applying Lemma 2.1b we see that $G^{(n)}(x)$ has at most n changes of sign.

Since $\lim_{x\to+\infty} |G(x)| = \lim_{x\to-\infty} |G(x)| = 0$, $G^{(1)}(t)$ has at least one change of sign (by Rolle's theorem). This together with $\lim_{x\to+\infty} |G^{(1)}(x)| = \lim_{x\to-\infty} |G^{(1)}(x)| = 0$ implies that $G^{(2)}(x)$ has at least two changes of sign, etc. Thus $G^{(n)}(x)$ has at least n changes of sign.

This result for $n = 0, 1, 2$ shows that the graph of $G(t)$ is bell-shaped.

5.2. It is possible to give an alternative proof of Theorem 5.1 which is of some interest.

We begin with the case in which $G(t)$ is of finite degree N. We may suppose that $N > 1$. Let $t_o = b + \sum_1^N a_k^{-1}$. By Theorem 8.2 of Chapter II $G(t)$ can be written in the form

$$\begin{aligned} G(t) &= Q_1(t) \qquad & t \leqq t_o, \\ &= Q_2(t) \qquad & t \geqq t_o, \end{aligned}$$

where $Q_1(t)$ and $Q_2(t)$ are exponential polynomials of degrees d_1 and d_2, respectively, and

$$d_1 = -1 + \sum_{a_k>0} 1 \qquad d_2 = -1 + \sum_{a_k<0} 1.$$

If $G^{(k)}(t)$ had more than k changes of sign then $G^{(N-1)}(t)$ would, by successive applications of Rolle's theorem, have more than $N - 1$ changes of sign. We shall show that this is not possible. We must distinguish three cases: (i) $d_1 \geqq 0$, $d_2 \geqq 0$; (ii) $d_1 \geqq 0$, $d_2 = -1$; (iii) $d_1 = -1$, $d_2 \geqq 0$.

(i) $G^{(N-1)}(t)$ has not more than d_1 zeros for $t < t_o$, not more than d_2 zeros for $t > t_o$, and not more than 1 zero at $t = t_o$. Thus $G^{(N-1)}(t)$ has at most $d_1 + d_2 + 1 = N - 1$ zeros.

(ii) Here $G(t) = 0$ for $t \geqq t_o$. $G^{(N-1)}(t)$ has no changes of sign at $t = t_o$, and has at most $d_1 = N - 1$ zeros for $t < t_o$.

(iii) Here $G(t) = 0$ for $t \leqq t_o$, etc. . . .

If $G(t)$ is of infinite degree then by §§ 4.1, 8.6 of Chapter III there exists a sequence of kernels of finite degree $G_r(t)$ $r = 1, 2, \cdots$, such that

$$\lim_{r\to\infty} G_r^{(k)}(t) = G^{(k)}(t).$$

It follows from Lemma 2.1b that $G^{(k)}(t)$ has at most k changes of sign.

5.3. If a continuous function $f(t)$ has m changes of sign in $(-\infty, \infty)$ then there exist points $t_1, t_2, \cdots, t_m$ such that $f(t)$ is of alternate sign in the successive intervals $(-\infty, t_1)$, (t_1, t_2), $\cdots$, (t_{m-1}, t_m), $(t_m, +\infty)$. The points $t_1, \cdots, t_m$ are zeros of $f(t)$; however they need not be of any definite order and indeed (since $f(t)$ can vanish in an interval) they need not be uniquely determined.

THEOREM 5.3. *If $G(t)$ defined by* 1.1 (1) *is of degree N then the points associated with the changes of sign of $G^{(k)}(t)$ $(k = 1, \cdots, N - 3)$ are simple zeros of $G^{(k)}(t)$.*

We recall that x is a simple zero of $f(t)$ if $f(x) = 0$ while $f'(x)$ exists and is not zero. Consider the case $k = 1$. Let t_1 be associated with the one change of sign of $G'(t)$. If our theorem were not true then t_1 would be

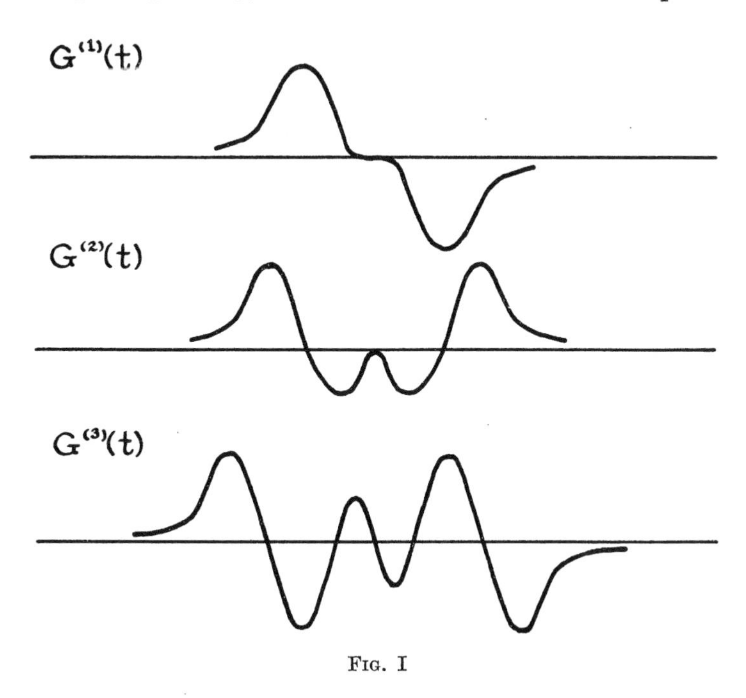

FIG. I

a zero of $G'(t)$ of order greater than 1, or $G'(t)$ might conceivably vanish in an interval containing t_1. It is quite easy to see by drawing pictures that this would constrain $G^{(3)}(t)$ to have 5 or more changes of sign, which however we know to be impossible. Similarly if the changes of sign of $G^{(k)}(t)$ are not effected by simple zeros $G^{(k+2)}(t)$ would have at least $k + 4$ changes of sign, etc.

6. INTERSECTION PROPERTIES

6.1. Let

$$E_1(s) = e^{-c's^2+b's}\prod_k \left(1 - \frac{s}{a'_k}\right)e^{s/a'_k},$$

$$E_2(s) = e^{-c''s^2+b''s}\prod_k \left(1 - \frac{s}{a''_k}\right)e^{s/a''_k},$$

where c', c'', b', b'', a'_k, a''_k, are real and $c' \geqq 0$, $c'' \geqq 0$, $\sum_k (a'_k)^{-2} < \infty$, $\sum_k (a''_k)^{-2} < \infty$, and let

$$H_1(t) = (2\pi i)^{-1} \int_{-i\infty}^{i\infty} [E_1(s)]^{-1} e^{st}\, ds,\ H_2(t) = (2\pi i)^{-1} \int_{-i\infty}^{i\infty} [E_2(s)]^{-1} e^{st}\, ds.$$

If $E(s)$ and $G(t)$ are defined by 1.1 (1) then the relations

$$E(s) = E_1(s)E_2(s) \tag{1}$$

and

$$G(t) = H_1(t) * H_2(t) \tag{2}$$

are equivalent. It is convenient and natural to express these relations by writing

$$H_1(t) \mid G(t), \qquad H_2(t) \mid G(t). \tag{3}$$

THEOREM 6.1a. *If*

1. *$G(t)$ is defined by* 1.1 (1)
2. $H_1(t) \mid G(t)$

then for any a, $-\infty < a < \infty$, $G(t) - aH_1(t)$ has at most two changes of sign.

Let us set $\Delta(t) = 0$ for $|t| \geqq 1$, $\Delta(t) = 1 - |t|$ for $|t| < 1$. It is easy to verify that

$$H_1(x) = \lim_{h \to 0+} \int_{-\infty}^{\infty} H_1(x - t)\, \{h^{-1}\Delta(t/h)\}\, dt.$$

If $H_2(t)$ is defined as above then

$$G(x) = \int_{-\infty}^{\infty} H_1(x - t) H_2(t)\, dt.$$

Thus

$$G(x) - aH_1(x) = \lim_{h \to 0+} \int_{-\infty}^{\infty} H_1(x - t)\, \{H_2(t) - ah^{-1}\Delta(t/h)\}\, dt.$$

For all sufficiently small h the function

$$H_2(t) - ah^{-1}\Delta(t/h)$$

has at most two changes of sign, and therefore since H_1 is variation diminishing the function

$$\int_{-\infty}^{\infty} H_1(x - t)\, \{H_2(t) - ah^{-1}\Delta(t/h)\}\, dt$$

has (for small h) at most two changes of sign. Appealing to Lemma 2.1b our theorem is established.

Theorem 6.1b. *If*

1. $G(t)$ *is defined by* 1.1 (1),

2. $H_1(t) \mid G(t)$,

then $H_1(t)/G(t)$ *has at most two changes of trend.*

If $G(t) = 0$, then $H_1(t)/G(t)$ must be defined (possibly as $+\infty$) by continuity considerations. It is easy to see by drawing pictures that if $H_1(t)/G(t)$ had more than two changes of trend, then there would exist a constant a such that $G(t) - aH_1(t)$ would have at least three changes of sign, and this is not possible.

7. GENERATION OF TOTALLY POSITIVE FUNCTIONS

7.1. The remaining sections of this chapter are devoted to the study of totally positive functions. We begin by showing that $G(t)$ defined by 1.1 (1) is totally positive.

Lemma 7.1a. Let $g(t)$ be defined as 2.1 (1). If a is a non-zero real constant then $|a| g(at)$ is a totally positive frequency function.

It is evident that if $\varphi(t)$ is a totally positive frequency function and if a is a real non-zero constant then $|a|\varphi(at)$ is a totally positive frequency function too. Thus it is sufficient to consider the case $a = 1$. If

$$x_1 < x_2 < \cdots < x_n, \qquad t_1 < t_2 < \cdots < t_n, \tag{1}$$

then $D_n = \det [g(x_i - t_j)]_n$ is of the form $\exp [\sum_1^n x_i - \sum_1^n t_j]E_n$ where

$$E_n = \begin{vmatrix} \frac{1}{2} & 1 & 1 \cdots 1 & 1 \\ 0 & 1 & 1 \cdots 1 & 1 \\ 0 & 0 & \frac{1}{2} \cdots 1 & 1 \\ \cdot & & & \\ \cdot & & & \\ \cdot & & & \\ 0 & 0 & 0 \cdots 0 & 1 \end{vmatrix},$$

the $i - j$th entry being 0, $\frac{1}{2}$ or 1 as $x_i - t_j$ is positive, zero, or negative. It is easy to see that $E_n = 0$ unless

$$x_1 \leqq t_1 \leqq x_2 \leqq \cdots \leqq x_n \leqq t_n \tag{2}$$

in which case

$$E_n = (\tfrac{1}{2})^\nu,$$

ν being the number of equalities in (2).

LEMMA 7.1b. If $\varphi(t),\ \psi(t) \in L^1(-\infty, \infty)$ and if $\chi(t) = \varphi(t) * \psi(t)$, then

$$\det [\chi(x_i - y_j)]_n = \frac{1}{n!} \int_{-\infty}^{\infty} \cdots \int_{-\infty}^{\infty} \det [\varphi(x_i - t_k)]_n \det [\psi(t_k - y_j)]_n \, dt_1 \cdots dt_n.$$

By the multiplication theorem for determinants we have

$$\det [\varphi(x_i - t_k)]_n \det [\psi(t_k - y_j)]_n = \det [\sum_{k=1}^{n} \varphi(x_i - t_k)\psi(t_k - y_j)]_n.$$

We recall that

$$\begin{vmatrix} a_{11} \cdots a_{1j-1} & a'_{1j} + a''_{1j} & a_{1j+1} \cdots a_{1n} \\ a_{21} \cdots a_{2j-1} & a'_{2j} + a''_{2j} & a_{2j+1} \cdots a_{2n} \\ \cdot & & \\ \cdot & & \\ \cdot & & \\ a_{n1} \cdots a_{nj-1} & a'_{nj} + a''_{nj} & a_{nj+1} \cdots a_{nn} \end{vmatrix}$$

is equal to

$$\begin{vmatrix} a_{11} \cdots a_{1j-1} & a'_{1j} & a_{1j+1} \cdots a_{1n} \\ a_{21} \cdots a_{2j-1} & a'_{2j} & a_{2j+1} \cdots a_{2n} \\ \cdot & & \\ \cdot & & \\ \cdot & & \\ a_{n1} \cdots a_{nj-1} & a'_{nj} & a_{nj+1} \cdots a_{nn} \end{vmatrix} + \begin{vmatrix} a_{11} \cdots a_{1j-1} & a''_{1j} & a_{1j+1} \cdots a_{1n} \\ a_{21} \cdots a_{2j-1} & a''_{2j} & a_{2j+1} \cdots a_{2n} \\ \cdot & & \\ \cdot & & \\ \cdot & & \\ a_{n1} \cdots a_{nj-1} & a''_{nj} & a_{nj+1} \cdots a_{nn} \end{vmatrix}.$$

Let N be the class of functions $k(j)$ from $1, \cdots, n$ to $1, \cdots, n$. Repeated application of the above identity gives

$$\det [\sum_{k=1}^{n} \varphi(x_i - t_k)\psi(t_k - y_j)]_n = \sum_{k(j) \in N} \det [\varphi(x_i - t_{k(j)})\psi(t_{k(j)} - y_j)]_n.$$

Let N' be the subclass of N consisting of those functions $k(j)$ such that $k(j_1) \neq k(j_2)$ if $j_1 \neq j_2$, and let $N'' = N - N'$. If $k(j) \in N''$ then two (or more) rows of the determinant

$$\det [\varphi(x_i - t_{k(j)})\psi(t_{k(j)} - y_j)]_n$$

are proportional, so that it is zero. If $k(j) \in N'$ then

$$\int_{-\infty}^{\infty} \cdots \int_{-\infty}^{\infty} \det [\varphi(x_i - t_{k(j)})\psi(t_{k(j)} - y_j)]_n \, dt_1 \cdots dt_n = \det [\chi(x_i - y_j)]_n.$$

The class N' contains $n!$ functions $k(j)$. Our lemma is a consequence of these results.

LEMMA 7.1c. If $\{\phi_r(t)\}_{r=1}^{R}$ are totally positive frequency functions and if $\varphi = \varphi_1 * \varphi_2 \cdots * \varphi_R$ then φ is a totally positive frequency function.

It is enough to prove this for $\varphi = \varphi_1 * \varphi_2$, the convolution of 2 functions, since it then can be extended by induction to the convolution of any finite number of functions. By Lemma 7.1b we have

$$(3)\quad \det[\varphi(x_i - t_j)]_n = \frac{1}{n!}\int_{-\infty}^{\infty}\cdots\int_{-\infty}^{\infty} \det[\varphi_1(x_i - u_k)]_n \det[\varphi_2(u_k - t_j)]_n \, du_1 \cdots du_n.$$

Let $x_1 < x_2 < \cdots < x_n$, $t_1 < t_2 < \cdots < t_n$. If any two u_k's are equal then $\det[\varphi_1(x_i - u_k)]_n$ and $\det[\varphi_2(u_k - t_j)]_n$ are both zero. If the u_k's are distinct then $\det[\varphi_1(x_i - u_k)]_n$ and $\det[\varphi_2(u_k - t_j)]_n$ are both non-negative if a permutation of even order is needed to arrange $u_1, \cdots, u_k$ in increasing order, and are both non-positive if a permutation of odd order is needed. Thus the integrand on the right side of equation (3) is non-negative. Our lemma follows.

THEOREM 7.1. *If $G(t)$ is defined by* 1.1 (1) *then $G(t)$ is a totally positive frequency function.*

If $G(t)$ is a finite kernel then our conclusion follows from Lemmas 7.1a and 7.1c. If $G(t)$ is not a finite kernel then we can choose a sequence $G_r(t)$ of finite kernels such that

$$\lim_{r\to\infty} G_r(t) = G(t) \qquad -\infty < t < \infty.$$

See § 8, Chapter III. This implies that

$$\det[G(x_i - t_j)]_n = \lim_{r\to\infty} \det[G_r(x_i - t_j)]_n.$$

If the inequalities (1) are satisfied then $\det[G_r(x_i - t_j)]_n \geqq 0$, $r = 1, 2, \cdots$, and thus $\det[G(x_i - t_j)]_n \geqq 0$.

8. MATRIX TRANSFORMATIONS

8.1. In order to establish the converse of Theorem 7.1 we must show that every totally positive frequency function is also variation diminishing. This necessitates a preliminary study of matrix transformations and it is to this that the present section is devoted. The results which we shall prove are due to Schoenberg [1930], Motzkin [1936], and Schoenberg and Whitney [1951].

DEFINITION 8.1a. A real matrix $A = [a(i,j)]$, $(i = 1, \cdots, m,$ $j = 1, \cdots, n)$, is said to be variation diminishing if

$$y(i) = \sum_{j=1}^{n} a(i,j)x(j) \qquad (i = 1, \cdots, m)$$

implies that $\mathscr{V}[y(1), \cdots, y(m)] \leqq \mathscr{V}[x(1), \cdots, x(n)]$.

Consider the following matrix operations:

(i) Multiplication of a row or column by a non-negative constant;
(ii) Addition of a row or column to an adjacent row or column;
(iii) Omission of a row or column.

It may be verified that these operations applied to a variation diminishing matrix yield a variation diminishing matrix. The unit $n \times n$ matrix is variation diminishing. By repeated application of the operations i and ii, we may, for example, deduce that the $n \times n$ matrices

$$\left|\begin{matrix} 1 & 0 & 0\cdots 0 \\ 1 & 1 & 0\cdots 0 \\ & \cdot & \\ & \cdot & \\ & \cdot & \\ 1 & 1 & 1\cdots 1 \end{matrix}\right|, \qquad \left|\begin{matrix} 1 & 0 & 0\cdots 0 & \\ 1 & 1 & 0\cdots 0 & \\ \cdot & & & \\ \cdot & & & \\ \cdot & & & \\ C_0^{n-1} & C_1^{n-1} & C_2^{n-1} & \cdots\ C_{n-1}^{n-1} \end{matrix}\right|, \tag{1}$$

are variation diminishing. Here $C_j^n = n!/[(n-j)!j!]$.

Let us put

$$A\begin{pmatrix} i_1, \cdots, i_k \\ j_1, \cdots, j_k \end{pmatrix} = \left|\begin{matrix} a(i_1, j_1) & a(i_1, j_2) \cdots a(i_1, j_k) \\ a(i_2, j_1) & a(i_2, j_2) \cdots a(i_2, j_k) \\ \cdot & \\ \cdot & \\ \cdot & \\ a(i_k, j_1) & a(i_k, j_2) \cdots a(i_k, j_k) \end{matrix}\right|$$

where $1 \leqq i_1 < i_2 < \cdots < i_k \leqq m$, $1 \leqq j_1 < j_2 < \cdots < j_k \leqq n$. $A\begin{pmatrix} i_1, \cdots, i_k \\ j_1, \cdots, j_k \end{pmatrix}$ is the minor of A formed from the rows $i_1, \cdots, i_k$ and the columns $j_1, \cdots, j_k$. The order of this minor is k. The rank $r(A)$ of the matrix A is the largest value of k for which there exists a minor of order k which is not zero.

DEFINITION 8.1b. A real matrix $A = [a(i, j)]$, $(i = 1, \cdots, m,$ $j = 1, \cdots, n)$, is said to be minor definite if minors of the same order $R < r(A)$ have the same sign and if minors of the same order $R = r(A)$ have the same sign if they belong to the same combination of columns.

The operations i and ii and iii applied to a minor definite matrix yield a minor definite matrix. The unit matrix is evidently minor definite; thus the matrices (1) are minor definite. These examples suggest that the properties of being variation diminishing and of being minor definite may be equivalent. This is indeed the case. It is convenient to first prove an intermediate result.

Given a sequence $x(1), \cdots, x(n)$ of real numbers we define $\mathscr{H}[x(1), \cdots, x(n)]$ as the number of $x(i)$'s which are not zero.

DEFINITION 8.1c. A real matrix $A = [a(i, j)]$, $(i = 1, \cdots, m$, $j = 1, \cdots, n)$, is said to be variation limiting if

$$y(i) = \sum_{j=1}^{n} a(i, j)x(j) \qquad (i = 1, \cdots, m)$$

implies that

$$\mathscr{V}[y(1), \cdots, y(m)] \leqq \mathscr{H}[x(1), \cdots, x(n)] - 1.$$

DEFINITION 8.1d. A real matrix $A = [a(i, j)]$, $(i = 1, \cdots, m$, $j = 1, \cdots, n)$ is said to be column definite if minors of order ν belonging to the same combination of columns are of the same sign.

THEOREM 8.1. *A real matrix $A = [a(i, j)]$ is variation limiting if and only if it is column definite.*

Let us prove that if A is column definite then it is variation limiting. We first note that it is no restriction to prove this under the assumption that no $x_k = 0$, or, what is equivalent, that $\mathscr{H}[x(1), \cdots, x(n)] = n$. If some $x_k = 0$ we set

$$x'(j) = \begin{cases} x(j) & j < k \\ x(j+1) & j \geqq k \end{cases} \qquad (j = 1, \cdots, n-1)$$
$$a'(i, j) = \begin{cases} a(i, j) & j < k \\ a(i, j+1) & j \geqq k . \end{cases}$$

We have

$$y(i) = \sum_{j=1}^{n-1} a'(i, j)x'(j).$$

The matrix $[a'(i, j)]$ is column definite, and if we can show that $\mathscr{V}[y(1), \cdots, y(m)] \leqq \mathscr{H}[x'(1), \cdots, x'(n-1)] - 1$ then since $\mathscr{H}[x(1), \cdots, x(n)] = \mathscr{H}[x'(1), \cdots, x'(n-1)]$ it will follow that $\mathscr{V}[y(1), \cdots, y(m)] \leqq \mathscr{H}[x(1), \cdots, x(n)] - 1$. Repeating this argument, it follows that we may assume no x_k is zero.

We next assert that if A is column definite then $\mathscr{V}[y(1), \cdots, y(m)] \leqq r(A) - 1$. We proceed by induction on $r(A)$. The result is clearly true if $r(A) = 0$ or 1. Suppose that it has been established for $r(A) = 0, 1, \cdots, R-1$; we will show that it is true for R. We may suppose that $m > R$ since if $m = R$ then $\mathscr{V}[y(1), \cdots, y(m)] \leqq R - 1$ trivially. It is enough to show that if $1 \leqq i_0 < i_1 < \cdots < i_R \leqq m$ then $\mathscr{V}[y(i_0), \cdots, y(i_R)] \leqq R - 1$. If A' is the submatrix of A consisting of the rows $i_0, i_1, \cdots, i_R$, then A' is column definite. There are two cases: $r(A') < R$ and $r(A') = R$. If the first case is realized then $\mathscr{V}[y(i_0), \cdots, y(i_R)] \leqq r(A') - 1$ by our induction assumption and $r(A') - 1 < R - 1$.

Suppose the second case holds and let $j_1, \cdots, j_R$ be a selection of columns of A'. We set

$$\alpha(k) = A\begin{pmatrix} i_0, \cdots, i_{k-1}, & i_{k+1}, \cdots, i_R \\ j_1, \cdots, & \cdots, j_R \end{pmatrix} \qquad (k = 0, \cdots, R).$$

Since A' is minor definite no two α_k's are of opposite sign, and because $r(A') = R$ we can so choose $j_1, \cdots, j_R$ that not all $\alpha(k)$'s are zero. We have

$$\alpha(0)y(i_0) - \alpha(1)y(i_1) + \cdots + (-1)^R\alpha(R)y(i_R) = 0. \tag{2}$$

It is easily seen that $\mathscr{V}[y(i_0), \cdots, y(i_R)] = R$ is not compatible with the identity (2) so that in this case too, $\mathscr{V}[y(i_0), \cdots, y(i_R)] \leqq R - 1$.

We can complete our proof. We suppose, as we may, that $\mathscr{H}[x(1), \cdots, x(n)] = n$. We have $\mathscr{V}[y(1), \cdots, y(m)] \leqq r(A) - 1$. Since $r(A) \leqq n$, it follows that $\mathscr{V}[y(1), \cdots, y(m)] \leqq \mathscr{H}[x(1), \cdots, x(n)] - 1$.

We shall now show that if $A = [a(i, j)], (i = 1, \cdots, m; j = 1, \cdots, n)$ is variation limiting then it is column definite. Consider the non-zero minors of order R contained in the columns $j_1, \cdots, j_R$. Two such minors α' and α'' are said to be adjacent if they contain $R - 1$ rows in common. We begin by proving that adjacent non-zero minors are of the same sign. Both α' and α'' are contained in $R + 1$ rows $i_0, \cdots, i_R$. Let

$$\alpha(k) = A\begin{pmatrix} i_0, \cdots, i_{k-1}, & i_{k+1}, \cdots, i_R \\ j_1, \cdots, & \cdots, j_R \end{pmatrix};$$

then α' and α'' are among $\alpha(0), \cdots, \alpha(R)$. A necessary and sufficient condition that $y(i_0), \cdots, y(i_R)$ be represented in the form

$$y(i_\mu) = \sum_{\nu=1}^{R} a(i_\mu, j_\nu)x(j_\nu) \qquad \mu = 0, \cdots, R,$$

is that

$$\alpha(0)y(i_0) - \alpha(1)y(i_1) + \cdots + (-1)^R\alpha(R)y(i_R) = 0. \tag{3}$$

If $\operatorname{sgn}(\alpha' \cdot \alpha'') = -1$, then it is possible to choose $y(i_0), \cdots, y(i_R)$ none of which are zero, which alternate in sign, and which satisfy the relation (3). We would then have $\mathscr{V}[y(i_0), \cdots, y(i_R)] = R > \mathscr{H}[x(j_1), \cdots, x(j_R)] - 1$, a contradiction.

Let α' and α'' be two non-zero minors of order R contained in the columns $j_1, \cdots, j_R$. We define the distance $\Delta(\alpha', \alpha'')$ between α' and α'' to be the number of rows in which they both lie less R. If $\Delta(\alpha', \alpha'') = 0$ then $\alpha' = \alpha''$; if $\Delta(\alpha', \alpha'') = 1$, then α' and α'' are adjacent. We assert that if $\Delta(\alpha', \alpha'') = d$, $d > 1$, then there exists a non-zero minor α (contained in the same combination of columns) such that $\Delta(\alpha, \alpha') = 1$,

$\Delta(\alpha, \alpha'') = d - 1$. Let α' lie in the rows $i_1, \cdots, i_R$, where the rows $i_1, i_2, \cdots, i_{R-d}$ also belong to α''. Let i represent the vector whose components are $[a(i, j_1), \cdots, a(i, j_R)]$. Since α' is not zero the vectors $i_1, \cdots, i_R$ are linearly independent. Choose i such that the row i is contained in α'' but not in α' and consider the set of vectors $i, i_1, \cdots, i_R$. Proceeding from left to right we strike out the first vector which is linearly dependent on its predecessors. We obtain in this fashion a linearly independent set of vectors and from the corresponding set of rows a non-zero minor α. It is clear that $\Delta(\alpha, \alpha') = 1$, $\Delta(\alpha, \alpha'') = d - 1$.

We may now show that if α' and α'' are non-zero minors of order R contained in the columns $j_1, \cdots, j_R$ then α' and α'' are of the same sign. We proceed by induction on $\Delta(\alpha', \alpha'')$. We know our result to be true if $\Delta(\alpha', \alpha'') = 1$; suppose that it has been established for $\Delta(\alpha', \alpha'') = 1, \cdots, d - 1$; we will prove it for $\Delta(\alpha', \alpha'') = d$. We have shown that there exists a non-zero minor α such that $\Delta(\alpha, \alpha'') = d - 1$, $\Delta(\alpha, \alpha') = 1$. By our induction assumption both α and α' and α and α'' have the same sign and so α' and α'' have the same sign.

8.2. We now proceed to the demonstration of the conjecture made at the beginning of the preceding section.

THEOREM 8.2. *A real matrix* $A = [a(i, j)]$, $(i = 1, \cdots, m;$ $j = 1, \cdots, n)$, *is variation diminishing if and only if it is minor definite.*

Let us prove that if A is minor definite then it is variation diminishing. We assert that it is sufficient to demonstrate this under the assumption that $\mathscr{V}[x(1), \cdots, x(n)] = \mathscr{H}[x(1), \cdots, x(n)] - 1$. If any $x(k) = 0$ we set

$$x'(j) = \begin{cases} x(j) & j < k \\ x(j+1) & j \geqq k \end{cases} \qquad (j = 1, \cdots, n-1)$$

$$a'(i, j) = \begin{cases} a(i, j) & j < k \\ a(i, j+1) & j \geqq k \end{cases} \qquad (j = 1, \cdots, n-1)$$

then we have

$$y(i) = \sum_{j=1}^{n-1} a'(i, j) x'(j).$$

The matrix $[a'(i, j)]$ is minor definite and if we can show $\mathscr{V}[y(1), \cdots, y(m)] \leqq \mathscr{V}[x'(1), \cdots, x'(n-1)]$ then since $\mathscr{V}[x'(1), \cdots, x'(n-1)] = \mathscr{V}[x(1), \cdots, x(n)]$ it will follow that $\mathscr{V}[y(1), \cdots, y(m)] \leqq \mathscr{V}[x(1), \cdots, x(n)]$. Repeated application of this algorithm shows that we may assume that no $x(k)$ is zero. Next we may suppose that $x(k)$ and $x(k+1)$ are of opposite sign. If this is not true then there exists $\lambda > 0$ such that $x(k+1) = \lambda x(k)$. If we set

$$x'(j) = \begin{cases} x(j) & j < k \\ & \\ x(j+1) & j \geq k \end{cases} \qquad (j = 1, \cdots, n-1)$$

$$a'(i, j) = \begin{cases} a(i, j) & j < k \\ a(i, k) + \lambda a(i, k+1) & j = k \\ a(i, j+1) & j > k \end{cases} \qquad (j = 1, \cdots, n-1)$$

then we have

$$y(i) = \sum_{j=1}^{n-1} a'(i, j) x'(j).$$

The matrix $[a'(i, j)]$ is minor definite, and if we can show that $\mathscr{V}[y(1), \cdots, y(m)] \leqq \mathscr{V}[x'(1), \cdots, x'(n-1)]$ then since $\mathscr{V}[x(1), \cdots, x(n)] = \mathscr{V}[x'(1), \cdots, x'(n-1)]$ it will follow that $\mathscr{V}[y(1), \cdots, y(m)] \leqq \mathscr{V}[x(1), \cdots, x(n)]$. This shows that we may assume $\mathscr{V}[x(1), \cdots, x(n)] = \mathscr{H}[x(1), \cdots, x(n)] - 1$. Since a minor definite matrix is a fortiori column definite we have from Theorem 8.1 that $\mathscr{V}[y(1), \cdots, y(m)] \leqq \mathscr{H}[x(1), \cdots, x(n)] - 1$, and our desired result follows.

Let us prove that if A is variation diminishing then it is minor definite. We know from Theorem 8.1 that minors of order R are of the same sign if they lie in the same combination of columns. It remains to show that if $R < r(A)$ then minors of order R are of the same sign without restriction as to position. Let α' and α'' be two non-zero minors of order R. Let us suppose that α' and α'' lie in the columns $j_0, \cdots, j_R$ and the rows $i_0, \cdots, i_R$ and that if

$$\alpha = A \begin{pmatrix} i_0, \cdots, i_R \\ j_0, \cdots, j_R \end{pmatrix},$$

then $\alpha \neq 0$. We set

$$\alpha(\mu, \nu) = A \begin{pmatrix} i_0, \cdots, i_{\mu-1}, i_{\mu+1}, \cdots, i_R \\ j_0, \cdots, j_{\nu-1}, j_{\nu+1}, \cdots, j_R \end{pmatrix};$$

then $\alpha' = \alpha(\mu_1, \nu_1)$ and $\alpha'' = \alpha(\mu_2, \nu_2)$. We know that $\alpha(\mu, \nu)$, $(\mu = 0, \cdots, R)$, are of the same sign. Thus we may suppose that $\nu_1 \neq \nu_2$. Let us solve the system of equations

$$(-1)^{\mu} = \sum_{\nu=0}^{R} a(i_{\mu}, j_{\nu}) x(\nu) \qquad \mu = 0, \cdots, R.$$

We have

$$\alpha x(\nu_1) = (-1)^{\nu_1} \sum_{\mu=0}^{R} \alpha(\mu, \nu_1)$$

$$\alpha x(\nu_2) = (-1)^{\nu_2} \sum_{\mu=0}^{R} \alpha(\mu, \nu_2)$$

and thus $\operatorname{sgn}[x(\nu_1)x(\nu_2)] = (-1)^{\nu_1-\nu_2} \operatorname{sgn}[\alpha' \cdot \alpha'']$. Since A is variation diminishing we must have $\mathscr{V}[x(0), \cdots, x(R)] = R$ and hence $\operatorname{sgn}[x(\nu_1)x(\nu_2)] = (-1)^{\nu_1-\nu_2}$. This shows that α' and α'' have the same sign.

Now suppose that α' and α'' are contained in $R + 1$ columns $J = (j_0, \cdots, j_R)$ of A and that these columns considered as vectors are linearly independent. We can then choose $R + 1$ rows $I = (i_0, \cdots, i_R)$ so that the minor

$$A\begin{pmatrix} i_0, & \cdots, & i_R \\ j_0, & \cdots, & j_R \end{pmatrix} \neq 0.$$

Let us choose a non-zero minor β' of order R contained in the same combination of columns as α' and in the rows I. Similarly let β'' of order R be contained in the same combination of columns as α'' and in the rows I. We know that α' and β' and also α'' and β'' are of the same sign. We have just proved that β' and β'' are of the same sign; it follows that α' and α'' are of the same sign.

Let $\delta(\alpha', \alpha'')$ be defined as the number of columns containing α' and α'' less R. Suppose first that $\delta(\alpha', \alpha'') = 1$. Let J' be the set of columns of A containing α' and J'' the set of columns of A containing α''. If $J' \cup J''$ are linearly independent then we have shown above that α' and α'' are of the same sign. If $J' \cup J''$ are linearly dependent then there is a column j such that $J' \cup j$ and $J'' \cup j$ are linearly independent. Let us choose a non-zero minor β contained in $(J' \cap J'') \cup j$. We have shown above that α' and β and also α'' and β are of the same sign ; it follows that α' and α'' are of the same sign. Our proof may now be completed by an induction on $\delta(\alpha', \alpha'')$.

9. TOTALLY POSITIVE FREQUENCY FUNCTIONS

9.1. We can now establish the converse of Theorem 7.1.

THEOREM 9.1a. *If $\varphi(t)$ is a totally positive frequency function then $\varphi(t)$ is variation diminishing.*

Let us first prove our theorem under the additional assumption that $\varphi(t) \in B \cdot C(-\infty, \infty)$. We must show that if $g(t) \in B \cdot C(-\infty, \infty)$ and if

$$f(x) = \int_{-\infty}^{\infty} \varphi(x - t)g(t)\, dt \tag{1}$$

then $\mathscr{V}[f(x)] \leqq \mathscr{V}[g(t)]$. If $\mathscr{V}[f(x)] \geqq m$ then we can choose $m+1$ points $x_0 < x_1 < \cdots < x_m$ so that $\mathscr{V}[f(x_0), \cdots, f(x_m)] = m$. We can then take T so large that if

$$f_1(x_i) = \int_{-T}^{T} \varphi(x_i - t)g(t)\,dt \qquad (i = 0, \cdots, m)$$

then

$$\operatorname{sgn}[f(x_i)f_1(x_i)] = 1 \qquad (i = 0, \cdots, m)$$

and thus $\mathscr{V}[f_1(x_0), \cdots, f_1(x_m)] = m$. Let us now subdivide the interval $[-T, T]$ into n equal parts each of length $h = 2T/n$, $t_j = -T + jh$ being the points of subdivision. Consider the Riemann sums

$$f_2(x_i) = h\sum_{j=1}^{n} \varphi(x_i - t_j)g(t_j) \qquad (i = 0, \cdots, m).$$

If n is large enough then

$$\operatorname{sgn}[f_1(x_i)f_2(x_i)] = 1 \qquad (i = 0, \cdots, m)$$

and thus $\mathscr{V}[f_2(x_0), \cdots, f_2(x_m)] = m$. By Theorem 8.2

$$m = \mathscr{V}[f_2(x_0), \cdots, f_2(x_m)] \leqq \mathscr{V}[g(t_1), \cdots, g(t_n)] \leqq \mathscr{V}[g(t)],$$

and our theorem is established for $\varphi(t) \in B \cdot C(-\infty, \infty)$.

In the general case we set $\varphi_\epsilon(t) = \varphi(t) * \pi^{-1/2}\epsilon^{-1}e^{-(t/\epsilon)^2}$. Since $\pi^{-1/2}\epsilon^{-1}e^{-(t/\epsilon)^2}$ is totally positive so is $\varphi_\epsilon(t)$, and $\varphi_\epsilon(t)$ is bounded and continuous for each $\epsilon > 0$. Thus if $g(t) \in B \cdot C(-\infty, \infty)$ and if

$$f_\epsilon(x) = \int_{-\infty}^{\infty} \varphi_\epsilon(x - t)g(t)\,dt$$

then $\mathscr{V}[f_\epsilon(x)] \leqq \mathscr{V}[g(t)]$. We assert that if $f(x)$ is given by (1) then

$$f(x) = \lim_{\epsilon \to 0+} \int_{-\infty}^{\infty} \varphi_\epsilon(x - t)g(t)\,dt.$$

It is easy to see that

$$\int_{-\infty}^{\infty} \varphi_\epsilon(x - t)g(t)\,dt = \int_{-\infty}^{\infty} \varphi(x - t)g_\epsilon(t)\,dt$$

where $g_\epsilon(t) = g(t) * \pi^{-1/2}\epsilon^{-1}e^{-(t/\epsilon)^2}$. We have, making use of Lemma 3.1b,

$$\lim_{\epsilon \to 0+} g_\epsilon(t)\varphi(x - t) = g(t)\varphi(x - t) \qquad (-\infty < t < \infty);$$

$$|g_\epsilon(t)\,\varphi(x - t)| \leqq ||g(t)||_\infty\,|\varphi(x - t)| \quad (-\infty < t < \infty).$$

Applying Lebesgue's limit theorem we obtain our desired result. By Lemma 2.1b $\mathscr{V}[f(x)] \leqq \varliminf_{\epsilon \to 0+} \mathscr{V}[f_\epsilon(x)] \leqq \mathscr{V}[g(t)]$ and our theorem is proved.

We require the following result which was first proved by Sierpinski [1920]. If A is any measurable set on the line then we denote by mA the Lebesgue measure of A.

THEOREM 9.1b. *If*

1. $-\infty < f(x) \leqq +\infty \qquad a < x < b,$
2. *$f(x)$ is measurable, and finite almost everywhere,*
3. $2f\left(\frac{x_1 + x_2}{2}\right) \leqq f(x_1) + f(x_2) \qquad a < x_1, \quad x_2 < b,$

then $f(x)$ is continuous for $a < x < b$.

We first assert that f is everywhere finite. Suppose to the contrary that $f(x) = \infty$. Let $(x - c, x + c)$ be a neighborhood of x contained in (a, b). If x' is any point of this neighborhood and if $x'' = 2x - x'$ then x'' also belongs to this neighborhood. Since

$$2f(x) \leqq f(x') + f(x'')$$

one of the relations

$$f(x') = \infty, \qquad f(x'') = \infty,$$

must hold. It follows that if E is the set of points where $f = \infty$ then $mE \geqq c$; but this is contrary to assumption 2. Thus $f(x)$ must be finite.

We next assert that if f is not continuous at x then

$$\overline{\lim_{y \to x}} f(y) = +\infty.$$

f discontinuous at x implies that there exists $\delta > 0$ such that in every neighborhood of x there is a point y for which $|f(x) - f(y)| > \delta$. Let $(x - c, x + c)$ be a neighborhood of x contained in (a, b). There exists a point x' in this neighborhood such that either $f(x') \geqq f(x) + \delta$ or $f(x') \leqq f(x) - \delta$. If the second inequality holds let $x'' = 2x - x'$. We have $x'' \in (x - c, x + c)$ and using assumption 3 it is easily seen that $f(x'') \geqq f(x) + \delta$. Thus $(x - c, x + c)$ contains a point x_1 such that

$$f(x_1) \geqq f(x) + \delta.$$

Let x_1' be a point in $(x - \frac{1}{2}c, x + \frac{1}{2}c)$ for which $f(x_1') \geqq f(x) + \delta$, and let $x_2 = 2x_1' - x$. It is clear that $x_2 \in (x - c, x + c)$; further

$$2f(x_1') \leqq f(x) + f(x_2),$$
$$f(x_2) \geqq f(x) + 2[f(x_1') - f(x)],$$
$$f(x_2) \geqq f(x) + 2\delta.$$

Repeating this argument n times we find that there is a point $x_n \in (x - c, x + c)$ such that

$$f(x_n) \geqq f(x) + 2^{n-1}\delta.$$

We have thus verified our assertion.

If f is not continuous at x then, as we have seen, there exist a sequence of points ξ_k contained in $(x - \frac{1}{2}c, x + \frac{1}{2}c)$ such that $f(\xi_k) \geqq k$. If x' is any point of the interval $(x - \frac{1}{2}c \cdot x + \frac{1}{2}c)$ and if $x'' = 2\xi_k - x'$ then x' and x'' belong to $(x - c, x + c)$. Since

$$2f(\xi_k) \leqq f(x') + f(x'')$$

at least one of the inequalities

$$f(x') \geqq k, \qquad f(x'') \geqq k$$

must hold. It follows that if E_k is the set of $y \in (x - c, x + c)$ for which $f(y) > k$ then $mE_k > \frac{1}{2}c$. Since this is incompatible with assumption 2, f must be continuous at x.

THEOREM 9.1c. *If $\varphi(t)$ is a totally positive frequency function, then there exists a kernel $G(t)$ of the form* 1.1 (1) *such that if $G(t)$ is of order greater than* 1 *then $\varphi(t) = G(t)$ everywhere, and if $G(t)$ is of order* 1 *then $\varphi(t) = G(t)$ except possibly at the discontinuity ξ of G where we have only that*

$$0 \leqq \varphi(\xi) \leqq \overline{\lim_{t \to \xi}}\, G(t).$$

It follows from the fact that $\varphi(t)$ is totally positive that

$$\varphi(t) \geqq 0, \qquad \varphi(t)^2 \geqq \varphi(t + h)\varphi(t - h).$$

The second of these relations may also be written in the form

$$2\left[-\log \varphi\left(\frac{t_1 + t_2}{2}\right)\right] \leqq [-\log \varphi(t_1)] + [-\log \varphi(t_2)].$$

Theorems 4.1 and 9.1a show that there exists a kernel $G(t)$ such that $\varphi(t) = G(t)$ almost everywhere. There are three cases:

(i) $G(t) > 0 \qquad (-\infty < t < \infty)$;

(ii) $G(t) > 0 \qquad (-\infty < t < \xi), \qquad G(t) = 0 \qquad (\xi < t < \infty)$;

(iii) $G(t) > 0 \qquad (\xi < t < \infty), \qquad G(t) = 0 \qquad (-\infty < t < \xi)$.

Let us consider case (iii). We first assert that $\varphi(t) = 0$ for $-\infty < t < \xi$. If this were not true then there would exist $\eta < \xi$ such that $\varphi(\eta) > 0$. For all x and $\eta < x$ we have $\varphi(x)^2 \geqq \varphi(\eta)\varphi(2x - \eta)$. From the fact that $\varphi(2x - \eta) > 0$ almost everywhere for $\frac{1}{2}(\xi + \eta) < x$, we would have $\varphi(x) > 0$ almost everywhere for $\frac{1}{2}(\xi + \eta) < x < \xi$ which is impossible. Thus $\varphi(t)$ and $G(t)$ coincide for $-\infty < t < \xi$. Theorem 9.1b applied to $-\log \varphi(t)$ shows that $-\log \varphi(t)$ is continuous for $\xi < t < \infty$.

Since both $\varphi(t)$ and $G(t)$ are continuous for $\xi < t < \infty$ and since they are equal almost everywhere, they coincide for $\xi < t < \infty$. It remains to determine $\varphi(\xi)$. We have $\varphi(\xi) \leqq \varphi(\xi + h)^2/\varphi(\xi + 2h) = G(\xi + h)^2/G(\xi + 2h)$. If $G(x)$ is not of order 1 then $G(\xi) = 0$ and $G(t)$ is continuous and non-decreasing in a neighborhood of ξ. Thus $G(\xi + h)^2/G(\xi + 2h) \leqq G(\xi + h) = o(1)$ as $h \to 0+$, and $\varphi(\xi) = G(\xi) = 0$. If $G(x)$ is of order 1 then $G(x) = ae^{-a(x-\xi)} (x > \xi)$ for some $a > 0$ so that $G(\xi + h)^2/G(\xi + 2h) \to a$ as $h \to 0+$, and thus $0 \leqq \varphi(\xi) \leqq a$. Cases (i) and (ii) may be dealt with similarly.

10. SUMMARY

10.1. In this chapter we have characterized (in several ways) the class of kernels $G(t)$ and in this process we have established many qualitative properties. These qualitative properties in conjunction with the quantitative estimates of Chapter V constitute weapons more than sufficient to enable us to attack many problems concerning the transforms $G * \varphi$.

CHAPTER V

Asymptotic Behaviour of Kernels

1. INTRODUCTION

1.1. This chapter is devoted to a detailed study of the behaviour of $G(t)$ as $t \to \pm\infty$ In the case that $G(t)$ is a finite kernel,

$$\int_{-\infty}^{\infty} G(t)e^{-st}\,dt = 1/e^{bs} \prod_1^n \left(1 - \frac{s}{a_k}\right)e^{s/a_k},$$

$G(t)$ can be expressed as one exponential polynomial for $t \geq b + \sum_1^n a_k^{-1}$ and as another for $t \leq b + \sum_1^n a_k^{-1}$, and from these expressions the behaviour of $G(t)$ at $\pm\infty$ can be easily read off. In the general case where

$$\int_{-\infty}^{\infty} G(t)e^{-st}\,dt = 1/E(s), \quad E(s) = e^{-cs^2+bs} \prod_k \left(1 - \frac{s}{a_k}\right)e^{s/a_k}, \qquad E(s) \in E,$$

the behaviour of $G(t)$ at $\pm\infty$ may be more difficult to determine.

2. ASYMPTOTIC ESTIMATES

2.1. Let us recall that

$$\alpha_1 = \max_{a_k<0}\,[a_k, -\infty], \qquad \alpha_2 = \min_{a_k>0}\,[a_k, +\infty].$$

We further define $\mu_1 + 1$ as the multiplicity of $s - \alpha_1$ as a zero of $E(s)$ and $\mu_2 + 1$ as the multiplicity of $s - \alpha_2$ as a zero of $E(s)$. As we shall see the behaviour of $G(t)$ at $+\infty(-\infty)$ is largely determined by $\alpha_1(\alpha_2)$.

THEOREM 2.1. *If $G(t)$ is non-finite then:*

A. $\alpha_1 > -\infty$ *implies*

$$G^{(n)}(t) = [e^{\alpha_1 t}p(t)]^{(n)} + O(e^{kt}) \qquad t \to +\infty,$$

$n = 0, 1, \cdots,$ *where k is a real number $<\alpha_1$ and $p(t)$ is a real polynomial of degree μ_1;*

B. $\alpha_1 = -\infty$ *implies*

$$G^{(n)}(t) = O(e^{kt}) \qquad t \to +\infty,$$

$n = 0, 1, \cdots$, *where* k *is an arbitrary (negative) real number;*

C. $\alpha_2 < +\infty$ *implies*

$$G^{(n)}(t) = [e^{\alpha_2 t}q(t)]^{(n)} + O(e^{kt}) \qquad t \to -\infty,$$

$n = 0, 1, \cdots$, *where* k *is a real number* $> \alpha_2$ *and* $q(t)$ *is a real polynomial of degree* μ_2;

D. $\alpha_2 = +\infty$ *implies*

$$G^{(n)}(t) = O(e^{kt}) \qquad t \to -\infty,$$

$n = 0, 1, \cdots$, *where* k *is an arbitrary (positive) real number.*

We have

$$\left(\frac{d}{dt}\right)^n G(t) = \frac{1}{2\pi i}\int_{-i\infty}^{i\infty} \frac{s^n e^{st}}{E(s)}\, ds \qquad (-\infty < t < \infty).$$

Suppose first that $\alpha_1 > -\infty$. Choose a real number k smaller than α_1 but greater than any other negative zero of $E(s)$. Let $T > 0$ and define D as the rectangular contour with vertices at $\pm iT$, $k \pm iT$. Integrals about D proceed counterclockwise. The integral

$$\frac{1}{2\pi i}\int_D \frac{s^n e^{st}}{E(s)}\, ds = \left(\frac{d}{dt}\right)^n \frac{1}{2\pi i}\int_D \frac{e^{st}}{E(s)}\, ds$$

is by Cauchy's residue theorem equal to the nth derivative of the residue of $e^{st}/E(s)$ at $s = \alpha_1$. Let $E(s) = (s - \alpha_1)^{\mu_1+1}E_1(s)$. The expansion

$$1/E_1(s) = \sum_{j=0}^{\infty} A(j)(s - \alpha_1)^j \qquad A(0) \neq 0,$$

is valid in some circle about α_1. Thus

$$e^{st}/E(s) = (s - \alpha_1)^{-\mu_1-1}\Big[\sum_{j=0}^{\infty} A(j)(s - \alpha_1)^j\Big]\Big[e^{\alpha_1 t}\sum_{j=0}^{\infty}\frac{t^j}{j!}(s - \alpha_1)^j\Big].$$

It follows that the residue of $e^{st}/E(s)$ at α_1 is

$$e^{\alpha_1 t}\sum_{j=0}^{\mu_1}\frac{A(\mu_1 - j)}{j!}\, t^j = e^{\alpha_1 t}p(t).$$

We write

$$\int_D \frac{s^n e^{st}}{E(s)}\, ds = \int_{-iT}^{iT} + \int_{k+iT}^{k-iT} + \int_{iT}^{k+iT} + \int_{k-iT}^{-iT},$$

$$= I_1 + I_2 + I_3 + I_4.$$

By Theorem 5.3 of Chapter III we have

$$\lim_{T\to\infty} I_3 = \lim_{T\to\infty} I_4 = 0.$$

Hence

$$\left(\frac{d}{dt}\right)^n G(t) = \left(\frac{d}{dt}\right)^n (e^{\alpha_1 t}p(t)) + \frac{1}{2\pi i}\int_{k-i\infty}^{k+i\infty}\frac{s^n e^{st}}{E(s)}\, ds.$$

A second application of Theorem 5.3 of III gives

$$\frac{1}{2\pi i}\int_{k-i\infty}^{k+i\infty}\frac{s^n e^{st}}{E(s)}\,ds = O(e^{kt}) \qquad t\to+\infty.$$

We have thus established conclusion A. To prove B our argument proceeds in the same fashion except that k is chosen as any negative number. Here, of course, there is no residue. Similar arguments serve to establish conclusions C and D.

2.2. The argument used in the proof of Theorem 2.1 can be made to yield additional information. Let $0 > A_1 > A_2 > \cdots$ be the distinct negative zeros of $E(s)$, the multiplicity of A_i being $M_i + 1$. If in the demonstration of Theorem 2.1 we choose k, $A_{m+1} < k < A_m$, then

$$G(t) = \sum_{i=1}^{m} P_i(t)e^{A_i t} + O(e^{kt}) \qquad t\to+\infty \tag{1}$$

where $P_i(t)$ is an (ordinary) polynomial in t of degree M_i for $i = 1, \cdots, m$. Similarly, if $0 < B_1 < B_2 < \cdots$ are the distinct positive zeros of $E(s)$, the multiplicity of B_i being $N_i + 1$ and if $B_m < k < B_{m+1}$ we may show that

$$G(t) = \sum_{i=1}^{m} Q_i(t)e^{B_i t} + O(e^{kt}) \qquad t\to-\infty \tag{2}$$

where $Q_i(t)$ is a polynomial of degree N_i. The formulas (1) and (2) are analogous to the expression, obtained in § 8 of Chapter II, for a finite kernel as two exponential polynomials joined together.

2.3. Let us consider several examples referring for the necessary information to the table of Chapter III. If $E(s) = \cos \pi s$ and if $G(t)$ is the corresponding kernel then we find from Theorem 2.1 that

$$G(t) = \frac{1}{\pi}e^{-t/2} + O(e^{kt}) \qquad t\to+\infty$$

where k is any fixed number greater than $-3/2$. Since $G(t) = \dfrac{1}{2\pi}\operatorname{sech}\left(\dfrac{t}{2}\right)$ this relation may be verified directly.

If $E(s) = 1/\Gamma(1-s)$ and if $G(t)$ is the corresponding kernel then, applying Theorem 2.1 again, we find that

$$G(t) = O(e^{kt}) \qquad t\to+\infty$$

for every (negative) k. Since $G(t) = e^t e^{-e^t}$ it is evident that this relation is correct.

Let $E(s) = \prod_1^\infty \left(1 - \dfrac{s}{k^2}\right)$ and let $G(t)$ be the corresponding kernel.

Theorem 2.1 implies that

$$G(t) = O(e^{kt}) \qquad t \to +\infty$$

for every (negative) k; however, we know that $G(t) = 0$ for $t \geqq 0$.

Let $E(s) = e^{-cs^2}$ and let $G(t)$ be the corresponding kernel. Theorem 2.1 shows that

$$G(t) = O(e^{kt}) \qquad t \to +\infty$$

for every k. Here $G(t) = \dfrac{1}{\sqrt{4\pi c}} e^{-t^2/4c}$.

3. ASYMPTOTIC ESTIMATES CONTINUED

3.1. Theorem 2.1 fails in the cases $\alpha_1 = -\infty$ $(\alpha_2 = +\infty)$ to give precise information as to the behaviour of $G(t)$ as $t \to +\infty$ $(t \to -\infty)$. It tells us only that $G(t) = O(e^{kt})$, $t \to +\infty$ $(t \to -\infty)$ for every k, which leaves open a wide range of possibilities as to the actual behaviour of $G(t)$, and as we have seen, a very great range of behaviour does take place. In this section we shall obtain precise information for these cases. Let

$$\lambda(r) = 2cr + b + \sum_k \frac{r}{a_k(a_k + r)},$$

$$\sigma(r) = \left[2c + \sum_k \frac{1}{(a_k + r)^2}\right]^{1/2}, \tag{1}$$

$$\Lambda(r) = e^{-r\lambda(r)}/\sigma(r)E(-r).$$

THEOREM 3.1a. *If $G(t)$ is a non-finite kernel then:*

A. $\alpha_1 = -\infty$ *implies that*

$$G^{(n)}[\lambda(r)] \sim (2\pi)^{-1/2}(-r)^n\Lambda(r) \qquad r \to +\infty$$

for $n = 0, 1, \cdots$;

B. $\alpha_2 = +\infty$ *implies that*

$$G^{(n)}[\lambda(r)] \sim (2\pi)^{-1/2}(-r)^n\Lambda(r) \qquad r \to -\infty$$

for $n = 0, 1, \cdots$.

Let us consider first the case $\alpha_1 = -\infty$. We have

$$G^{(n)}(u) = \frac{1}{2\pi i}\int_{-i\infty}^{i\infty} \frac{s^n e^{su}}{E(s)}\, ds.$$

Making the change of variable $s = s' - r$ we obtain

$$G^{(n)}(u) = \frac{1}{2\pi i}\int_{r-i\infty}^{r+i\infty} (s' - r)^n e^{(s'-r)u}[E(s' - r)]^{-1}\, ds'.$$

It is evident from Theorem 5.3 of Chapter III that we can deform the path of integration back to the imaginary axis. Let us make the change of variable $s' = s/\sigma(r)$. We now have

$$G^{(n)}(u) = \frac{e^{-ru}(-r)^n}{\sigma(r)} \frac{1}{2\pi i} \int_{-i\infty}^{i\infty} \left(1 - \frac{s}{r\sigma(r)}\right)^n e^{su/\sigma(r)} \left[E\left(\frac{s}{\sigma(r)} - r\right)\right]^{-1} ds.$$

If $A_k(r) = \sigma(r)\,[a_k + r]$, $C(r) = c/\sigma(r)^2$,

$$E_r(s) = e^{-C(r)s^2} \prod_k \left[1 - \frac{s}{A_k(r)}\right] e^{s/A_k(r)},$$

then it may be verified that

$$E\left(\frac{s}{\sigma(r)} - r\right) = E(-r)E_r(s)e^{\lambda(r)s/\sigma(r)}.$$

See § 10.2 of III for a similar computation. Thus

$$G^{(n)}(u) = \frac{e^{-ru}(-r)^n}{E(-r)\sigma(r)} \frac{1}{2\pi i} \int_{-i\infty}^{i\infty} \left(1 - \frac{s}{r\sigma(r)}\right)^n E_r(s)^{-1} \exp\left(\frac{s}{\sigma(r)}[u - \lambda(r)]\right) ds. \tag{2}$$

If we set $u = \lambda(r)$ we find that

$$G^{(n)}(\lambda(r)) = (-r)^n \Lambda(r) I_r,$$

where

$$I_r = (2\pi i)^{-1} \int_{-i\infty}^{i\infty} \left[1 - \frac{s}{r\sigma(r)}\right]^n E_r(s)^{-1}\, ds.$$

We note that

$$2C(r) + \sum_k A_k^{-2}(r) = 1, \qquad 0 \leq r < \infty, \tag{3'}$$

$$\lim_{r \to +\infty} A_k(r) = \infty, \qquad k = 1, 2, \cdots. \tag{3''}$$

The first of these relations is obvious. The second follows from the fact that $r\sigma(r) \uparrow \infty$ as $r \to +\infty$. Equivalently one may show that $r^2\sigma(r)^2 \uparrow \infty$ as $r \to +\infty$. We have

$$r^2\sigma(r)^2 = 2cr^2 + \sum_k r^2(a_k + r)^{-2}.$$

Clearly $r^2\sigma(r)^2 \in \uparrow$, and since either $c > 0$, or there are infinitely many a_k's, or both, $r^2\sigma(r)^2 \uparrow \infty$ as $r \to +\infty$.

We assert that

$$\lim_{r \to +\infty} E_r(s) = e^{-s^2/2} \qquad s = \sigma + i\tau, \tag{4}$$

for all s, uniformly for s in any compact set. By a familiar inequality, see E. C. Titchmarsh [1939; 246],

$$|\log [(1-s)e^{s+(s^2/2)}]| \leq 2|s|^3 \qquad |s| \leq 1/2.$$

Here that branch of the logarithm is taken for which $\log 1 = 0$. It follows that if $|s| \leq \frac{1}{2}A(r)$, where

$$A(r) = \sigma(r)\,(r + \alpha_2)$$

then

$$|\log E_r(s)e^{s^2/2}| \leq 2\;|s|^3 \sum_k A_k^{-3}(r)$$

since, by equation (3′),

$$\log E_r(s)e^{s^2/2} = \sum_k \log \left\{\left[1 - \frac{s}{A_k(r)}\right] \exp\,[sA_k(r)^{-1} + \frac{1}{2}\,s^2A_k(r)^{-2}]\right\}.$$

We have

$$\lim_{r\to\infty} \sum_k A_k(r)^{-3} = 0;$$

for

$$\sum_k A_k^{-3}(r) \leq A(r)^{-1} \sum_k A_k(r)^{-2} \leqq A(r)^{-1},$$

and $A(r) \to \infty$ as $r \to +\infty$. Thus equation (4) is verified. By Theorems 2.3 and 8.6 of III

$$\lim_{r\to+\infty} (2\pi i)^{-1} \int_{-i\infty}^{i\infty} \frac{s^m}{E_r(s)}\,ds = \left(\frac{d}{du}\right)^m [(2\pi)^{-1/2}e^{-u^2/2}]_{u=0}.$$

Since

$$I_r = (2\pi i)^{-1} \sum_{m=0}^{n} (-r\sigma(r))^{-m} \binom{n}{m} \int_{-i\infty}^{i\infty} s^m E_r(s)^{-1}\,ds.$$

and since $r\sigma(r) \to \infty$ as $r \to +\infty$ it follows that

$$\lim_{r\to\infty} I_r = (2\pi)^{-1/2}.$$

The proof of our theorem is thus completed in the case $\alpha_1 = -\infty$. The case $\alpha_2 = +\infty$ can be dealt with similarly.

The reader will recognize that the mechanism of the latter part of the proof is exactly that of the central limit theorem of probability.

If $\alpha_1 = -\infty$ and if $c = 0$, $\sum_{k=1}^{\infty} a_k^{-1} < \infty$, then we know that $G(t) = 0$ for $t \geq b + \sum_1^{\infty} a_k^{-1}$. In this case conclusion A tells us the behaviour of $G(t)$ as $t \to b + \sum_1^{\infty} a_k^{-1}-$. Similarly if $\alpha_2 = +\infty$ and if $c = 0$, $\sum_1^{\infty} a_k^{-1} > -\infty$, then $G(t) = 0$ for $t \leq b + \sum_1^{\infty} a_k^{-1}$. In this case conclusion B tells us the behaviour of $G(t)$ as $t \to b + \sum_1^{\infty} a_k^{-1}+$.

COROLLARY 3.1. If $G(t)$ is non-finite and if $c = 0$, $\sum_1^\infty a_k^{-1}$ is finite, then:

A. $\alpha_1 = -\infty$ implies that

$$G(t) > 0 \qquad -\infty < t < b + \sum_1^\infty a_k^{-1},$$

$$G(t) = 0 \qquad b + \sum_1^\infty a_k^{-1} \leq t < \infty;$$

B. $\alpha_2 = +\infty$ implies that

$$G(t) > 0 \qquad b + \sum_1^\infty a_k^{-1} < t < \infty,$$

$$G(t) = 0 \qquad -\infty < t < b + \sum_1^\infty a_k^{-1}.$$

This follows from Theorem 3.1 and the fact that $G'(t)$ has one and $G(t)$ no change of sign for $-\infty < t < \infty$. See § 5 of IV.

3.2. Let us consider as an example the kernel $e^t e^{-e^t}$ corresponding to $E(s) = 1/\Gamma(1-s)$. We have, see Titchmarsh [1939; 150],

$$\begin{aligned} \lambda(r) &= -\gamma + \sum_1^\infty \frac{r}{k(k+r)}, \\ &= \frac{d}{dr} \log \Gamma(1+r), \\ &= \log r + \frac{1}{2r} + O(r^{-2}) \qquad (r \to +\infty), \end{aligned}$$

so that

$$[e^t e^{-e^t}]_{t=\lambda(r)} \sim r e^{-r-1/2} \qquad (r \to +\infty).$$

Further

$$\begin{aligned} \sigma(r)^2 &= \sum_1^\infty \frac{1}{(k+r)^2}, \\ &= \left(\frac{d}{dr}\right)^2 \log \Gamma(1+r), \\ &= \frac{1}{r} + O(r^{-2}) \qquad (r \to +\infty). \end{aligned}$$

Finally

$$\begin{aligned} E(-r) &= 1/\Gamma(1+r), \\ &\sim (2\pi)^{-1/2} e^r r^{-r-1/2}. \end{aligned}$$

It is now easily seen that

$$(2\pi)^{-1/2} e^{-r\lambda(r)} [\sigma(r) E(-r)]^{-1} \sim r e^{-r-1/2} \qquad (r \to +\infty),$$

thus verifying, in this special case, Theorem 3.1.

3.3. Let $G(t) = e^{-\chi(t)}$; then Theorem 3.1 implies that

$$(1) \qquad \begin{aligned} \chi'[\lambda(r)] &\sim r \qquad r \to +\infty \quad (\alpha_1 = -\infty), \\ \chi'[\lambda(r)] &\sim r \qquad r \to -\infty \quad (\alpha_2 = +\infty). \end{aligned}$$

These relations are simpler in that $\Lambda(r)$ no longer appears. The above formulas can be rewritten in a more advantageous form. If $c = 0$ and $\sum_k \frac{1}{a_k}$ is finite, we define M as a function of t by the equation

$$(2) \qquad t = \sum_k \frac{1}{M + a_k} \qquad \begin{array}{l} 0 < t < \infty \text{ if } \alpha_1 = -\infty \\ -\infty < t < 0 \text{ if } \alpha_2 = +\infty. \end{array}$$

If $c > 0$ or if $\sum_k \frac{1}{a_k}$ is infinite we define L as a function of t by the equation

$$(3) \qquad t = 2cL + b + \sum_k \frac{L}{a_k(a_k + L)} \qquad \begin{array}{l} b < t < \infty \text{ if } \alpha_1 = -\infty \\ -\infty < t < b \text{ if } \alpha_2 = +\infty. \end{array}$$

THEOREM 3.3. *If $G(t)$ is non-finite, then:*

A. $\alpha_1 = -\infty$, $c = 0$ *and* $\sum_k \frac{1}{a_k}$ *finite implies that*

$$\chi'[b + \sum_1^\infty \frac{1}{a_k} - t] \sim M(t) \qquad t \to 0+;$$

B. $\alpha_1 = -\infty$, $c > 0$ *or* $\sum_k \frac{1}{a_k}$ *infinite implies that*

$$\chi'[t] \sim L(t) \qquad t \to +\infty;$$

C. $\alpha_2 = +\infty$, $c = 0$ *and* $\sum_k \frac{1}{a_k}$ *finite implies that*

$$\chi'\left[b + \sum_1^\infty \frac{1}{a_k} - t\right] \sim M(t) \qquad t \to 0-;$$

D. $\alpha_2 = +\infty$, $c > 0$ *or* $\sum_k \frac{1}{a_k}$ *infinite implies that*

$$\chi'[t] \sim L(t) \qquad t \to -\infty.$$

Let us prove A, since B, C, and D are entirely similar. We suppose that $\alpha_1 = -\infty$, $c = 0$ and $\sum_1^\infty a_k^{-1} < \infty$. Using (2) we have

$$\begin{aligned} b + \sum_1^\infty a_k^{-1} - t &= b + \sum_{k=1}^\infty \frac{M(t)}{a_k(a_k + M(t))}, \\ &= \lambda(M(t)). \end{aligned}$$

It follows from (1) that

$$\chi'[\lambda(M(t))] \sim M(t) \qquad M \to +\infty,$$

$$\chi'[b + \sum_1^\infty a_k^{-1} - t] \sim M(t) \qquad t \to 0+,$$

which is what we wished to show.

3.4. A result very similar to Theorem 3.3 may be obtained by a different argument. Actually it is the theorem of the present section which we shall require in subsequent chapters.

THEOREM 3.4. *If* $\chi(t)$ *is defined as above and* $L(t)$ *as in* 3.3 (3) *then:*

A. $\alpha_1 = -\infty$, $c > 0$ *or* $\sum_k a_k^{-1} = \infty$ *implies that*

$$\chi'(t) = L[t + o(1)] \qquad t \to +\infty;$$

B. $\alpha_2 = +\infty$, $c > 0$ *or* $\sum_k a_k^{-1} = -\infty$ *implies that*

$$\chi'(t) = L[t + o(1)] \qquad t \to -\infty.$$

Let us establish conclusion A. If in equation 3.1 (2) we set $n = 0$ and multiply through by e^{ru} we obtain

$$e^{ru-\chi(u)} = [E(-r)\sigma(r)]^{-1}H_r\left[\frac{u - \lambda(r)}{\sigma(r)}\right],$$

where

$$H_r(u) = (2\pi i)^{-1}\int_{-i\infty}^{i\infty}[E_r(s)]^{-1}e^{su}\,ds.$$

Differentiating with respect to u we have

$$[r - \chi'(u)]e^{ru-\chi(u)} = [E(-r)\sigma(r)^2]^{-1}H'_r\left[\frac{u - \lambda(r)}{\sigma(r)}\right]. \tag{1}$$

Let $z(r)$ be the zero of H'_r associated with its one change of sign; since $H''_r(z(r)) \neq 0$, $z(r)$ is uniquely determined; see § 5 of IV. By Theorems 2.3 and 8.6 of III if $0 \leq r_0 < \infty$

$$\lim_{r\to r_0} \| H_{r_0}^{(n)}(u) - H_r^{(n)}(u) \|_\infty = 0 \qquad n = 0, 1, \cdots \tag{2}$$

We assert that

$$\lim_{r\to r_0} z(r) = z(r_0), \tag{3}$$

that is, $z(r)$ is a continuous function of r. This is an immediate consequence of (2) and the fact that $H''(z(r_0)) \neq 0$. A second application of Theorems 2.3 and 8.6 of III yields

$$\lim_{r\to+\infty} \| H_\infty^{(n)}(u) - H_r^{(n)}(u) \|_\infty = 0 \tag{4}$$

where $H_\infty(u)$ is $(2\pi)^{-1/2}e^{-u^2/2}$. Since $z(\infty) = 0$, $H''_\infty(0) \neq 0$, it follows, as above, that

$$\lim_{r\to+\infty} z(r) = 0.$$

Setting

$$u = \lambda(r) + \sigma(r)z(r)$$

in (1) we obtain

$$\chi'[\lambda(r) + \sigma(r)z(r)] = r.$$

Since $z(r)$ is continuous every large positive value of t can be represented in the form

$$t = \lambda(r) + \sigma(r)z(r), \tag{5}$$

possibly in several ways. Let $r(t)$ be for each large value of t some one solution of (5); then

$$\chi'[t] = r(t).$$

Note that $r(t) \to +\infty$ as $t \to +\infty$. Now

$$t = \lambda(r(t)) + \sigma(r(t))z(r(t)),$$
$$r(t) = L[t - \sigma(r(t))z(r(t))],$$

and thus

$$\chi'(t) = L[t + o(1)] \qquad t \to +\infty,$$

as desired.

Theorems 3.3 and 3.4 are so similar that one would think it possible to deduce one from the other. This does not however seem to be the case.

3.5. In this section we shall consider in detail the asymptotic behaviour of the kernels $K_\alpha(t)$ corresponding to the products

$$E_\alpha(s) = \prod_{k=1}^{\infty}\left(1 - \frac{s}{k^\alpha}\right)e^{s/k^\alpha}.$$

The parameter α may be either in the range $\frac{1}{2} < \alpha < 1$ (class II) or $1 < \alpha < \infty$ (class III).

Let us first suppose that $\frac{1}{2} < \alpha < 1$. We consider the integral

$$I(r) = \frac{1}{2\pi i}\int_{c-i\infty}^{c+i\infty} \frac{\zeta(z\alpha)}{\sin \pi z}\, r^z\, dz$$

where $c < 2$, $c\alpha > 1$. Here $\zeta(z)$ is the Riemann zeta function. We require here various elementary properties of $\zeta(z)$: $\zeta(z)$ is analytic in the complex plane except at $z = 1$ where it has a pole of order 1 and residue 1; $\log^+ |\zeta(x + iy)| = o(|y|)$ as $y \to \pm\infty$ uniformly for x in any finite interval; $\zeta(z) = \sum_1^\infty n^{-z}$ the series converging absolutely, boundedly, and uniformly in every half plane $\mathrm{Rl}\, z \geq 1 + \epsilon$. These results may all be found in Titchmarsh [1951]. We suppose that $0 < r < \infty$. Thus $r^{x+iy} = O(1)$ as $y \to \pm\infty$ uniformly for x in any finite interval. Further $1/|\sin \pi(x+iy)| = O(e^{-\pi|y|})$ $y \to \pm\infty$ uniformly for x in any finite interval. Applying the Lebesgue limit theorem we find that

$$I(r) = \sum_{k=1}^{\infty} (2\pi i)^{-1} \int_{c-i\infty}^{c+i\infty} \frac{r^z}{\sin \pi z} \frac{dz}{k^{z\alpha}} = \sum_{k=1}^{\infty} I_k(r).$$

It is easy to see that

$$I_k(r) = (2\pi i)^{-1} \int_{(3/2)-i\infty}^{(3/2)+i\infty} \frac{e^{zv}}{\sin \pi z}\, dz;$$

here $v = \log r - \alpha \log k$. If we set $z = (3/2) + w$ we obtain

$$I_k(r) = -e^{3v/2}(2\pi i)^{-1} \int_{-i\infty}^{i\infty} \frac{e^{wv}}{\cos \pi w}\, dw.$$

By the results of § 9.5 of III

$$I_k(r) = -e^{3v/2}(2\pi)^{-1} \operatorname{sech} \frac{v}{2},$$

$$= -\frac{r}{\pi} \frac{r}{k^\alpha(k^\alpha + r)}.$$

Thus

$$I(r) = \sum_{k=1}^{\infty} -\frac{r}{\pi} \frac{r}{k^\alpha(k^\alpha + r)}.$$

Let us now deform the line of integration in the integral defining $I(r)$ from $\mathrm{Rl}\, z = c$ to $\mathrm{Rl}\, z = -n - \frac{1}{2}$. We obtain

$$I(r) = \frac{r^{1/\alpha}}{\alpha \sin \dfrac{\pi}{\alpha}} - \frac{\zeta(\alpha) r}{\pi} + \sum_{m=0}^{n} (-1)^m \frac{1}{\pi} \zeta(-m\alpha) r^{-m} + O(r^{-n-1/2}).$$

Combining these relations we find that

$$(1)\quad \sum_{k=1}^{\infty} \frac{r}{k^\alpha(k^\alpha + r)} \sim -\frac{\pi}{\alpha \sin \dfrac{\pi}{\alpha}} r^{\frac{1}{\alpha}-1} + \zeta(\alpha) + \sum_{m=0}^{\infty} (-1)^{m+1} \zeta(-m\alpha) r^{-m-1}$$

$$(\tfrac{1}{2} < \alpha < 1).$$

In the case $1 < \alpha < \infty$ we choose c so that $\alpha c > 1$, $c < 1$ and proceed as before. Here we obtain

$$(2)\quad \sum_{1}^{\infty} \frac{1}{(k^\alpha + r)} \sim \frac{\pi}{\alpha \sin \dfrac{\pi}{\alpha}} r^{\frac{1}{\alpha}-1} + \sum_{m=0}^{\infty} (-1)^m \zeta(-m\alpha) r^{-m-1} \quad (1 < \alpha < \infty).$$

If we define $L_\alpha(t)$ and $M_\alpha(t)$ by the equations

$$t = \sum_1^\infty \frac{L_\alpha}{k^\alpha(k^\alpha + L_\alpha)} \qquad (\tfrac{1}{2} < \alpha < 1),$$

$$t = \sum_1^\infty \frac{1}{(k^\alpha + M_\alpha)} \qquad (1 < \alpha < \infty),$$

the asymptotic developments (1) and (2) imply that

$$L_\alpha(t) \sim \left[-\frac{\alpha}{\pi} t \sin\left(\frac{\pi}{\alpha}\right)\right]^{\frac{\alpha}{1-\alpha}} \qquad t \to +\infty,$$

$$M_\alpha(t) \sim \left[\frac{\alpha}{\pi} t \sin\left(\frac{\pi}{\alpha}\right)\right]^{\frac{-\alpha}{\alpha-1}} \qquad t \to 0^+.$$

(More precise estimates could, of course, be obtained.) Applying Theorem 3.3, we have

$$\frac{d}{dt} \log K_\alpha(t) \sim -\left[-\frac{\alpha}{\pi} t \sin\left(\frac{\pi}{\alpha}\right)\right]^{\frac{\alpha}{1-\alpha}} \qquad (t \to +\infty,\ \tfrac{1}{2} < \alpha < 1),$$

$$\frac{d}{dt} \log K_\alpha(\zeta(\alpha) - t) \sim \left[\frac{\alpha}{\pi} t \sin\left(\frac{\pi}{\alpha}\right)\right]^{\frac{-\alpha}{\alpha-1}} \qquad (t \to 0^+,\ 1 < \alpha < \infty).$$

Integrating these relations we find that

$$\log K_\alpha(t) \sim -(1-\alpha)\left[-\frac{\alpha}{\pi} \sin\left(\frac{\pi}{\alpha}\right)\right]^{\frac{\alpha}{1-\alpha}} t^{\frac{1}{1-\alpha}} \qquad (t \to +\infty,\ \tfrac{1}{2} < \alpha < 1),$$

$$\log K_\alpha(\zeta(\alpha) - t) \sim -(\alpha - 1)\left[+\frac{\alpha}{\pi} \sin\left(\frac{\pi}{\alpha}\right)\right]^{\frac{-\alpha}{\alpha-1}} t^{\frac{-1}{\alpha-1}}$$

$$(t \to 0^+,\ 1 < \alpha < \infty).$$

4. SUMMARY

4.1. In the present chapter we have studied the asymptotic behaviour of $G(t)$ for large values of t. This behaviour varies widely according to the structure of $G(t)$ and this foreshadows the separation of kernels into classes, described in I, which will prove necessary in later chapters.

CHAPTER VI

Real Inversion Theory

1. INTRODUCTION

1.1. In the present chapter we shall make use of the results of Chapters IV and V in order to obtain in its sharpest form the inversion theory for the kernels $G(t)$,

$$(1) \qquad G(t) = (2\pi i)^{-1} \int_{-i\infty}^{i\infty} \left[e^{bs} \prod_k \left(1 - \frac{s}{a_k}\right) e^{s/a_k} \right]^{-1} e^{st}\, ds.$$

As one might anticipate from the results of Chapter V it is necessary to classify these kernels on the basis of the properties of the a_k's. See § 8 of Chapter I.

Definition 1.1. Let $G(t)$ be defined by (1). $G(t)$ is said to belong to class I if there are both positive and negative a_k's; to class II if there are only positive a_k's and $\sum_k a_k^{-1} = \infty$; and to class III if there are only positive a_k's and $\sum_k a_k^{-1} < \infty$.

Either $G(t)$ or $G(-t)$ belongs to one of these classes. It is necessary to treat separately the kernels belonging to class I, class II, and class III. However, although the details differ from class to class, the main outline is in every case the same.

2. SOME PRELIMINARY RESULTS

2.1. We develop here several elementary lemmas which it will be convenient to have at our disposal.

Lemma 2.1a. Let $\varphi(t)$ be continuous and $\alpha(t)$ of bounded variation on every finite subinterval of $a \leqq t < \infty$. If

1. $\displaystyle\int_a^\infty \varphi(t)\, d\alpha(t)$ converges (conditionally),

2. $\psi(x, t)/\varphi(t)$ is continuous and has no change of trend $a \leqq t < \infty$ for each $x \in I$,

3. $|\psi(x, t)/\varphi(t)| \leqq M$ for $a \leqq t < \infty$, $x \in I$,

then the integral

$$\int_a^\infty \psi(x, t)d\alpha(t)$$

converges uniformly for $x \in I$.

By the mean value theorem

$$\int_t^{t_2} \psi(x, t)d\alpha(t) = \int_{t_1}^{t_2} \frac{\psi(x, t)}{\varphi(t)} \varphi(t)d\alpha(t)$$

$$= \frac{\psi(x, t_1)}{\varphi(t_1)} \int_{t_1}^{\xi} \varphi(t)d\alpha(t) + \frac{\psi(x, t_2)}{\varphi(t_2)} \int_{\xi}^{t_2} \varphi(t)d\alpha(t),$$

where $t_1 < \xi < t_2$. It is evident from our assumptions that

$$\lim_{t_1, t_2 \to \infty} \int_{t_1}^{t_2} \psi(x, t)\, d\alpha(t) = 0$$

uniformly for $x \in I$, which is equivalent to our assertion.

LEMMA 2.1b. Let $\varphi(t)$ be continuous and $\alpha(t)$ of bounded variation in every finite interval of $a \leqq t < \infty$. If

1. $\int_a^\infty \varphi(t)\, d\alpha(t)$ converges (conditionally),

2. $\psi(t)/\varphi(t)$ is continuous and has only a finite number of changes of trend,

3. $|\psi(t)/\varphi(t)| \leqq M$ for $a \leqq t < \infty$,

then the integral

$$\int_a^\infty \psi(t)\, d\alpha(t)$$

converges.

It is only necessary to note that $\psi(t)/\varphi(t)$ is monotonic for t sufficiently large and then proceed as in the proof of Lemma 2.1a.

Lemma 2.1c. Let $\varphi(t)$ be continuous and $\alpha(t)$ of bounded variation in every finite subinterval of $a \leqq t < \infty$. If

1. $\varphi(t)$ is positive and monotonic,

2. $\int_a^\infty \varphi(t)d\alpha(t)$ converges,

then:

A. $\lim_{t\to\infty} \varphi(t) = 0$ implies that

$$\alpha(t) = o\left[\frac{1}{\varphi(t)}\right] \qquad t \to +\infty;$$

B. $0 < \lim_{t\to\infty} \varphi(t) < \infty$ implies that $\alpha(+\infty)$ exists and that

$$\alpha(+\infty) - \alpha(t) = o(1) \qquad t \to +\infty;$$

C. $\lim_{t\to\infty} \varphi(t) = \infty$ implies that $\alpha(+\infty)$ exists and that

$$\alpha(+\infty) - \alpha(t) = o\left[\frac{1}{\varphi(t)}\right] \qquad t \to +\infty.$$

Let us demonstrate only conclusion C; A and B may be established in a similar manner. By the mean value theorem

$$\alpha(t_2) - \alpha(t_1) = \int_{t_1}^{t_2} \frac{1}{\varphi(t)} \varphi(t)\, d\alpha(t),$$

$$\alpha(t_2) - \alpha(t_1) = \frac{1}{\varphi(t_1)} \int_{t_1}^{\xi} \varphi(t)\, d\alpha(t), \tag{1}$$

where $t_1 < \xi < t_2$. Thus if $t_2 > t_1$

$$\alpha(t_2) - \alpha(t_1) = o\left[\frac{1}{\varphi(t_1)}\right].$$

Since $1/\varphi(t) \to 0$ as $t \to \infty$ we have

$$\lim_{t_1, t_2 \to \infty} [\alpha(t_2) - \alpha(t_1)] = 0,$$

from which it follows that $\alpha(+\infty)$ exists. Just as we established (1) we may show that

$$\alpha(+\infty) - \alpha(t) = \frac{1}{\varphi(t)} \int_t^{\xi} \varphi(t)\, d\alpha(t)$$

where $t < \xi < \infty$. It follows that

$$\alpha(+\infty) - \alpha(t) = o\left[\frac{1}{\varphi(t)}\right] \qquad t \to +\infty,$$

as desired.

LEMMA 2.1d. If

1. $k(t)$ belongs to L on every finite interval,
2. $r(t) \in \downarrow, \qquad r(+\infty) = 0,$
3. $k(x_0 - t)r(t) \in L(-\infty, \infty)$ for some x_0,

then

$$\lim_{x\to+\infty} \int_{-\infty}^{\infty} k(x - t)r(t)\, dt = 0.$$

We have

$$\int_{-\infty}^{\infty} k(x-t)r(t)\,dt = \int_{-\infty}^{\infty} k(x_0-u)r(x-x_0+u)\,du,$$

$$= \int_{-\infty}^{\infty} [k(x_0-u)r(u)]\,[r(x-x_0+u)/r(u)]\,du \tag{2}$$

where we suppose $x \geqq x_0$. (We define 0/0 as 0.) We have

$$0 \leqq r(x-x_0+u)/r(u) \leqq 1, \qquad \lim_{x\to+\infty} r(x-x_0+u)/r(u) = 0.$$

Applying Lebesgue's limit theorem in (2) we obtain our desired result.

3. CONVERGENCE

3.1. We can now determine the convergence behaviour of the transform $\int_{-\infty}^{\infty} G(x-t)\,d\alpha(t)$.

THEOREM 3.1. *If*

1. $G(t) \in$ class I,
2. *$\alpha(t)$ is of bounded variation in every finite interval,*
3. $\int_{-\infty}^{\infty} G(x_0-t)\,d\alpha(t)$ *converges (conditionally),*

then

$$\int_{-\infty}^{\infty} G(x-t)\,d\alpha(t)$$

converges uniformly for x in any infinite interval; i.e. for

$$-\infty < x_1 \leqq x \leqq x_2 < \infty.$$

Let I be the interval $[x_1, x_2]$. By Theorem 3.1 of Chapter IV $G(x-t)/G(x_0-t)$ is, as a function of t, non-increasing for $x \leqq x_0$, and non-decreasing for $x \geqq x_0$. Thus $G(x-t)/G(x_0-t)$ has no change of trend for each $x \in I$. This together with Theorem 2.1 of Chapter V shows that $0 \leqq G(x-t)/G(x_0-t) \leqq M$ for $x \in I$, where

$$M = \max\,[e^{\alpha_2(x_2-x_0)}, e^{\alpha_1(x_1-x_0)}].$$

Appealing to Lemma 2.1a our theorem is proved.

As a consequence of this theorem we see that if $G(t) \in$ class I then $G \# \alpha$ may be referred to as either convergent or divergent. The reader will recognize here the familiar convergence behaviour of the Stieltjes transform.

3.2.

THEOREM 3.2. *If*

1. $G(t) \in$ class II,
2. $\alpha(t)$ *is of bounded variation in every finite interval,*
3. $\int_{-\infty}^{\infty} G(x_0 - t)\, d\alpha(t)$ *converges (conditionally),*

then

$$\int_{-\infty}^{\infty} G(x - t)\, d\alpha(t)$$

converges uniformly for x in any finite interval bounded on the left by x_0; i.e. for $x_0 \leqq x \leqq x_1 < \infty$.

Let I be the interval $[x_0, x_1]$. By Theorem 3.1 of Chapter IV $G(x - t)/G(x_0 - t)$ is non-decreasing as a function of t for $x \in I$. This together with Theorem 2.1 of Chapter V shows that $0 \leqq G(x-t)/G(x_0-t) \leqq M$ for $x \in I$ where

$$M = e^{\alpha_2(x_1 - x_0)}.$$

Applying Lemma 2.1a our theorem follows.

As a consequence of Theorem 3.2 we see that if $G(t) \in$ class II then there exists a constant γ_c, which may be $+\infty$ or $-\infty$, depending on $\alpha(t)$ such that the transform $G(t) \# \alpha(t)$ converges for $x > \gamma_c$ and diverges for $x < \gamma_c$. For $x = \gamma_c$ it may converge or diverge. The number γ_c is called the abscissa of convergence. The reader will recognize here the convergence behaviour of the Laplace transform.

3.3. If $G(t) \in$ class III then $G(t) = 0$ for $t > b + \sum_k a_k^{-1}$. It follows that we need not suppose $\alpha(t)$ defined for all t; it is enough to assume $\alpha(t)$ defined for $T < t < \infty$ and of bounded variation in every subinterval provided that we consider only those x for which $x > T + b + \sum_k a_k^{-1}$.

THEOREM 3.3. *If*

1. $G(x) \in$ class III,
2. $\alpha(t)$ *is defined and of bounded variation in every interval*

$$T < t_1 \leqq t \leqq t_2 < \infty,$$

3. $\int_{-\infty}^{\infty} G(x_0 - t)\, d\alpha(t)$ *converges,* $x_0 > T + b + \sum_k a_k^{-1}$,

then

$$\int_{-\infty}^{\infty} G(x - t)\, d\alpha(t)$$

converges uniformly for x in any finite interval lying to the right of $T + b + \sum_k a_k^{-1}$; i.e. for

$$T + b + \sum_k a_k^{-1} < x_1 \leqq x \leqq x_2 < \infty.$$

Let I be the interval $[x_1, x_2]$. Choose u

$$u > x_0 - b - \sum_k a_k^{-1},$$

and let

$$J_1(x) = \int_{-\infty}^{u} G(x - t)\, d\alpha(t), \qquad J_2(x) = \int_{u}^{\infty} G(x - t)\, d\alpha(t).$$

Since $G(t)$ vanishes for $t \geqq b + \sum_a a_k^{-1}$, $J_1(x)$ is uniformly convergent for $x \in I$. On the other hand $G(x - t)/G(x_0 - t)$ has as a function of t no changes of trend in $u \leqq t < \infty$ for each $x \in I$. In addition $G(x - t)/G(x_0 - t)$ is easily seen to be uniformly bounded for $u \leqq t < \infty$, $x \in I$. Appealing to Lemma 2.1a we see that $J_2(x)$ is uniformly convergent.

4. THE SEQUENCE OF KERNELS

4.1. We suppose that we are given a sequence $\{b_m\}_0^\infty$ of real numbers such that $b_0 = b$, $\lim_{m\to\infty} b_m = 0$. As in Chapter III we set

$$P_m(D) = e^{(b-b_m)D} \prod_{k=1}^{m} \left(1 - \frac{D}{a_k}\right) e^{D/a_k},$$

$$E_m(s) = e^{b_m s} \prod_{m+1}^{\infty} \left(1 - \frac{s}{a_k}\right) e^{s/a_k},$$

$$G_m(t) = (2\pi i)^{-1} \int_{-i\infty}^{i\infty} [E_m(s)]^{-1} e^{st}\, ds,$$

$$S_m = \sum_{m+1}^{\infty} a_k^{-2}.$$

We put

$$\alpha_1(m) = \underset{\substack{k>m \\ a_k<0}}{\text{l.u.b.}} [a_k, -\infty], \qquad \alpha_2(m) = \underset{\substack{k>m \\ a_k>0}}{\text{g.l.b.}} [a_k, +\infty],$$

and we define $\mu_1(m) + 1 (\mu_2(m) + 1)$ as the multiplicity of $\alpha_1(m)$ $(\alpha_2(m))$ as a zero of $E_m(s)$. The unique zero of $G_m(t)'$ we denote by ζ_m.

The $\{G_m(t)\}_0^\infty$ form a sequence of frequency functions. The mean of $G_m(t)$ is b_m and the variance is S_m. Each $G_m(t)$ is bell-shaped and two different $G_m(t)$'s cut in at most two points. Since as $m \to \infty$ b_m converges to the origin and S_m decreases to zero, it seems natural to suppose that the $G_m(t)$ appear somewhat as in the following illustration.

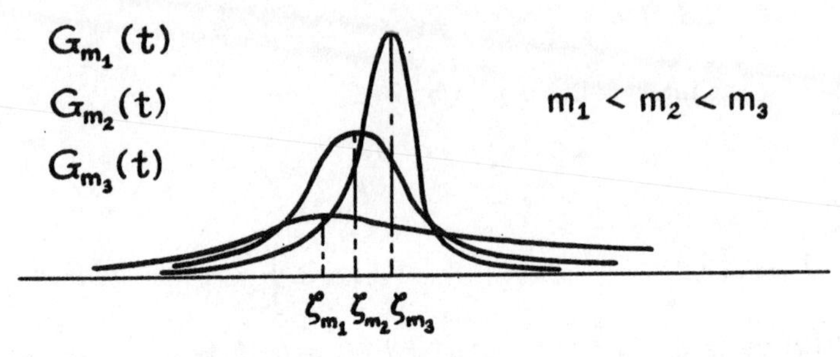

FIG. II

Further verification of this picture is contained in the following result.

LEMMA 4.1. If $\{G_m(t)\}_0^\infty$, ζ_m, and S_m are defined as above then:

A. $|\zeta_m - b_m| \leqq 8\, S_m^{1/2}$;

B. $\lim_{m\to\infty} G_m(t) = 0 \qquad (0 < |t| < \infty)$.

We have

$$\int_{|t|\leqq 2S_m^{1/2}} G_m\,(t + b_m)\,dt = 1 - \int_{|t|\geqq 2S_m^{1/2}} G_m(t + b_m)\,dt$$

$$\geqq 1 - \frac{1}{4\,S_m}\int_{-\infty}^{\infty} t^2 G_m(t + b_m)\,dt = \frac{3}{4}.$$

From this it follows that

$$\operatorname*{l.u.b.}_{|t|\leqq 2S_m^{1/2}} G_m(t + b_m) \geqq \tfrac{3}{16}\, S_m^{-1/2}.$$

We will prove that $|\zeta_m - b_m| \leqq 8\, S_m^{1/2}$. Suppose the contrary. Then since $G_m(t)$ has a unique maximum at ζ_m it is monotonic in the interval from b_m to ζ_m. If I is the interval consisting of those points between b_m and ζ_m which lie at a distance $2\, S_m^{1/2}$ or more from b_m then the length of I is at least $6\, S_m^{1/2}$. In I $G_m(t)$ is not less than $\frac{3}{16}\, S_m^{-1/2}$. This would imply that

$$\int_{-\infty}^{\infty} G_m(t)dt \geqq 9/8,$$

a contradiction. Thus conclusion A is valid.

Let t_0 be an arbitrary number different from zero and choose m so large that $|b_m| < |t_0|/2$. Then

$$\left| \int_{t_0}^{t_0/2} G_m(t)dt \right| \leqq \int_{|t| \geqq |t_0|/2} G_m(t)\, dt \leqq \left(\frac{|t_0|}{2} - |b_m| \right)^{-2} \int_{-\infty}^{\infty} (t - b_m)^2 G_m(t)\, dt,$$

$$\leqq \left(\frac{|t_0|}{2} - |b_m| \right)^{-2} S_m.$$

Hence

$$\lim_{m \to \infty} \left| \int_{t_0}^{t_0/2} G_m(t)\, dt \right| = 0.$$

But if m is so large that $|\zeta_m| < |t_0|/2$, then $G_m(t)$ is monotonic over the range of this integral and takes its smallest value at t_0; i.e.

$$G_m(t_0)\tfrac{1}{2}|t_0| \leqq \left| \int_{t_0}^{t_0/2} G_m(t)\, dt \right|.$$

From this inequality conclusion B is immediate.

5. THE INVERSION THEOREM

5.1. In this section we shall treat the case $G(t) \in$ class I. We begin by considering the application of the differential operator.

THEOREM 5.1a. *If*

1. $G(t) \in$ class I,

2. $P_m(D)$, $G_m(t)$ *are defined as in* § 4,

3. $\alpha(t)$ *is of bounded variation in every finite interval,*

4. $\int_{-\infty}^{\infty} G(x - t)\, d\alpha(t) = f(x)$ *converges,*

then

$$P_m(D)f(x) = \int_{-\infty}^{\infty} G_m(x - t)\, d\alpha(t),$$

the integral converging uniformly for x in any finite interval.

Let us choose numbers x' and x''. The function $G_m(x'' - t)/G(x' - t)$ has a function of t at most two changes of trend. We have either $\alpha_2(m) > \alpha_2$ or $\alpha_2(m) = \alpha_2$, $\mu_2(m) \leqq \mu_2$. It follows from Theorem 2.1 of Chapter V that

$$\overline{\lim_{t \to +\infty}}\, G_m(x'' - t)/G(x' - t) < \infty.$$

Similarly since either $\alpha_1(m) < \alpha_1$ or $\alpha_1(m) = \alpha_1$, $\mu_1(m) \leq \mu_1$, we have

$$\overline{\lim_{t \to -\infty}} \; G_m(x'' - t)/G(x' - t) < \infty.$$

Applying Lemma 2.1b we find that the integral

$$\int_{-\infty}^{\infty} G_m(x - t)\, d\alpha(t) \tag{1}$$

converges for $x = x''$. If $G_m(t) \in$ class I then (1) is by Theorem 3.1 uniformly convergent for x in any finite interval; if $G_m(t) \in$ class II then (1) is by Theorem 3.2 uniformly convergent for x in any finite interval bounded on the left by x'' and since x'' is arbitrary (1) is uniformly convergent for x in any finite interval; if $G_m(t) \in$ class III then by Theorem 3.3 (1) is uniformly convergent for x in any finite interval. It remains to verify that

$$P_m(D) \int_{-\infty}^{\infty} G(x - t)\, d\alpha(t) = \int_{-\infty}^{\infty} G_m(x - t)\, d\alpha(t).$$

This follows from the fact that a linear differential operator may be applied under the integral sign if the resulting integral is uniformly convergent in the small.

THEOREM 5.1b. *If*

1. $G(t) \in$ class I, $G(t)$ *is non-finite,*
2. $\varphi(t)$ *is integrable on every finite interval,*
3. $f(x) = \int_{-\infty}^{\infty} G(x - t)\varphi(t)\, dt$ *converges,*
4. $P_m(D)$ *is defined as in* § 4,
5. $\varphi(t)$ *is continuous at* x,

then

$$\lim_{m \to \infty} P_m(D) f(x) = \varphi(x).$$

By Theorem 5.1a we have

$$P_m(D) f(x) = \int_{-\infty}^{\infty} G_m(x - t)\varphi(t)\, dt.$$

Since $G_m(t)$ is a frequency function

$$P_m(D) f(x) - \varphi(x) = \int_{-\infty}^{\infty} G_m(x - t)\, [\varphi(t) - \varphi(x)]\, dt.$$

It is therefore enough to show that

$$\lim_{m \to \infty} \int_{-\infty}^{\infty} G_m(x - t)\, [\varphi(t) - \varphi(x)]\, dt = 0.$$

Given $\epsilon > 0$ let us choose $\delta > 0$ so small that

$$|\varphi(t) - \varphi(x)| \leq \epsilon \qquad\qquad |t - x| \leq \delta.$$

We set

$$\int_{-\infty}^{\infty} G_m(x-t)\,[\varphi(t)-\varphi(x)]\,dt = I_1 + I_2 + I_3,$$

corresponding to the ranges of integration $(-\infty, x-\delta)$, $(x-\delta, x+\delta)$, and $(x+\delta, +\infty)$. We have

$$I_2 = \int_{x-\delta}^{x+\delta} G_m(x-t)\,[\varphi(t)-\varphi(x)]\,dt,$$

$$|I_2| \leqq \epsilon \int_{x-\delta}^{x+\delta} G_m(x-t)\,dt \leqq \epsilon.$$

We have

$$I_1 = \int_{-\infty}^{x-\delta} G_m(x-t)\,[\varphi(t)-\varphi(x)]\,dt$$

$$= \int_{-\infty}^{x-\delta} [G_m(x-t)/G(x_1-t)]\,[G(x_1-t)\,\{\varphi(t)-\varphi(x)\}]\,dt,$$

where x_1 is any real number. By Theorem 6.1b of Chapter IV $G_m(x-t)/G(x_1-t)$ has at most two changes of trend. If m is sufficiently large then $G_m(x-t)/G(x_1-t)$ is increasing for t near $-\infty$ and decreasing for t near $+\infty$. Thus $G_m(x-t)/G(x_1-t)$ has an odd number of changes of trend and hence one change of trend. By Lemma 4.1 this must lie within $(x-\delta, x+\delta)$ if m is sufficiently large. It follows that $G_m(x-t)/G(x_1-t)$ is increasing for $-\infty < t \leqq x-\delta$ if m is large. By the mean value theorem.

$$I_1 = [G_m(\delta)/G(x_1-x+\delta)] \int_{\xi}^{x-\delta} G(x_1-t)\,[\varphi(t)-\varphi(x)]\,dt$$

where $-\infty < \xi < x-\delta$. By Lemma 4.1 we have $\lim_{m\to\infty} I_1 = 0$. Similarly we may show that $\lim_{m\to\infty} I_3 = 0$. Thus

$$\overline{\lim_{m\to\infty}} \left| \int_{-\infty}^{\infty} G_m(x-t)\,[\varphi(t)-\varphi(x)]\,dt \right| \leqq \epsilon.$$

Since ϵ is arbitrary our theorem is proved.

5.2. In this section we consider the case $G(t) \in$ class II.

THEOREM 5.2a. *If*

1. $G(t) \in$ class II,
2. $P_m(D)$, $G_m(t)$ *are defined as in* § 4,
3. $\alpha(t)$ *is of bounded variation in every finite interval,*
4. $\int_{-\infty}^{\infty} G(x-t)\,d\alpha(t) = f(x)$ *converges for* $x > \gamma_c$,

then

$$P_m(D)f(x) = \int_{-\infty}^{\infty} G_m(x-t)\,d\alpha(t), \tag{1}$$

the integral converging uniformly for x in any finite interval

$$\gamma_c - b + b_m - \sum_1^m a_k^{-1} < x_1 \leqq x \leqq x_2 < \infty.$$

Let us choose numbers x', x'' such that $x' > \gamma_c$, $x'' > x' - b + b_m - \sum_1^m a_k^{-1}$. The function $G_m(x'' - t)/G(x' - t)$ has, as a function of t, at most two changes of trend. We have either $\alpha_2(m) > \alpha_2$ or $\alpha_2(m) = \alpha_2$ and $\mu_2(m) \leqq \mu_2$. It follows from Theorem 2.1 of Chapter V that

$$\overline{\lim_{t \to +\infty}} G_m(x'' - t)/G(x' - t) < \infty.$$

From Theorem 3.4 of Chapter V we have

$$\frac{d}{dt} \log G_m(x'' - t)/G(x' - t) = L_m(x'' - t + o(1)) - L(x' - t + o(1)) \qquad t \to -\infty,$$

where $L(t)$ and $L_m(t)$ are defined by the equations

$$t = b + \sum_1^\infty \frac{L(t)}{a_k(a_k + L(t))}, \qquad t = b_m + \sum_{m+1}^\infty \frac{L_m(t)}{a_k(a_k + L_m(t))}.$$

We have

$$t + b - b_m + \sum_1^m \frac{L_m(t)}{a_k(a_k + L_m(t))} = b + \sum_1^\infty \frac{L_m(t)}{a_k(a_k + L_m(t))}$$

from which we obtain

$$L_m(t) = L\left[t + b - b_m + \sum_1^m \frac{L_m(t)}{a_k(a_k + L_m(t))}\right],$$

$$= L\left[t + b - b_m + \sum_1^m \frac{1}{a_k} + o(1)\right] \qquad t \to -\infty.$$

Thus

$$\frac{d}{dt} \log G_m(x'' - t)/G(x' - t) = L(x'' - t + b - b_m + \sum_1^m a_k^{-1} + o(1)) - L(x' - t + o(1)).$$

Since $L(t)$ is non-decreasing it follows that $\log G_m(x'' - t)/G(x' - t)$ is non-increasing as $t \to -\infty$ and thus that

$$\overline{\lim_{t \to -\infty}} G_m(x'' - t)/G(x' - t) < \infty.$$

Applying Lemma 2.1b we see that

$$\int_{-\infty}^{\infty} G_m(x - t)\, d\alpha(t)$$

is convergent for $x = x''$. By Theorem 3.2 the integral is uniformly convergent for x in any finite interval bounded on the left by x'', etc.

THEOREM 5.2b. *If*

1. $G(t) \in$ class II, $G(t)$ *is non-finite,*
2. $\varphi(t)$ *is integrable on every finite interval,*
3. $f(x) = \int_{-\infty}^{\infty} G(x-t)\varphi(t)\, dt$ *converges for* $x > \gamma_c$,
4. $P_m(D)$ *is defined as in* § 4,
5. $\varphi(t)$ *is continuous for* $t = x$,

then

$$\lim_{m\to\infty} P_m(D)f(x) = \varphi(x).$$

As in the proof of Theorem 5.1b it is enough to show that

$$\lim_{m\to\infty} \int_{-\infty}^{\infty} G_m(x-t)\,[\varphi(t) - \varphi(x)]\, dt = 0.$$

Given $\epsilon > 0$ we choose $\delta > 0$ so small that

$$|\varphi(t) - \varphi(x)| \leqq \epsilon \qquad\qquad |t - x| \leqq \delta,$$

and we set, as before,

$$\int_{-\infty}^{\infty} G_m(x-t)\,[\varphi(t) - \varphi(x)]\, dt = I_1 + I_2 + I_3,$$

corresponding to the ranges of integration $(-\infty, x-\delta)$, $(x-\delta, x+\delta)$, and $(x+\delta, \infty)$. Arguing as before we find that $|I_2| \leqq \epsilon$. We have

$$I_1 = \int_{-\infty}^{x-\delta} G_m(x-t)\,[\varphi(t) - \varphi(x)]\, dt$$

$$= \int_{-\infty}^{x-\delta} [G_m(x-t)/G(x_1-t)]\,[G(x_1-t)\,\{\varphi(t) - \varphi(x)\}]\, dt$$

where x_1 is any number $> \gamma_c$. By Theorem 6.1b of Chapter IV $G_m(x-t)/G(x_1-t)$ has at most two changes of trend. If m is sufficiently large then $G_m(x-t)/G(x_1-t)$ is increasing for t near $-\infty$ and decreasing for t near $+\infty$. (See the argument used in the proof of Theorem 5.2a.) Thus $G_m(x-t)/G(x_1-t)$ has one change of trend and this must lie within $[x-\delta, x+\delta]$ if m is sufficiently large. It follows that $G_m(x-t)/G(x_1-t)$ is increasing for $-\infty < t \leqq x - \delta$ if m is sufficiently large. By the mean value theorem

$$I_1 = G_m(\delta)/G(x_1 - x + \delta) \int_{\xi}^{x-\delta} G(x_1-t)\,[\varphi(t) - \varphi(x)]\, dt$$

where $-\infty < \xi < x - \delta$. By Lemma 4.1 we have $\lim_{m\to\infty} I_1 = 0$. Similarly we may show that $\lim_{m\to\infty} I_3 = 0$, etc.

5.3. We now turn to the case $G(t) \in$ class III. The demonstrations of the following results follow in the pattern of sections 5.1 and 5.2.

THEOREM 5.3a. *If*

1. $G(t) \in$ class III,

2. $P_m(D)$, $G_m(t)$ *are defined as in* § 4,

3. $\alpha(t)$ *is defined for* $T < t < \infty$ *and is of bounded variation in every interval* $T < t_1 \leqq t \leqq t_2 < \infty$,

4. $\int_{-\infty}^{\infty} G(x - t)\, d\alpha(t) = f(x)$ *converges for* $x > T + b + \sum_1^\infty a_k^{-1}$,

then

$$P_m(D)f(x) = \int_{-\infty}^{\infty} G_m(x - t)\, d\alpha(t)$$

the integral converging uniformly for x in any finite interval of the form $T + b_m + \sum_{m+1}^{\infty} a_k^{-1} < x_1 \leqq x \leqq x_2 < \infty$.

THEOREM 5.3b. *If*

1. $G(t) \in$ class III, $G(t)$ *is non-finite,*

2. $\varphi(t)$ *is defined for* $T < t < \infty$ *and is integrable over every interval*

$$T < t_1 \leqq t \leqq t_2 < \infty,$$

3. $f(x) = \int_{-\infty}^{\infty} G(x - t)\varphi(t)\, dt$ *converges for* $x > T + b + \sum_1^\infty a_k^{-1}$,

4. $P_m(D)$ *is defined as in* § 4,

5. $\varphi(t)$ *is continuous at* $t = x (x > T)$,

then

$$\lim_{m \to \infty} P_m(D)f(x) = \varphi(x).$$

6. STIELTJES INTEGRALS

6.1. In these sections we shall consider the inversion formula for the Stieltjes convolution $G \# \alpha$ rather than the Lebesgue convolution $G * \varphi$. We begin with $G(t) \in$ class I.

THEOREM 6.1a. *If*

1. $G(t) \in$ class I,

2. $P_m(D)$, $G_m(t)$ *are defined as in* § 4,

3. $\alpha(t)$ *is of bounded variation in every finite interval,*

4. $f(x) = \int_{-\infty}^{\infty} G(x - t)\, e^{ct} d\alpha(t)$ *converges,*

then for m sufficiently large:

A. $\alpha_1 < c < \alpha_2$ *implies that*

$$\int_{x_1}^{x_2} e^{-cx} P_m(D) f(x)\, dx = \int_{-\infty}^{\infty} G_m(x_2 - t) e^{-c(x_2-t)} \alpha(t)\, dt - \int_{-\infty}^{\infty} G_m(x_1 - t) e^{-c(x_2-t)} \alpha(t)\, dt;$$

B. $c \geqq \alpha_2$ *implies that* $\alpha(+\infty)$ *exists and that*

$$\int_{x_1}^{\infty} e^{-cx} P_m(D) f(x)\, dx = \int_{-\infty}^{\infty} G_m(x_1 - t) e^{-c(x_1-t)} [\alpha(+\infty) - \alpha(t)]\, dt;$$

C. $c \leqq \alpha_1$ *implies that* $\alpha(-\infty)$ *exists and that*

$$\int_{-\infty}^{x_2} e^{-cx} P_m(D) f(x)\, dx = \int_{-\infty}^{\infty} G_m(x_2 - t) e^{-c(x_2-t)} [\alpha(t) - \alpha(-\infty)]\, dt.$$

The function $G(x - t)e^{ct}$ has as a function of t either one or no change of trend. It is therefore monotonic near $+\infty$ and near $-\infty$. By Lemma 2.1c if x_0 is any real number we have

$$\alpha(t) = o[G(x_0 - t)e^{ct}]^{-1} \qquad t \to \pm\infty, \qquad \alpha_1 < c < \alpha_2, \tag{1}$$

$$\alpha(+\infty) - \alpha(t) = o[G(x_0 - t)e^{ct}]^{-1} \qquad t \to \pm\infty, \qquad c \geqq \alpha_2, \tag{2}$$

$$\alpha(t) - \alpha(-\infty) = o[G(x_0 - t)e^{ct}]^{-1} \qquad t \to \pm\infty, \qquad c \leqq \alpha_1. \tag{3}$$

CONCLUSION A. By Theorem 5.1a we have

$$P_m(D) f(x) = \int_{-\infty}^{\infty} G_m(x - t) e^{ct}\, d\alpha(t) \quad (m = 0, 1, \cdots),$$

$$e^{-cx} P_m(D) f(x) = \int_{-\infty}^{\infty} G_m(x - t) e^{-c(x-t)}\, d\alpha(t),$$

the integral converging uniformly for x in any finite interval. Integrating by parts we obtain

$$e^{-cx} P_m(D) f(x) = \Big[G_m(x - t) e^{-c(x-t)} \alpha(t) \Big]_{-\infty}^{+\infty} - \int_{-\infty}^{\infty} \left[\frac{\partial}{\partial t} G_m(x - t) e^{-c(x-t)} \right] \alpha(t)\, dt.$$

The estimates given above and Theorem 2.1 of Chapter V show that the integrated term vanishes uniformly for $x_1 \leqq x \leqq x_2$. Thus

$$e^{-cx} P_m(D) f(x) = -\int_{-\infty}^{\infty} \left[\frac{\partial}{\partial t} G_m(x - t) e^{-c(x-t)} \right] \alpha(t)\, dt,$$

$$= +\int_{-\infty}^{\infty} \left[\frac{\partial}{\partial x} G_m(x - t) e^{-c(x-t)} \right] \alpha(t)\, dt,$$

the integral converging uniformly for $x_1 \leqq x \leqq x_2$. We have

$$\int_{x_1}^{x_2} e^{-cx} P_m(D) f(x)\, dx = \int_{x_1}^{x_2} dx \int_{-\infty}^{\infty} \left[\frac{\partial}{\partial x} G_m(x-t)e^{-c(x-t)}\right] \alpha(t)\, dt.$$

Because of the uniform convergence of the inner integral the order of the integrations may be inverted to give

$$\int_{x_1}^{x_2} e^{-xc} P_m(D) f(x)\, dx = \int_{-\infty}^{\infty} [G_m(x_2-t)e^{-c(x_2-t)} - G_m(x_1-t)e^{-c(x_1-t)}]\alpha(t)\, dt.$$

Using Theorem 2.1 of Chapter V and the estimate (1) we see that if m is sufficiently large the integrals

$$\int_{-\infty}^{\infty} G_m(x_2-t)e^{-c(x_2-t)}\alpha(t)\, dt, \quad \int_{-\infty}^{\infty} G_m(x_1-t)e^{-c(x_1-t)}\alpha(t)\, dt,$$

will converge absolutely. Q.E.D.

CONCLUSION B. Proceeding exactly as in the previous case we obtain

$$\int_{x_1}^{x_2} e^{-cx} P_m(D) f(x)\, dx = -\int_{-\infty}^{\infty} G_m(x_2-t)e^{-c(x_2-t)}[\alpha(+\infty)-\alpha(t)]\, dt$$
$$+ \int_{-\infty}^{\infty} G_m(x_1-t)e^{-c(x_1-t)}[\alpha(+\infty)-\alpha(t)]\, dt.$$

We must show that

$$\lim_{x_2 \to +\infty} \int_{-\infty}^{\infty} G_m(x_2-t)e^{-c(x_2-t)}[\alpha(+\infty)-\alpha(t)] = 0. \tag{4}$$

We define

$$r(t) = [G(x_0-t)e^{ct}]^{-1},$$

$$k(t) = G_m(t)e^{-ct}.$$

We have, using (2), that

$$\overline{\lim_{x \to +\infty}} \left| \int_{-\infty}^{\infty} G_m(x-t)e^{-c(x-t)}[\alpha(+\infty)-\alpha(t)]\, dt \right|$$
$$\leqq O(1) \overline{\lim_{x \to +\infty}} \int_{-\infty}^{\infty} k(x-t) r(t)\, dt.$$

The assumptions of Lemma 2.1d are satisfied so that

$$\lim_{x \to +\infty} \int_{-\infty}^{\infty} k(x-t) r(t)\, dt = 0.$$

Thus equation (4) has been established.

THEOREM 6.1b. *If*

1. $G(t) \in$ class I, $G(t)$ *is non-finite,*
2. $\alpha(t)$ *is of bounded variation in every finite interval,*
3. $f(x) = \int_{-\infty}^{\infty} G(x-t)e^{ct}d\alpha(t)$ *converges,*
4. $P_m(D)$ *is defined as in* § 4,
5. $\alpha(t)$ *is continuous at* x_1, x_2,

then:

A. $\alpha_1 < c < \alpha_2$ *implies that*

$$\lim_{m\to\infty} \int_{x_1}^{x_2} e^{-cx}P_m(D)f(x)\,dx = \alpha(x_2) - \alpha(x_1);$$

B. $c \geqq \alpha_2$ *implies that*

$$\lim_{m\to\infty} \int_{x_1}^{\infty} e^{-cx}P_m(D)f(x)\,dx = \alpha(+\infty) - \alpha(x_1);$$

C. $c \leqq \alpha_1$ *implies that*

$$\lim_{m\to\infty} \int_{-\infty}^{x_2} e^{-cx}P_m(D)f(x)\,dx = \alpha(x_2) - \alpha(-\infty).$$

Let us demonstrate conclusion B. The other conclusions follow from similar arguments. We have shown in Theorem 6.1a that there exists an integer m_0 such that if $m \geqq m_0$ then

$$\int_{x_1}^{\infty} e^{-cx}P_m(D)f(x)\,dx = \int_{-\infty}^{\infty} G_m(x_1-t)e^{-c(x_1-t)}[\alpha(+\infty)-\alpha(t)]\,dt.$$

Thus if

$$f^*(x) = \int_{-\infty}^{\infty} G_{m_0}(x-t)e^{ct}[\alpha(+\infty)-\alpha(t)]\,dt$$

then

$$\int_{x_1}^{\infty} e^{-cx}P_m(D)f(x)\,dx = e^{-cx}P_m^*(D)f^*(x)\,\Big|_{x=x_1},$$

where

$$P_m^*(D) = e^{(b_m-b_{m_0})D}\prod_{m_0+1}^{m}\left(1-\frac{D}{a_k}\right)e^{D/a_k}.$$

Appealing to Theorem 5.1b, 5.2b, or 5.3b as $G_{m_0}(t)$ belongs to class I, class II, or class III we obtain our desired result.

6.2. $G(t) \in$ class II. Since only small changes in argument are necessary for this case we shall forgo detailed demonstrations of our results.

THEOREM 6.2a. *If*

1. $G(t) \in$ class II,

2. *$P_m(D)$, $G_m(t)$ are defined as in* § 4,

3. *$\alpha(t)$ is of bounded variation in every finite interval,*

4. $f(x) = \int_{-\infty}^{\infty} G(x-t)e^{ct}\,d\alpha(t)$ *converges for* $x > \gamma_c$,

then for m sufficiently large:

A. $c < \alpha_2$ *implies that*

$$\int_{x_1}^{x_2} e^{-cx}P_m(D)f(x)\,dx = \int_{-\infty}^{\infty} G_m(x_2-t)e^{-c(x_2-t)}\alpha(t)\,dt - \int_{-\infty}^{\infty} G_m(x_1-t)e^{-c(x_1-t)}\alpha(t)\,dt;$$

B. $c \geqq \alpha_2$ *implies that* $\alpha(+\infty)$ *exists and that*

$$\int_{x_1}^{\infty} e^{-cx}P_m(D)f(x)\,dx = \int_{-\infty}^{\infty} G_m(x_1-t)e^{-c(x_1-t)}[\alpha(+\infty)-\alpha(t)]\,dt.$$

THEOREM 6.2b. *If*

1. $G(t) \in$ class II, *$G(t)$ is non-finite,*

2. *$\alpha(t)$ is of bounded variation in every finite interval,*

3. $f(x) = \int_{-\infty}^{\infty} G(x-t)e^{ct}\,d\alpha(t)$ *converges for* $x > \gamma_c$,

4. *$P_m(D)$ is defined as in* § 4,

5. *$\alpha(t)$ is continuous at x_1, x_2,*

then:

A. $c < \alpha_2$ *implies that*

$$\lim_{m\to\infty} \int_{x_1}^{x_2} e^{-cx}P_m(D)f(x)\,dx = \alpha(x_2) - \alpha(x_1);$$

B. $c \geqq \alpha_2$ *implies that*

$$\lim_{m\to\infty} \int_{x_1}^{\infty} e^{-cx}P_m(D)f(x)\,dx = \alpha(+\infty) - \alpha(x_1).$$

6.3. $G(t) \in$ class III. As in the preceding section it is left to the reader to supply proofs.

THEOREM 6.3a. *If*

1. $G(t) \in$ class III,

2. $\alpha(t)$ *is defined for* $T < t < \infty$ *and is of bounded variation in every finite interval* $T < t_1 \leqq t \leqq t_2 < \infty$,

3. $P_m(D)$, $G_m(t)$ *are defined as in* § 4,

4. $f(x) = \int_{-\infty}^{\infty} G(x-t)e^{ct}\, d\alpha(t)$ *converges for* $x > T + b + \sum_1^\infty a_k^{-1}$,

then for $x_1, x_2 > T$ *and m sufficiently large:*

A. $c < \alpha_2$ *implies that*

$$\int_{x_1}^{x_2} e^{-cx} P_m(D) f(x)\, dx = \int_{-\infty}^{\infty} G_m(x_2 - t)e^{-c(x_2-t)}\alpha(t)\, dt - \int_{-\infty}^{\infty} G_m(x_1 - t)e^{-c(x_1-t)}\alpha(t)\, dt;$$

B. $c \geqq \alpha_2$ *implies that* $\alpha(+\infty)$ *exists and that*

$$\int_{x_1}^{\infty} e^{-cx} P_m(D) f(x)\, dx = \int_{-\infty}^{\infty} G_m(x_1 - t)e^{-c(x_1-t)}[\alpha(+\infty) - \alpha(t)]\, dt.$$

THEOREM 6.3b. *If*

1. $G(t) \in$ class III, $G(t)$ *is non-finite,*

2. $\alpha(t)$ *is defined for* $T < t < \infty$ *and is of bounded variation in every finite interval* $T < t_1 \leqq t \leqq t_2$,

3. $f(x) = \int_{-\infty}^{\infty} G(x-t)e^{ct}\, d\alpha(t)$ *converges for* $x > T + b + \sum_1^\infty a_k^{-1}$,

4. $P_m(D)$ *is defined as in* § 4,

5. $\alpha(t)$ *is continuous at* x_1, x_2 $\quad (x_1, x_2 > T)$,

then:

A. $c < \alpha_2$ *implies that*

$$\lim_{m\to\infty} \int_{x_1}^{x_2} e^{-cx} P_m(D) f(x)\, dx = \alpha(x_2) - \alpha(x_1);$$

B. $c \geqq \alpha_2$ *implies that*

$$\lim_{m\to\infty} \int_{x_1}^{\infty} e^{-cx} P_m(D) f(x)\, dx = \alpha(+\infty) - \alpha(x_1).$$

7. RELAXATION OF CONTINUITY CONDITIONS

7.1. It is natural to suppose from the known examples of our theory, e.g. the Laplace and Stieltjes transforms, that if

$$f(x) = \int_{-\infty}^{\infty} G(x-t)\varphi(t)\,dt$$

converges then

$$\lim_{m\to\infty} P_m(D)f(x) = \varphi(x)$$

almost everywhere. We shall show that this is true if the constants b_m approach zero not too slowly.

LEMMA 7.1. If $\{G_m(t)\}_0^\infty$, S_m are defined as in § 4 then

$$G_m(t) \leqq (2/S_m)^{1/2} \qquad (-\infty < t < \infty;\quad m = 0, 1, \cdots).$$

There are two cases we must consider:

A. $S_m \geqq 2 \max (a_{m+1}^{-2}, a_{m+2}^{-2}, \cdots)$,

B. $S_m < 2 \max (a_{m+1}^{-2}, a_{,m+2}^{-2} \cdots)$.

If case A obtains we have

$$G_m(t) \leqq (2\pi)^{-1} \int_{-\infty}^{\infty} \left[\prod_{m+1}^{\infty}\left(1+\frac{\tau^2}{a_k^2}\right)\right]^{-1/2} d\tau.$$

Now

$$\prod_{m+1}^{\infty}\left(1+\frac{\tau^2}{a_k^2}\right) = 1 + \tau^2 S_m + \tfrac{1}{2}\tau^4 \sum_{m+1}^{\infty}\frac{1}{a_k^2}\left(S_m - \frac{1}{a_k^2}\right) + \cdots,$$

the higher terms having positive coefficients. Since $(S_m - a_k^{-2}) \geqq \frac{1}{2}S_m$ for $k = m+1, m+2, \cdots$ we have

$$\prod_{m+1}^{\infty}\left(1+\frac{\tau^2}{a_k^2}\right) \geqq 1 + \tau^2 S_m + \tfrac{1}{4}\tau^4 S_m^2.$$

Hence

$$G_m(t) \leqq (2\pi)^{-1}\int_{-\infty}^{\infty} [1 + \tfrac{1}{2}\tau^2 S_m]^{-1}\,d\tau,$$

$$\leqq 1/(2S_m)^{1/2}.$$

If case B obtains there exists $k_m > m$ such that $2/a_{k_m}^2 > S_m$, $|a_{k_m}| < (2/S_m)^{1/2}$. If we set

$$E_m^*(s) = e^{s/a_{k_m}} e^{b_m s} \prod_{\substack{k=m+1\\ k\neq k_m}}^{\infty} \left(1 - \frac{s}{a_k}\right) e^{s/a_k},\quad G_m^*(t) = \frac{1}{2\pi i}\int_{-i\infty}^{i\infty} [E_m^*(s)]^{-1} e^{st}\,dt,$$

then we have

$$G_m(t) = \int_{-\infty}^{\infty} g(a_{k_m}, u) G_m^*(t-u)\,du.$$

It follows that

$$G_m(t) \leqq [\text{l.u.b.}_u\ g(a_{k_m}, u)] \int_{-\infty}^{\infty} G_m^*(t - u)\, du,$$

$$G_m(t) \leqq |a_{k_m}| < (2/S_m)^{1/2}.$$

It is interesting to note that $S_m^{-1/2}$ is the true order of the maximum of $G_m(t)$, since it was shown in § 4 that the maximum exceeds $AS_m^{-1/2}$ where A is an absolute constant.

THEOREM 7.1a. *If in Theorems* 5.1b, 5.2b, *and* 5.3b *the assumption that $\varphi(t)$ is continuous at x is replaced by the weaker assumption*

$$\int_x^{x+h} [\varphi(t) - \varphi(x)]\, dt = o(h) \qquad h \to 0$$

and if it is assumed that

$$b_m = O(S_m^{1/2}) \qquad m \to \infty,$$

then the conclusions of these theorems, that

$$\lim_{m\to\infty} P_m(D)f(x) = \varphi(x), \tag{1}$$

still hold. In particular equation (1) *holds at all points of the Lebesgue set of $\varphi(t)$ and therefore almost everywhere.*

We must show that given $\epsilon > 0$ there exists $\delta > 0$ such that if

$$I_m = \int_{-\delta}^{\delta} G_m(t)\, [\varphi(x - t) - \varphi(x)]\, dt,$$

then

$$\overline{\lim_{m\to\infty}}\, |I_m| \leqq \epsilon.$$

If ζ_m is the point where $G_m(t)'$ changes sign then by Lemma 4.1

$$|\zeta_m - b_m| \leqq AS_m^{1/2}$$

where A is an absolute constant; thus

$$|\zeta_m| = O(S_m^{1/2}) \qquad m \to \infty.$$

Let $M = \text{l.u.b.}_{m=0,1,\ldots} |\zeta_m|\, G_m(\zeta_m)$. This is finite because of Lemma 7.1. We now choose δ so small that

$$|\psi(t)| \leqq \epsilon\, |t|\, (2M + 1)^{-1} \qquad |t| \leqq \delta$$

where

$$\psi(t) = \int_0^t [\varphi(x - u) - \varphi(x)]\, du.$$

Integrating by parts and using Lemma 4.1 we have

$$I_m = [G_m(t)\psi(t)]_{-\delta}^{\delta} - \int_{-\delta}^{\delta} G_m(t)'\psi(t)\,dt,$$

$$|I_m| \leqq o(1) + \epsilon(2M+1)^{-1} \int_{-\infty}^{\infty} |G_m(t)'t|\,dt.$$

Now $|t| \leqq |t - \zeta_m| + |\zeta_m|$ from which it follows that

$$\int_{-\delta}^{\delta} |G_m(t)'t|\,dt \leqq \int_{-\delta}^{\delta} G_m(t)'(\zeta_m - t)\,dt + |\zeta_m| \int_{-\delta}^{\delta} |G_m(t)'|\,dt.$$

We have

$$\int_{-\delta}^{\delta} G_m(t)'(\zeta_m - t)\,dt \leqq \int_{-\infty}^{\infty} G_m(t)'(\zeta_m - t)\,dt$$

$$\leqq [(\zeta_m - t)G_m(t)]_{-\infty}^{\infty} + \int_{-\infty}^{\infty} G_m(t)\,dt = 1.$$

Further

$$|\zeta_m| \int_{-\delta}^{\delta} |G_m(t)'|\,dt \leqq |\zeta_m| \int_{-\infty}^{\zeta_m} G_m(t)'\,dt - |\zeta_m| \int_{\zeta_m}^{\infty} G_m(t)'\,dt,$$

$$\leqq 2\,|\zeta_m|\,G(\zeta_m) \leqq 2M.$$

Combining these results we obtain

$$\overline{\lim_{m\to\infty}}\,|I_m| \leqq \epsilon.$$

It is interesting to note that Theorem 7.1b depends essentially upon information concerning the mode of $G_m(t)$.

7.2. In this section we shall discuss the conditions under which the inversion formula for the Stieltjes convolution is valid at points of discontinuity of the integrator function $\alpha(t)$.

THEOREM 7.2a. *Let* $C_m = \sum_{m+1}^{\infty} |a_k|^{-3}$. *If* $b_m = o(S_m)^{1/2}$, $C_m = o(S_m)^{3/2}$ *then*

$$\lim_{m\to\infty} S_m^{1/2} G_m(S_m^{1/2}t) = (2\pi)^{-1/2} e^{-t^2/2},$$

uniformly for t in any finite interval.

Let $A_k(m) = a_k S_m^{1/2}$ and let

$$F_m(z) = e^{zb_m S_m^{-1/2}} \prod_{m+1}^{\infty} \left(1 - \frac{z}{A_k(m)}\right) e^{z/A_k(m)}.$$

If in the formula

$$G_m(t) = (2\pi i)^{-1} \int_{-i\infty}^{i\infty} [E_m(s)]^{-1} e^{st}\,ds$$

we set $s = zS_m^{-1/2}$ and $t = uS_m^{1/2}$ we obtain

$$S_m^{1/2}G(S_m^{1/2}u) = (2\pi i)^{-1}\int_{-i\infty}^{i\infty}[F_m(z)]^{-1}e^{uz}\,dz.$$

We will show that

$$\lim_{m\to\infty} F_m(z) = e^{-z^2/2} \tag{1}$$

uniformly for $|z| \leqq R$ for every $R < \infty$. Let us first note that

$$\sum_{m+1}^{\infty} A_k(m)^{-2} = 1, \qquad \sum_{m+1}^{\infty} |A_k(m)|^{-3} = C_m S_m^{-3/2} = o(1).$$

If R is given then we will have $|A_k(m)| \geqq 2R$ if m is sufficiently large. The following inequality is well known

$$|\log(1-z)e^z e^{z^2/2}| \leqq 2|z|^3 \qquad (|z| \leqq \tfrac{1}{2});$$

here that branch of $\log z$ is taken for which $\log 1 = 0$. See Titchmarsh [1939; 246]. Thus R being given and m being sufficiently large we have

$$\left|\log\left(1 - \frac{z}{A_k(m)}\right) e^{z/A_k(m)} e^{z^2/2A_k^2(m)}\right| \leqq 2R^3 |A_k(m)|^{-3}.$$

It follows that

$$\begin{aligned}|\log F_m(z)e^{z^2/2}| &\leqq R|b_m|S_m^{-1/2} + 2R^3\sum_{m+1}^{\infty}|A_k(m)|^{-3},\\ &= o(1) \qquad (m\to\infty).\end{aligned}$$

We have thus established (1). Applying Theorem 8.6 of Chapter III we obtain our desired result.

THEOREM 7.2b. *If in Theorems* 6.1b, 6.2b, *and* 6.3b, *we make the additional assumption that* $C_m = o(S_m)^{3/2}$, $b_m = o(S_m)^{1/2}$, *and if* $\alpha(t) = \frac{1}{2}[\alpha(t+) + \alpha(t-)]$ *then the conclusions of these theorems hold for all* x_1, x_2.

We have

$$\int_0^{\infty} G_m(t)\,dt = \int_0^{\infty} S_m^{1/2}G_m(tS_m^{1/2})\,dt.$$

By Fatou's lemma and Theorem 7.2a we have

$$\varliminf_{m\to\infty}\int_0^{\infty} G_m(t)\,dt \geqq (2\pi)^{-1/2}\int_0^{\infty} e^{-t^2/2}\,dt = \tfrac{1}{2}. \tag{2}$$

Similarly

$$\varliminf_{m\to\infty}\int_{-\infty}^{0} G_m(t)\,dt \geqq (2\pi)^{-1/2}\int_{-\infty}^{0} e^{-t^2/2}\,dt = \tfrac{1}{2}. \tag{3}$$

Since

$$\int_0^\infty G_m(t)\,dt + \int_{-\infty}^0 G_m(t)\,dt = 1,$$

we have

$$\overline{\lim_{m\to\infty}} \int_0^\infty G_m(t)\,dt + \underline{\lim}_{m\to\infty} \int_{-\infty}^0 G_m(t)\,dt = 1, \tag{4}$$

$$\underline{\lim}_{m\to\infty} \int_0^\infty G_m(t)\,dt + \overline{\lim_{m\to\infty}} \int_{-\infty}^0 G_m(t)\,dt = 1. \tag{4'}$$

The relations (2), (3), (4), (4′) imply that

$$\lim_{m\to\infty} \int_0^\infty G_m(t)\,dt = \tfrac{1}{2}, \qquad \lim_{m\to\infty} \int_{-\infty}^0 G_m(t)\,dt = \tfrac{1}{2},$$

and our theorem follows.

It is to be noted that this result depends upon information concerning the median of $G_m(t)$.

8. FACTORIZATION

8.1. Let $E(s) = F_n(s)E_n(s)$ where

$$F_n(s) = \prod_{k=1}^n \left(1 - \frac{s}{a_k}\right), \ E_n(s) = e^{sa_1^{-1}+\dots+sa_n^{-1}+sb} \prod_{k=n+1}^\infty \left((1 - \frac{s}{a_k}\right) e^{s/a_k},$$

and let $H_n(t)$ and $G_n(t)$ be the kernels corresponding to $F_n(s)$ and $E_n(s)$ respectively, so that $G = G_n * H_n$. If φ is sufficiently restricted, for example if $\varphi \in L^1(-\infty, \infty)$, then Fubini's theorem may be applied to show that

$$G * \varphi = G_n * (H_n * \varphi).$$

In this section we will prove that if it is merely supposed that $G * \varphi$ is defined, then $G_n * (H_n * \varphi)$ is defined also and they are equal. The converse is false; $G_n * (H_n * \varphi)$ may be defined even though $G * \varphi$ is not. This result, which is related to the material of § 3, is needed in later chapters.

THEOREM 8.1a. *Let $G(t) \in$ class I, let $G_n(t)$ and $H_n(t)$ be defined as above, and let $\alpha(t)$ be of bounded variation in every finite interval $-\infty < t_1 \leqq t \leqq t_2 < \infty$. If the integral*

$$f(x) = \int_{-\infty}^\infty G(x - t)\,d\alpha(t)$$

converges for $-\infty < x < \infty$, then

$$f(x) = \int_{-\infty}^\infty G_n(x - t)\,dt \int_{-\infty}^\infty H_n(t - u)\,d\alpha(u)$$

for $-\infty < x < \infty$.

Let us suppose, for the sake of definiteness that $a_1 > 0$. If x is fixed the function $e^{-a_1 t}/G(x-t)$ is non-increasing. It follows from Lemma 2.1b that the integral

$$A(t) = \int_t^\infty e^{-a_1 u} d\alpha(u)$$

is convergent. Employing the mean value theorem we find that

$$A(t) = \int_t^\infty [e^{-a_1 u}/G(x-u)]G(x-u)\, d\alpha(u),$$

$$= [e^{-a_1 t}/G(x-t)] \int_t^\xi G(x-u)\, d\alpha(u) \qquad t < \xi < \infty.$$

It follows from this that

$$A(t) = o[e^{-a_1 t}/G(x-t)] \qquad t \to +\infty. \tag{1}$$

Let t_1 be arbitrary and $t < t_1$. We have

$$A(t) = A(t_1) + \int_t^{t_1} [e^{-a_1 u}/G(x-u)]G(x-u)\, d\alpha(u),$$

$$= A(t_1) + [e^{-a_1 t}/G(x-t)] \int_t^\xi G(x-u)\, d\alpha(u), \quad t < \xi < t_1.$$

Given $\epsilon > 0$ we can choose t_1 so large and negative that

$$\left| \int_a^b G(x-u)\, d\alpha(u) \right| \leqq \epsilon$$

for $a, b \leqq t_1$. Thus $\overline{\lim}_{t \to -\infty} |A(t)G(x-t)e^{a_1 t}| \leqq \epsilon$, or since ϵ is arbitrary,

$$A(t) = o[e^{-a_1 t}/G(x-t)] \qquad t \to -\infty. \tag{2}$$

We have

$$\int_{-\infty}^\infty G(x-t)\, d\alpha(t) = \int_{-\infty}^\infty G(x-t)e^{a_1 t}e^{-a_1 t}\, d\alpha(t),$$

$$= \Big[-A(t)\, e^{a_1 t} G(x-t)\Big]_{-\infty}^{\infty}$$

$$+ \int_{-\infty}^\infty A(t) \left[\frac{d}{dt} G(x-t)\, e^{a_1 t}\right] dt.$$

By (1) and (2) the integrated term is zero and

$$\int_{-\infty}^\infty G(x-t)\, d\alpha(t) = \int_{-\infty}^\infty A(t) \left[\frac{d}{dt} G(x-t)\, e^{a_1 t}\right] dt,$$

$$= \int_{-\infty}^\infty G_1(x-t) dt \int_{-\infty}^\infty g_1(t-u)\, d\alpha(u).$$

Here $g_k(t)$ $k = 1, \cdots, n$ is defined as in § 6 of Chapter II. This has been established under the assumption that $a_1 > 0$, but it is of course true if $a_1 < 0$. We may apply this same argument to G_1 and a_2, and then again to G_2 and a_3, and so on until after n such steps we have

$$G \# \alpha = G_n * (g_n * (g_{n-1} * (\cdots (g_2 * (g_1 \# \alpha)) \cdots))). \tag{3}$$

If x, x_0 are fixed then $H_n(x_0 - t)/G(x - t)$ has at most two changes of trend and is bounded. It follows from Lemma 2.1b that

$$\int_{-\infty}^{\infty} H_n(x - t)\, d\alpha(t)$$

converges for $-\infty < x < \infty$. Applying the same argument to H_n that we formerly applied to G we find that

$$H_n \# \alpha = g_n * (g_{n-1} * (\cdots (g_2 * (g_1 \# \alpha)) \cdots)). \tag{4}$$

Combining (3) and (4) we have our desired result.

THEOREM 8.1b. *Let $G(t) \in$ class II, let $G_n(t)$ and $H_n(t)$ be defined as above, and let $\alpha(t)$ be of bounded variation in every finite interval $-\infty < t_1 \leqq t \leqq t_2 < \infty$. If the integral*

$$f(x) = \int_{-\infty}^{\infty} G(x - t)\, d\alpha(t)$$

converges for $\gamma_c < x < \infty$, then

$$f(x) = \int_{-\infty}^{\infty} G_n(x - t)\, dt \int_{-\infty}^{\infty} H_n(t - u)\, d\alpha(u)$$

for $\gamma_c < x < \infty$.

THEOREM 8.1c. *Let $G(t) \in$ class III, let $G_n(t)$ and $H_n(t)$ be defined as above, and let $\alpha(t)$ be of bounded variation in every interval $T < t_1 \leqq t \leqq t_2 < \infty$. If the integral*

$$f(x) = \int_{-\infty}^{\infty} G(x - t)\, d\alpha(t)$$

converges for $T + b + \sum_{1}^{\infty} \frac{1}{a_k} < x < \infty$, then

$$f(x) = \int_{-\infty}^{\infty} G_n(x - t)\, dt \int_{-\infty}^{\infty} H_n(t - u)\, d\alpha(u)$$

for $T + b + \sum_{1}^{\infty} a_k < x < \infty$.

These theorems are proved in the same manner as Theorem 8.1a.

9. SUMMARY

9.1. In this chapter a substantially complete development of the operational inversion theory of the convolution transforms with kernels

$$G(t) = (2\pi i)^{-1} \int_{-i\infty}^{i\infty} \left[e^{bs} \prod_k \left(1 - \frac{s}{a_k}\right) e^{s/a_k} \right]^{-1} e^{st} \, ds$$

has been effected. It is to be noted that the results proved here are no less precise than those obtained in special cases by particular methods.

CHAPTER VII

Representation Theory

1. INTRODUCTION

1.1. It has been shown, see Widder [1946; 310], that necessary and sufficient conditions for $F(x)$ to be represented as a Laplace transform

$$F(x) = \int_{0+}^{\infty} e^{-xt}\, dA(t) \qquad (0 \leqq x_0 < x < \infty)$$

with $A(t)$ non-decreasing are that

$$(-1)^k F^{(k)}(x) \geqq 0 \qquad (x_0 < x < \infty;\ k = 0, 1, \cdots),$$

$$F(x) = o(1) \qquad (x \to +\infty).$$

In this chapter we shall see how to associate such a theorem with each convolution transform

$$f(x) = \int_{-\infty}^{\infty} G(x - t)\, d\alpha(t) \tag{1}$$

with kernel $G(t)$,

$$G(t) = \frac{1}{2\pi i} \int_{-i\infty}^{i\infty} [E(s)]^{-1} e^{st}\, ds,$$

$$E(s) = e^{bs} \prod_{1}^{\infty} \left(1 - \frac{s}{a_k}\right) e^{s/a_k}.$$

If $G(t) \in$ class II then, as we shall show, $f(x)$ can be represented in the form (1) with $\alpha(t) \in \uparrow$ for $\gamma < x < \infty$ if and only if

$$\prod_{1}^{n} \left((1 - \frac{D}{a_k}\right) f(x) \geqq 0 \qquad (\gamma < x < \infty;\ n = 0, 1, \cdots), \tag{2}$$

$$f(x) = o(e^{\alpha_2 x}) \qquad (x \to +\infty).$$

If we put $E(s) = 1/\Gamma(1 - s)$ then we obtain as a special case the result concerning the Laplace transform which we quoted above.

2. BEHAVIOUR AT INFINITY

2.1. It is a familiar result that if the Stieltjes transform

$$f(x) = \int_{0+}^{\infty} \frac{d\varphi(t)}{x+t}$$

converges, then

$$\begin{aligned} f(x) &= o(x^{-1}) \qquad && (x \to 0+) \\ &= o(1) && (x \to +\infty). \end{aligned}$$

Similarly if the Laplace transform

$$f(x) = \int_{0+}^{\infty} e^{-xt}\, d\varphi(t)$$

converges for some value of x, and therefore for all sufficiently large x, then

$$f(x) = o(1) \qquad (x \to +\infty).$$

These results are special cases of a general theorem which plays an important role in the representation theory.

THEOREM 2.1. *If the transform*

$$f(x) = \int_{-\infty}^{\infty} G(x-t)\, d\alpha(t) \tag{1}$$

converges for some value of x then:

A. $G(t) \in$ class I *implies*

$$\begin{aligned} f(x) &= o(e^{\alpha_2 x}) \qquad && (x \to +\infty) \\ &= o(e^{\alpha_1 x}) && (x \to -\infty); \end{aligned}$$

B. $G(t) \in$ class II *or* III *implies*

$$f(x) = o(e^{\alpha_2 x}) \qquad (x \to +\infty).$$

Let us suppose for the sake of definiteness that $G(t) \in$ class II. Let x_0 be any value for which the integral (1) is convergent. We write

$$f(x) = I_1(\zeta, x) + I_2(\zeta, x)$$

where

$$I_1(\zeta, x) = \int_{-\infty}^{\zeta} G(x-t)\, d\alpha(t), \qquad I_2(\zeta, x) = \int_{\zeta}^{\infty} G(x - t\, d\alpha(t).$$

If $x > x_0$ then $G(x-t)/G(x_0-t)$ is non-decreasing as a function of t and we have

$$I_1(\zeta, x) = \int_{-\infty}^{\zeta} \frac{G(x-t)}{G(x_0-t)} G(x_0-t)\, d\alpha(t),$$

$$= \frac{G(x-\zeta)}{G(x_0-\zeta)} \int_{\xi}^{\zeta} G(x_0-t)\, d\alpha(t) \qquad (-\infty < \xi < \zeta),$$

by the mean value theorem. It follows from this that

$$|I_1(\zeta, x)| = O(1) \qquad x \to +\infty.$$

Similarly since $\lim\limits_{t\to\infty} G(x-t)/G(x_0-t) = e^{\alpha_2(x-x_0)}$ we have

$$I_2(\zeta, x) = e^{\alpha_2(x-x_0)} \int_{\xi}^{\infty} G(x_0-t) d\alpha(t) \qquad (\zeta < \xi < \infty).$$

Thus

$$\overline{\lim_{x\to\infty}}\, |I_2(x, \zeta)|/e^{\alpha_2(x-x_0)} \leqq \varepsilon$$

where

$$\varepsilon = \underset{\xi \geqq \zeta}{\text{l.u.b.}} \left| \int_{\xi}^{\infty} G(x_0-t)\, d\alpha(t) \right|.$$

Since ε can be made arbitrarily small by taking ζ large our theorem is established for $G(t) \in$ class II. The other cases may be dealt with similarly.

2.2. We require certain elementary Tauberian theorems which will be used in conjunction with the preceding result. See Boas [1937].

THEOREM 2.2. *If*

1. $f(x) \in C^2 \qquad (0 \leqq x < \infty),$
2. $f(x) = o(e^{\alpha x}) \qquad \alpha > 0 \qquad (x \to +\infty),$
3. $f''(x) \geqq O(e^{\alpha x}) \qquad (x \to +\infty),$

then

$$f'(x) = o(e^{\alpha x}) \qquad (x \to +\infty).$$

Let θ be any real constant not equal to zero. The identity,

$$f'(x) = \theta^{-1}[f(x+\theta) - f(x)] + \theta^{-1} \int_{x+\theta}^{x} (x+\theta-t) f''(t) dt, \tag{1}$$

may be verified using integration by parts. Assumption 3 implies that there exists a non-negative constant A such that

$$f''(x) \geqq -Ae^{\alpha x} \qquad (0 \leqq x < \infty).$$

Using equation (1) we may establish by elementary estimations that

$$f'(x) \leqq o(e^{\alpha x}) + \theta A e^{\alpha(x+\theta)} \qquad (x \to +\infty, \quad \theta > 0),$$

$$f'(x) \geqq o(e^{\alpha x}) + \theta A e^{\alpha x} \qquad (x \to +\infty, \quad \theta < 0).$$

Since θ may be chosen arbitrarily small these inequalities imply that

$$f'(x) = o(e^{\alpha x}) \qquad (x \to +\infty),$$

as desired.

THEOREM 2.2b. *Let $Q(D)$ be a linear differential operator of degree n with constant coefficients $Q(D) = q_n D^n + q_{n-1} D^{n-1} + \cdots + q_0 (q_n \neq 0)$. If*

1. $f(x) \in C^n \qquad (0 \leqq x < \infty)$,
2. $f(x) = o(e^{\alpha x}) \qquad \alpha > 0 \qquad (x \to +\infty)$,
3. $Q(D) f(x) \geqq O(e^{\alpha x}) \qquad (x \to +\infty)$,

then

$$f^{(k)}(x) = o(e^{\alpha x}) \qquad (x \to +\infty; \quad k = 1, \cdots, n-1).$$

It will be sufficient to prove our theorem for $k = 1$. For, suppose that it has been established in this case and that we have $f'(x) = o(e^{\alpha x})$ as $x \to +\infty$. Let $Q^{(1)}(D) = q_n D^{n-1} + q_{n-1} D^{n-2} + \cdots + q_1$. Since $Q^{(1)}(D) f'(x) = Q(D) f(x) + q_0 f(x)$ we have $Q^{(1)}(D) f'(x) \geqq O(e^{\alpha x})$ as $x \to +\infty$. Applying our theorem with $k = 1$ to $f'(x)$ and $Q^{(1)}(D)$ we find that $f''(x) = o(e^{\alpha x})$ as $x \to +\infty$. Proceeding in this way we may show that $f^{(k)}(x) = o(e^{\alpha x})$ as $x \to +\infty$ for $k = (2, 3, \cdots, n-1)$.

Let us establish our theorem for $k = 1$. We set

$$F(x) = \int_0^x (x-t)^{n-1} Q(D) f(t)\, dt.$$

Integrating by parts we may show that

$$F(x) = (n-1)! q_n f(x) + \sum_{j=0}^{n-1} \frac{\Gamma(n)}{\Gamma(n-j)} q_j \int_0^x (x-t)^{n-j-1} f(t)\, dt + \Omega(x)$$

where $\Omega(x)$ is a polynomial of degree $n-1$. Assumption 2 implies that

$$F(x) = o(e^{\alpha x}) \qquad (x \to +\infty).$$

On the other hand

$$F''(x) = (n-1)(n-2) \int_0^x (x-t)^{n-3} Q(D) f(t)\, dt;$$

hence by assumption 3

$$F''(x) \geqq O(e^{\alpha x}) \qquad (x \to +\infty).$$

Applying Theorem 2.2a to $F(x)$ we obtain

$$F'(x) = o(e^{\alpha x}) \qquad (x \to +\infty).$$

But

$$F'(x) = (n-1)!q_n f'(x) + o(e^{\alpha x}) \qquad (x \to +\infty),$$

and so since $q_n \neq 0$, we have

$$f'(x) = o(e^{\alpha x}) \qquad (x \to +\infty)$$

as desired.

3. AN ELEMENTARY REPRESENTATION THEOREM

3.1. Throughout the present chapter we shall write

$$g(t) = \begin{cases} e^{t-1} & (-\infty < t < 1) \\ \frac{1}{2} & (t = 1) \\ 0 & (1 < t < \infty), \end{cases} \tag{1}$$

$$g(a, t) = |a|\, g(at). \tag{2}$$

This represents a slight departure from the notation of previous chapters.

Let $\{a_k\}$ $k = 1, \cdots, n$ be real numbers not zero. We set

$$\alpha_1 = \underset{a_k<0}{\text{l.u.b.}}\, [a_k, -\infty], \qquad \alpha_2 = \underset{a_k>0}{\text{g.l.b.}}\, [a_k, +\infty]. \tag{3}$$

The following simple result lies at the foundation of our representation theory.

THEOREM 3.1. *Let α_1 and α_2 be defined as above. If*

1a. $\alpha_1 > -\infty$, $\alpha_2 < +\infty$,

2a. $f(x) \in C^n(-\infty < x < \infty)$,

3a. $f^{(k)}(x) = o(e^{\alpha_2 x}) \qquad (x \to +\infty; \quad k = 0, 1, \cdots, n-1)$

$\qquad = o(e^{\alpha_1 x}) \qquad (x \to -\infty; \quad k = 0, 1, \cdots, n-1)$,

then

A. $$f(x) = g(a_1, x) * \cdots * g(a_n, x) * \prod_{k=1}^{n} \left(1 - \frac{D}{a_k}\right) e^{D/a_k} f(x) \qquad (-\infty < x < \infty).$$

If

1b. $\alpha_1 = -\infty$, $\alpha_2 < +\infty$,

2b. $f(x) \in C^n(a < x < \infty)$,

3b. $f^{(k)}(x) = o(e^{\alpha_2 x}) \qquad (x \to +\infty; \quad k = 0, 1, \cdots, n-1)$,

then

B. $$f(x) = g(a_1, x) * \cdots * g(a_n, x) * \prod_{k=1}^{n} \left(1 - \frac{D}{a_k}\right) e^{D/a_k} f(x) \qquad (a < x < \infty).$$

If

1c. $\alpha_1 > -\infty$, $\alpha_2 = +\infty$,

2c. $f(x) \in C^n$ $(-\infty < x < a)$,

3c. $f^{(k)}(x) = o(e^{\alpha_1 x})$ $\qquad (x \to -\infty;\quad k = 0, 1, \cdots, n-1)$,

then

$$\text{C. } f(x) = g(a_1, x) * \cdots * g(a_n, x) * \prod_{k=1}^{n} \left(1 - \frac{D}{a_k}\right) e^{D/a_k} f(x). \qquad (-\infty < x < a).$$

Suppose that assumptions 1a, 2a, and 3a, are satisfied. Let us write

$$\prod_{1}^{m} \left(1 - \frac{D}{a_k}\right) e^{D/a_k} f(x) = f_m(x).$$

Integrating the linear differential equation

$$\left(1 - \frac{D}{a_n}\right) e^{D/a_n} f_{n-1}(x) = f_n(x)$$

we see that

$$f_{n-1}(x) = -a_n e^{a_n\left(x - \frac{1}{a_n}\right)} \int_0^{x - \frac{1}{a_n}} e^{-a_n t} f_n(t)\, dt + a_n C e^{a_n\left(x - \frac{1}{a_n}\right)}. \tag{4}$$

There are two possibilities, $a_n \geqq \alpha_2$ or $a_n \leqq \alpha_1$. We shall consider only the first of these which is typical. By assumption we have

$$f_{n-1}(x) = o(e^{\alpha_2 x}) \qquad (x \to +\infty). \tag{5}$$

Since $a_n \geqq \alpha_2$ equations (4) and (5) imply that

$$C = \int_0^{\infty} e^{-a_n t} f_n(t)\, dt,$$

this equation may be rewritten as

$$f_{n-1}(x) = a_n e^{a_n\left(x - \frac{1}{a_n}\right)} \int_{x - \frac{1}{a_n}}^{\infty} e^{-a_n t} f_n(t)\, dt,$$

$$f_{n-1}(x) = g(a_n, x) * f_n(x).$$

Repeating this argument n times we obtain conclusion A. The proofs of conclusions B and C are entirely similar.

4. DETERMINING FUNCTION IN L^p

4.1. Let $L^p(1 \leqq p < \infty)$ denote the class of functions $\varphi(t)$ such that $\|\varphi\|_p$ is finite where

$$\|\varphi\|_p = \left[\int_{-\infty}^{\infty} |\varphi(t)|^p \, dt\right]^{1/p} \qquad 1 \leqq p < \infty,$$

and let L^∞ denote the class of functions $\varphi(t)$ such that $\|\varphi\|_\infty$ is finite where

$$\|\varphi(t)\|_\infty = \underset{-\infty<t<\infty}{\text{essential upper bound}} |\varphi(t)|.$$

There corresponds to each index p a conjugate index q defined by the equation

$$\frac{1}{p} + \frac{1}{q} = 1\,.$$

With each kernel

$$G(t) = \frac{1}{2\pi i} \int_{-i\infty}^{i\infty} \frac{e^{st}}{E(s)} \, ds,$$

$$E(s) = e^{bs} \prod_1^\infty \left(1 - \frac{s}{a_k}\right) e^{s/a_k},$$

we associate a sequence of finite kernels

$$H_n(t) = \frac{1}{2\pi i} \int_{-i\infty}^{i\infty} \frac{e^{st}}{P_n(s)} \, ds,$$

$$P_n(s) = e^{(b-b_n)s} \prod_1^n \left(1 - \frac{s}{a_k}\right) e^{s/a_k}.$$

Here $\{b_n\}_1^\infty$ is any sequence of real numbers such that $\lim_{n\to\infty} b_n = 0$.

LEMMA 4.1. *If $G(t)$ and $H_n(t)$ are defined as above, then*

$$\lim_{n\to\infty} \|G(t) - H_n(t)\|_p = 0 \qquad 1 \leqq p \leqq \infty.$$

It was shown in § 8 of Chapter III that

$$\lim_{n\to\infty} \|G(t) - H_n(t)\|_\infty = 0.$$

We have

$$\int_{-\infty}^{\infty} G(t) \, dt = 1, \qquad \int_{-\infty}^{\infty} H_n(t) \, dt = 1.$$

It is now easy to see that

$$\lim_{n\to\infty} \|G(t) - H_n(t)\|_1 = 0.$$

Finally since

$$||G(t) - H_n(t)||_p \leqq ||G(t) - H_n(t)||_1^{\frac{1}{p}} ||G(t) - H_n(t)||_\infty^{1-\frac{1}{p}}$$

we have

$$\lim_{n\to\infty} ||G(t) - H_n(t)||_p = 0.$$

For a demonstration of the following well known result see Banach [1932; 130].

THEOREM 4.1. *Let* $1 < p \leqq \infty$ *and let* $\varphi_n(t) \in L^p\ (-\infty, \infty)$ $n = 1, 2, \cdots$. *If there exists a constant M independent of n such that*

$$||\varphi_n||_p \leqq M \qquad n = 1, 2, \cdots,$$

then there exists a subsequence $n_1 < n_2 < n_3 < \cdots$ *and a function* $\varphi(t) \in L^p$, $||\varphi||_p \leqq M$, *such that for every* $\psi(t) \in L^q$ *we have*

$$\lim_{i\to\infty} \int_{-\infty}^{\infty} \psi(t)\varphi_{n_i}(t)\,dt = \int_{-\infty}^{\infty} \psi(t)\varphi(t)\,dt.$$

The functions $\varphi_{n_1}(t), \varphi_{n_2}(t), \cdots$ are said to converge weakly to $\varphi(t)$ in L^p.

4.2. We are now in a position to establish our first representation theorem.

THEOREM 4.2a. *Necessary and sufficient conditions that*

$$f(x) = \int_{-\infty}^{\infty} G(x-t)\varphi(t)\,dt$$

with $||\varphi(t)||_p \leqq M$, $1 < p \leqq \infty$, *are:*

A. $f(x) \in C^\infty \qquad (-\infty < x < \infty);$

B. $||P_n(D)f(x)||_p \leqq M \qquad n = 0, 1, \cdots.$

The necessity of A is obvious. To establish the necessity of condition B we write

$$P_n(D)f(x) = \int_{-\infty}^{\infty} G_n(x-t)\varphi(t)\,dt,$$

$$= \int_{-\infty}^{\infty} [G_n(x-t)]^{1/q}[G_n(x-t)]^{1/p}\varphi(t)\,dt.$$

By Hölder's inequality

$$|P_n(D)f(x)|^p \leqq \left[\int_{-\infty}^{\infty} G_n(x-t)\,dt\right]^{\frac{p}{q}} \int_{-\infty}^{\infty} G(_n x-t)\,|\varphi(t)|^p\,dt,$$

$$\leqq \int_{-\infty}^{\infty} G_n(x-t)\,|\varphi(t)|^p\,dt.$$

Thus

$$\| P_n(D)f(x) \|_p^p \leq$$

$$\int_{-\infty}^{\infty} \int_{-\infty}^{\infty} G_n(x-t) \,|\varphi(t)|^p \,dt\,dx = \int_{-\infty}^{\infty} |\varphi(t)|^p \int_{-\infty}^{\infty} G_n(x-t)\,dx\,dt$$

from which we obtain

$$\| P_n(D)f(x) \|_p \leq M.$$

We now turn to the sufficiency of our conditions. A simple application of Hölder's inequality yields

$$\left| \int_0^x f(t)\,dt \right| \leq \|f\|_p \,|x|^{1/q}$$

which implies that

$$\int_0^x f(t)\,dt = o(e^{\alpha_2 x}) \qquad (x \to +\infty) \tag{1}$$

$$= o(e^{\alpha_1 x}) \qquad (x \to -\infty).$$

Since

$$P_n(D) \int_0^x f(t)\,dt = \int_0^x P_n(D)f(t)\,dt + c$$

where c is a constant and since by Hölder's inequality

$$\left| \int_0^x P_n(D)\,f(t)\,dt \right| \leq \| P_n(D)f(t) \|_p \,|x|^{1/q}$$

we have

$$P_n(D) \int_0^x f(t)\,dt = O(e^{\alpha_2 x}) \qquad x \to +\infty \tag{2}$$

$$= O(e^{\alpha_1 x}) \qquad x \to -\infty.$$

The relations (1) and (2) together with Theorem 2.2b imply that

$$f^{(k)}(x) = o(e^{\alpha_2 x}) \qquad x \to +\infty \tag{3}$$

$$= o(e^{\alpha_1 x}) \qquad x \to -\infty$$

for $k = 0, 1, \cdots, n-2$, or since n is arbitrary for all k. (If $\alpha_1 = -\infty$ or $\alpha_2 = +\infty$ the corresponding relations are to be interpreted vacuously.) By Theorem 3.1 we have

$$f(x) = g(a_1, x) * \cdots * g(a_n, x) * \prod_1^n \left(1 - \frac{D}{a_k}\right) e^{D/a_k} f(x).$$

Since $\prod_1^n \left(1 - \frac{D}{a_k}\right) e^{D/a_k} f(x) \in L^p$ by B this iterated integral is absolutely convergent and may be inverted to give, after an adjustment of the translations,

$$f(x) = \int_{-\infty}^{\infty} H_n(x-t)P_n(D)f(t)\,dt$$

for $n = 1, 2, \cdots$. Let $P_n(D)f(t) = \varphi_n(t)$. By B and Theorem 4.1 there exists a set of indices $n_1 < n_2 < n_3 < \cdots$ and a function $\varphi(t)$ with $\|\varphi\|_p \leqq M$, such that $\varphi_{n_1}(t), \varphi_{n_2}(t), \cdots$ converge weakly to $\varphi(t)$ in L^p. For fixed x $G(x-t)$ belongs to L^q and hence

$$\lim_{i\to\infty} \int_{-\infty}^{\infty} G(x-t)\varphi_{n_i}(t)\,dt = \int_{-\infty}^{\infty} G(x-t)\varphi(t)\,dt. \tag{4}$$

We have

$$f(x) - \int_{-\infty}^{\infty} G(x-t)\varphi_{n_i}(t)\,dt = \int_{-\infty}^{\infty} [H_{n_i}(x-t) - G(x-t)]\varphi_{n_i}(t)\,dt.$$

By Hölder's inequality

$$\left| f(x) - \int_{-\infty}^{\infty} G(x-t)\varphi_{n_i}(t)\,dt \right| \leqq M \, \| H_{n_i} - G \|_q.$$

Lemma 4.1 implies that

$$f(x) = \lim_{i\to\infty} \int_{-\infty}^{\infty} G(x-t)\varphi_{n_i}(t)\,dt. \tag{5}$$

Combining relations (4) and (5) we obtain our theorem.

It is clear that the preceding theorem would remain true if the conditions B were replaced by the weaker conditions

B_1. $\| P_{n_i}(D)f(x) \|_p \leqq M$ $\qquad 0 = n_0 < n_1 < n_2 < \cdots,$

or

B_2. $\| P_{n_i}(D)f(x) \|_p \leqq M$ $\qquad 0 \leqq n_0 < n_1 < \cdots,$

$$f(x) = o(e^{\alpha_2 x})\ (x \to +\infty), \qquad f(x) = o(e^{\alpha_1 x})\ (x \to -\infty).$$

A slight generalization of the preceding result is given in the following theorem.

THEOREM 4.2b. *Necessary and sufficient conditions that*

$$f(x) = \int_{-\infty}^{\infty} G(x-t)e^{ct}\varphi(t)\,dt$$

where $\alpha_1 < c < \alpha_2$, $\|\varphi(t)\|_p \leqq M$, $1 < p \leqq \infty$, *are:*

A. $f(x) \in C^\infty\,(-\infty, \infty)$;

B. $\| e^{-cx} P_n(D)f(x) \|_p \leqq M/E_n(c)$ $\qquad (n = 0, 1, \cdots).$

The necessity of condition A is obvious. We have

$$e^{-cx}P_n(D)f(x) = \int_{-\infty}^{\infty} G_n(x-t)e^{-c(x-t)}\varphi(t)\,dt.$$

Using Hölder's inequality we obtain as before

$$|e^{-cx}P_n(D)f(x)|^p$$
$$\leq \left[\int_{-\infty}^{\infty} G_n(x-t)e^{-c(x-t)}|\varphi(t)|^p\,dt\right]\left[\int_{-\infty}^{\infty} G_n(x-t)e^{-c(x-t)}\,dt\right]^{\frac{p}{q}},$$

$$|e^{-cx}P_n(D)f(x)|^p \leq E_n(c)^{-\frac{p}{q}}\int_{-\infty}^{\infty} G_n(x-t)e^{-c(x-t)}|\varphi(t)|^p\,dt,$$

$$\int_{-\infty}^{\infty}|e^{-cx}P_n(D)f(x)|^p\,dx \leq E_n(c)^{-\frac{p}{q}}\int_{-\infty}^{\infty}|\varphi(t)|^p\,dt$$
$$\left[\int_{-\infty}^{\infty} G_n(x-t)e^{-c(x-t)}\,dx\right],$$

$$\|e^{-cx}P_n(D)f(x)\|_p \leq M/E_n(c).$$

Thus the necessity of condition B is established.

The proof of the sufficiency goes exactly as before.

5. DETERMINING FUNCTIONS OF BOUNDED TOTAL VARIATION

5.1. The following theorem is essentially demonstrated in Widder [1946; 26].

THEOREM 5.1. *If $\beta_n(t)$, $n = 1, 2, \cdots$, are uniformly bounded and of total variation less than or equal to M, then there exists a set of indices $n_1 < n_2 < n_3 < \cdots$, a function $\beta(t)$ of total variation not exceeding M, and two constants B_1 and B_2 such that if $\varphi(t)$ is any continuous function of t for which $\varphi(+\infty)$ and $\varphi(-\infty)$ exist then*

$$\lim_{i\to\infty}\int_{-\infty}^{\infty}\varphi(t)\,d\beta_{n_i}(t) = B_1\varphi(-\infty) + B_2\varphi(+\infty) + \int_{-\infty}^{\infty}\varphi(t)\,d\beta(t).$$

5.2.

THEOREM 5.2a. *Necessary and sufficient conditions that*

$$f(x) = \int_{-\infty}^{\infty} G(x-t)\,d\beta(t)$$

with $\beta(t)$ of total variation not exceeding M are:

A. $f(x) \in C^{\infty}$ $\qquad (-\infty < x < \infty);$

B. $\|P_n(D)f(x)\|_1 \leq M$ $\qquad n = 0, 1, \cdots.$

It is clear that condition A is necessary. We have

$$P_n(D)f(x) = \int_{-\infty}^{\infty} G(x-t)\,d\beta(t)$$

$$|P_n(D)f(x)| \leqq \int_{-\infty}^{\infty} G(x-t)\,|d\beta(t)|$$

$$\int_{-\infty}^{\infty} |P_n(D)f(x)|\,dx \leqq \int_{-\infty}^{\infty} dx \int_{-\infty}^{\infty} G(x-t)\,|d\beta(t)|.$$

Since the integrand is non-negative the order of integrations may be inverted to give

$$\int_{-\infty}^{\infty} |P_n(D)f(x)|\,dx \leqq \int_{-\infty}^{\infty} |d\beta(t)| \int_{-\infty}^{\infty} G(x-t)\,dx,$$
$$\leqq M,$$

which establishes the necessity of condition B.

To establish the sufficiency of our conditions we begin by showing, just as in the proof of Theorem 4.2a, that

$$f^{(k)}(x) = o(e^{\alpha_2 x}) \qquad x \to +\infty,$$
$$= o(e^{\alpha_1 x}) \qquad x \to -\infty.$$

By Theorem 3.1 we have

$$f(x) = \int_{-\infty}^{\infty} H_n(x-t)P_n(D)f(t)\,dt \qquad (-\infty < x < \infty),$$

$$= \int_{-\infty}^{\infty} H_n(x-t)\,d\beta_n(t),$$

where

$$\beta_n(x) = \int_0^x P_n(D)f(t)\,dt.$$

Condition B implies that the $\beta_n(x)$ are uniformly bounded and of total variation not exceeding M. By Theorem 5.1 there exists $\beta(t)$ of total variation not exceeding M such that

$$\lim_{i\to\infty} \int_{-\infty}^{\infty} \varphi(t)\,d\beta_{n_i}(t) = \int_{-\infty}^{\infty} \varphi(t)\,d\beta(t)$$

for any continuous function $\varphi(t)$ vanishing at $\pm\infty$. Since for each x $G(x-t)$ is a continuous function of t and vanishing at $\pm\infty$ we have

$$\lim_{i\to\infty} \int_{-\infty}^{\infty} G(x-t)\,d\beta_{n_i}(t) = \int_{-\infty}^{\infty} G(x-t)\,d\beta(t). \tag{1}$$

Now

$$f(x) - \int_{-\infty}^{\infty} G(x-t)\, d\beta_{n_i}(t) = \int_{-\infty}^{\infty} [H_{n_i}(x-t) - G(x-t)]\, d\beta_{n_i}(t),$$

$$\left| f(x) - \int_{-\infty}^{\infty} G(x-t)\, d\beta_{n_i}(t) \right| \leqq M \, || H_{n_i} - G ||_\infty .$$

Lemma 4.1 implies that

$$f(x) = \lim_{i\to\infty} \int_{-\infty}^{\infty} G(x-t)\, d\beta_{n_i}(t). \tag{2}$$

Equations (1) and (2) yield our theorem.

Theorem 5.2a would remain true if conditions B were replaced by the weaker conditions

B_1. $|| P_{n_i}(D) f(x) ||_1 \leqq M \qquad 0 = n_0 < n_1 < \cdots,$

or

B_2. $|| P_{n_i}(D) f(x) ||_1 \leqq M \qquad 0 \leqq n_0 < n_1 < \cdots,$

$f(x) = o(e^{\alpha_2 x}) \quad (x \to +\infty), \qquad f(x) = o(e^{\alpha_1 x}) \quad (x \to -\infty).$

THEOREM 5.2b. *Necessary and sufficient conditions that*

$$f(x) = \int_{-\infty}^{\infty} G(x-t) e^{ct}\, d\beta(t)$$

where $\alpha_1 < c < \alpha_2$ *and where* $\beta(t)$ *is of total variation not exceeding* M *are*:

A. $f(x) \in C^\infty \qquad (-\infty < x < \infty);$

B. $|| e^{-cx} P_n(D) f(x) ||_1 \leqq M / E_n(c) \qquad (n = 0, 1, \cdots).$

The demonstration of this result is left to the reader.

6. DETERMINING FUNCTION NON-DECREASING

6.1. We must here vary our argument as $G(t)$ belongs to class I, II, or III. We begin with $G(t) \in$ class I.

THEOREM 6.1. *Let* $G(t) \in$ class I. *Necessary and sufficient conditions that*

$$f(x) = \int_{-\infty}^{\infty} G(x-t)\, d\beta(t)$$

where $\beta(t) \in \uparrow$ *are*:

A. $f(x) \in C^\infty \qquad (-\infty < x < \infty);$

B. $f(x) = o(e^{\alpha_2 x}) \qquad x \to +\infty$

$\quad = o(e^{\alpha_1 x}) \qquad x \to -\infty;$

C. $P_n(D) f(x) \geqq 0 \qquad (-\infty < x < \infty;\ n = 0, 1, 2, \cdots).$

The necessity of condition A is evident and the necessity of condition B follows from Theorem 2.1. The equation

$$P_n(D)f(x) = \int_{-\infty}^{\infty} G_n(x - t)\, d\beta(t)$$

and the fact that $\beta(t) \in \uparrow$ implies that

$$P_n(D)f(x) \geqq 0 \qquad (-\infty < x < \infty;\quad n = 0, 1, \cdots)$$

which establishes the necessity of condition C.

We now turn to the sufficiency of our conditions. Conditions B and C together with Theorem 2.2b imply that

$$\begin{aligned} f^{(k)}(x) &= o(e^{\alpha_2 x}) & x \to +\infty, \\ &= o(e^{\alpha_1 x}) & x \to -\infty, \end{aligned}$$

for every k. Using Theorem 3.1 we obtain

$$f(x) = g(a_1, x) * \cdots * g(a_n, x) * \prod_1^n \left(1 - \frac{D}{a_k}\right) e^{D/a_k} f(x).$$

Since $\prod_1^n \left(1 - \frac{D}{a_k}\right) f(x) \geqq 0$, this iterated integral is absolutely convergent and may be inverted to give, after an adjustment of the translations,

$$f(x) = \int_{-\infty}^{\infty} H_n(x - t) P_n(D) f(t)\, dt.$$

Take any x', $-\infty < x' < \infty$. We may rewrite the above formula as

$$\begin{aligned} f(x) &= \int_{-\infty}^{\infty} \frac{H_n(x - t)}{H_n(x' - t)} H_n(x' - t) P_n(D) f(t)\, dt, \\ &= \int_{-\infty}^{\infty} \psi_n(x, t)\, d\varphi_n(t), \end{aligned}$$

where

$$\varphi_n(t) = \int_0^t H_n(x' - u) P_n(D) f(u)\, du,$$

$$\psi_n(x, t) = H_n\,(x - t)/H_n(x' - t).$$

Let us set

$$\psi(x, t) = G(x - t)/G(x' - t).$$

Because of C $\varphi_n(t) \in \uparrow$, $n = 1, 2, \cdots$, and since

$$\int_{-\infty}^{\infty} d\varphi_n(t) = f(x') \qquad n = 1, 2, \cdots$$

it follows that the $\varphi_n(t)$ are uniformly bounded and of uniformly bounded total variation. For each x, $\psi(x, t)$ is a continuous function of t and

$$\psi(x, +\infty) = e^{\alpha_2(x-x')}, \qquad \psi(x, -\infty) = e^{\alpha_1(x-x')}.$$

It follows from Theorem 5.1 that there exist indices $n_1 < n_2 < n_3 < \cdots$, a function $\varphi(t)$ of bounded total variation, and constants p_1 and p_2 such that

$$(1) \qquad \lim_{i\to\infty} \int_{-\infty}^{\infty} \psi(x, t)\, d\varphi_{n_i}(t) = p_1 e^{\alpha_1(x-x')} + p_2 e^{\alpha_2(x-x')} + \int_{-\infty}^{\infty} \psi(x, t)\, d\varphi(t).$$

If n is sufficiently large then $\psi_n(x, t)$ is a continuous function of t and

$$\psi_n(x, -\infty) = e^{\alpha_1(x-x')}, \qquad \psi_n(x, +\infty) = e^{\alpha_2(x-x')}.$$

By Lemma 4.1

$$\lim_{n\to\infty} \psi_n(x, t) = \psi(x, t)$$

uniformly for t in every finite interval. The functions $\psi(x, t)$, $\psi_n(x, t)$, $n = 1, 2, \cdots$, are non-decreasing if $x \geqq x'$ and non-increasing if $x \leqq x'$. It follows that for each x

$$\lim_{n\to\infty} \| \psi(x, t) - \psi_n(x, t) \|_\infty = 0.$$

We have

$$f(x) - \int_{-\infty}^{\infty} \psi(x, t)\, d\varphi_{n_i}(t) = \int_{-\infty}^{\infty} [\psi_{n_i}(x, t) - \psi(x, t)]\, d\varphi_{n_i}(t),$$

$$\left| f(x) - \int_{-\infty}^{\infty} \psi(x, t)\, d\varphi_{n_i}(t) \right| \leqq f(x') \| \psi_{n_i}(x, t) - \psi(x, t) \|_\infty,$$

and from this we obtain

$$(2) \qquad f(x) = \lim_{i\to\infty} \int_{-\infty}^{\infty} \psi(x, t) \varphi_{n_i}(t).$$

Combining relations (1) and (2) we obtain

$$f(x) = p_1 e^{\alpha_1(x-x')} + p_2 e^{\alpha_2(x-x')} + \int_{-\infty}^{\infty} \psi(x, t)\, d\varphi(t),$$

$$= p_1 e^{\alpha_1(x-x')} + p_2 e^{\alpha_2(x-x')} + \int_{-\infty}^{\infty} G(x - t)\, d\beta(t),$$

where

$$\beta(t) = \int_0^t \frac{d\varphi(u)}{G(x' - u)}.$$

By Theorem 2.1

$$\int_{-\infty}^{\infty} G(x - t)\, d\beta(t) = o(e^{\alpha_2 x}) \qquad x \to +\infty,$$

$$= o(e^{\alpha_1 x}) \qquad x \to -\infty.$$

Making use of this and condition B we see that $p_1 = p_2 = 0$, so that we have

$$f(x) = \int_{-\infty}^{\infty} G(x-t)\, d\beta(t).$$

Theorem 6.1 remains true if condition C is replaced by the weaker condition

C$_1$. $P_{n_i}(D)f(x) \geqq 0 \qquad\qquad 0 \leqq n_0 < n_1 < \cdots.$

THEOREM 6.2. *Let* $G(t) \in$ class II. *Necessary and sufficient conditions that*

$$f(x) = \int_{-\infty}^{\infty} G(x-t)\, d\beta(t) \qquad (x > \gamma_c)$$

where $\beta(t) \in \uparrow$ *are:*

A. $f(x) \in C^\infty \qquad (\gamma_c < x < \infty)$;

B. $f(x) = o(e^{\alpha_2 x}) \qquad x \to +\infty$;

C. $P_n(D)f(x) \geqq 0\ (\gamma_c - b + b_n - \sum_1^n \frac{1}{a_k} < x < \infty,\ n = 0, 1, \cdots)$.

The necessity of our conditions may be established just as in the proof of the preceding theorem. In order to establish their sufficiency we show precisely as before that

$$f^{(k)}(x) = o(e^{\alpha_2 x}) \qquad (x \to +\infty)$$

for every k. There is of course no corresponding result as $x \to -\infty$. By Theorem 3.1

$$f(x) = g(a_1, x) * g(a_2, x) * \cdots * g(a_n, x) * \prod_{k=1}^{n} \left(1 - \frac{D}{a_k}\right) e^{D/a_k} f(x)$$

for $x > \gamma_c$. Since the functions $g(a_1, x), \cdots, g(a_n, x), \prod_1^n \left(1 - \frac{D}{a_k}\right) e^{D/a_k} f(x)$

are non-negative the order of the integrations may be inverted to give

$$f(x) = \int_{-\infty}^{\infty} H_n(x-t) P_n(D) f(t)\, dt \qquad x > \gamma_c.$$

Choose any $x' > \gamma_c$; for $x > x'$ we have

$$\begin{aligned} f(x) &= \int_{-\infty}^{\infty} \frac{H_n(x-t)}{H_n(x'-t)} H_n(x'-t) P_n(D) f(t)\, dt, \\ &= \int_{-\infty}^{\infty} \psi_n(x, t)\, d\varphi_n(t), \end{aligned}$$

where

$$\varphi_n(t) = \int_0^t H_n(x' - u)P_n(D)f(u)\,du,$$
$$\psi_n(x, t) = H_n(x - t)/H_n(x' - t).$$

Here 0/0 is defined as 0. The functions $\psi_n(x, t)$ are continuous and non-decreasing, and for n sufficiently large

$$\psi_n(x, -\infty) = 0, \qquad \psi_n(x, +\infty) = e^{\alpha_2(x-x')}.$$

If we set

$$\psi(x, t) = G(x - t)/G(x' - t),$$

then $\psi(x, t)$ is continuous and non-decreasing and

$$\psi(x, -\infty) = 0, \qquad \psi(x, +\infty) = e^{\alpha_2(x-x')}.$$

Further $\psi_n(x, t)$ converges to $\psi(x, t)$ as $n \to \infty$ uniformly in every finite interval by Lemma 4.1. It is now easy to deduce that

$$\lim_{n\to\infty} ||\psi(x, t) - \psi_n(x, t)||_\infty = 0.$$

The proof from this point on can be completed as before.

Theorem 6.2 is still true if we replace condition C by condition

C₁. $P_{n_i}(D)f(x) \geqq 0$

$$\left(\gamma_c - b + b_{n_i} - \sum_1^{n_i} \frac{1}{a_k} < x < \infty;\ \ 0 \leqq n_0 < n_1 < \cdots\right).$$

6.3.

THEOREM 6.3. *Let* $G(t) \in$ class III. *Necessary and sufficient conditions that*

$$f(x) = \int_{-\infty}^{\infty} G(x - t)\,d\beta(t) \quad \left(x > T + b + \sum_1^\infty \frac{1}{a_k}\right)$$

where $\beta(t) \in \uparrow$ $(T < t < \infty)$ *are:*

A. $f(x) \in C^\infty$ $\left(T + b + \sum_1^\infty \frac{1}{a_k} < x < \infty\right)$;

B. $f(x) = o(e^{\alpha_2 x})$ $(x \to +\infty)$;

C. $P_n(D)f(x) \geqq 0$

$$\left(T + \sum_{n+1}^\infty \frac{1}{a_k} + b_n < x < \infty;\ \ n = 0, 1, \cdots\right).$$

The proof of this theorem is left to the reader.

In Theorem 6.3 condition C could be replaced by the weaker condition

C₁. $P_{n_i}(D)f(x) \geqq 0$

$$\left(T + \sum_{n_i+1}^\infty \frac{1}{a_k} + b_{n_i} < x < \infty;\ \ 0 \leqq n_0 < n_1 < \cdots\right).$$

6.4. As we remarked in § 1, the familiar Hausdorff-Bernstein-Widder theorem is a particular case of the results of the present section. It is curious that the demonstration given here does not specialize to any of the known proofs of this theorem.

7. REPRESENTATIONS OF PRODUCTS

7.1. Let

$$E(s) = e^{bs} \prod_1^\infty \left(1 - \frac{s}{a_k}\right) e^{s/a_k},$$

$$E_j(s) = e^{b_j s} \prod_1^\infty \left(1 - \frac{s}{a_{jk}}\right) e^{s/a_{jk}} \qquad (j = 1, \cdots, n),$$

where we assume of course that $\sum a_k^{-2} < \infty$, $\sum a_{jk}^{-2} < \infty$ $(j = 1, \cdots, n)$, and let

$$G(t) = \frac{1}{2\pi i} \int_{-i\infty}^{i\infty} \frac{e^{st}}{E(s)}\, ds,$$

$$H_j(t) = \frac{1}{2\pi i} \int_{-i\infty}^{i\infty} \frac{e^{st}}{E_j(s)}\, ds \qquad (j = 1, \cdots, n).$$

We wish to give conditions which will insure that if, for example,

$$f_j(x) = \int_{-\infty}^{\infty} H_j(x - t)\, d\beta_j(t) \qquad (j = 1, \cdots, n)$$

with $\beta_j \in \uparrow$ $(j = 1, \cdots, n)$, and

$$f(x) = f_1(x) f_2(x) \cdots f_n(x)$$

then $f(x)$ will be representable in the form

$$f(x) = \int_{-\infty}^{\infty} G(x - t)\, d\beta(t)$$

with $\beta(t) \in \uparrow$. In order to do this we rewrite the products $E(s)$, $E_j(s)$ in the form

$$E(s) = e^{bs} \prod_{k=-\infty}^{\infty}{}' \left(1 - \frac{s}{A_k}\right) e^{s/A_k}$$

$$\cdots \leqq A_{-2} \leqq A_{-1} < 0 < A_1 \leqq A_2 \leqq \cdots,$$

$$E_j(s) = e^{b_j s} \prod_{k=-\infty}^{\infty}{}' \left(1 - \frac{s}{A_{jk}}\right) e^{s/A_{jk}}$$

$$\cdots \leqq A_{j,-2} \leqq A_{j,-1} < 0 < A_{j,1} \leqq A_{j,2} \leqq \cdots.$$

If necessary A_k and A_{jk} may be $+\infty$ or $-\infty$.

Definition 7.1. We write $G(t) \sim [H_1(t), \cdots, H_n(t)]$ if

$$\sum_{j=1}^n A_{j,\nu_j} \leqq A_\nu$$

for $\nu_j \geqq 1 (j = 1, \cdots, n)$, $\sum_{j=1}^n \nu_j \leqq \nu + n - 1$, and if

$$\sum_{j=1}^n A_{j,\nu_j} \geqq A_\nu$$

for $\nu_j \leqq -1 (j = 1, \cdots, n)$, $\sum_{j=1}^n \nu_j \geqq \nu - n + 1$.

LEMMA 7.1. If $G(t) \sim [H_1(t), \cdots, H_n(t)]$ and if

$$f(x) = f_1(x)f_2(x) \cdots f_n(x)$$

then

$$\prod_{k=q}^{p}{}' \left(1 - \frac{D}{A_k}\right) f(x) = \sum A(p, p_1, \cdots, p_n;\ \ q, q_1, \cdots, q_n)$$
$$\left[\prod_{q_1}^{p_1}{}' \left(1 - \frac{D}{A_{1k}}\right) f_1(x)\right] \cdots \left[\prod_{q_n}^{p_n}{}' \left(1 - \frac{D}{A_{nk}}\right) f_n(x)\right]$$

where $p \geqq 0$, $q \leqq 0$ and the summation extends over all values of the indices $p_1, \cdots, p_n, q_1, \cdots, q_n$ for which $p_j \geqq 0, q_j \leqq 0$ $(j = 1, \cdots, n)$, $p_1 + p_2 + \cdots + p_n \leqq p$, $q_1 + q_2 + \cdots + q_n \geqq q$. Moreover, we have

$$A(p, p_1, p_2, \cdots, p_n;\ \ q, q_1, \cdots, q_n) \geqq 0,$$
$$\sum A(p, p_1, p_2, \cdots, p_n;\ \ q, q_1, \cdots, q_n) = 1,$$

the range of indices in the summation being as above.

This may be established by induction making use of the identities

$$\text{(1)} \quad \left(1 - \frac{D}{A_{p+1}}\right) f(x) = f(x) \left[1 - \sum_{j=1}^{n} \frac{A_{j,p_j+1}}{A_{p+1}}\right] + \sum_{j=1}^{n} \frac{A_{j,p_j+1}}{A_{p+1}} \frac{f(x)}{f_j(x)} \left(1 - \frac{D}{A_{j,p_j+1}}\right) f_j(x),$$

$$\text{(2)} \quad \left(1 - \frac{D}{A_{q-1}}\right) f(x) = f(x) \left[1 - \sum_{j=1}^{n} \frac{A_{j,q_j-1}}{A_{q-1}}\right] + \sum_{j=1}^{n} \frac{A_{j,q_j-1}}{A_{q-1}} \frac{f(x)}{f_j(x)} \left(1 - \frac{D}{A_{j,q_j-1}}\right) f_j(x).$$

Note that if $p_1 + p_2 + \cdots + p_n \leqq p$, $q_1 + q_2 + \cdots + q_n \geqq q$, then

$$(p_1 + 1) + \cdots + (p_n + 1) \leqq (p + 1) + n - 1,$$
$$(q_1 - 1) + \cdots + (q_n - 1) \geqq (q - 1) - n + 1,$$

so that by our assumptions the coefficients in equations (1) and (2) are non-negative.

THEOREM 7.1. *If* $G(t) \sim [H_1(t), \cdots, H_n(t)]$, *if*

$$f_j(x) = \int_{-\infty}^{\infty} H_j(x - t)\varphi_j(t)dt \qquad \varphi_j(t) \in L^{r_j}(-\infty, \infty),$$

where $1 < r_j \leqq \infty$ $j = 1, \cdots, n$, *and if* $1 < r \leqq \infty$ *where*

$$1/r = (1/r_1) + \cdots + (1/r_n),$$

then

$$f_1(x)f_2(x) \cdots f_n(x) = \int_{-\infty}^{\infty} G(x - t)\varphi(t)\, dt$$

where $\varphi(t) \in L^r(-\infty, \infty)$. *More precisely we have*

$$||\varphi||_r \leqq ||\varphi_1||_{r_1} ||\varphi_2||_{r_2} \cdots ||\varphi_n||_{r_n}.$$

Let $f(x) = f_1(x)f_2(x) \cdots f_n(x)$. By Theorem 4.2a we have for every $p_j \geqq 0$, $q_j \leqq 0$

$$(3) \qquad ||\prod_{q_j}^{p_j}{}' \left(1 - \frac{D}{A_{jk}}\right) f_j(x) ||_{r_j} \leqq ||\varphi_j||_{r_j} \qquad (j = 1, \cdots, n)$$

By Lemma 7.1 we have

$$\prod_{q}^{p}{}' \left(1 - \frac{D}{A_k}\right) f(x) = \sum A(p, p_1, \cdots, p_n; \; q, q_1, \cdots, q_n) \left[\prod_{q_1}^{p_1}{}' \left(1 - \frac{D}{A_{1j}}\right) f_1(x)\right] \cdots \left[\prod_{q_n}^{p_n}{}' \left(1 - \frac{D}{A_{nj}}\right) f_n(x)\right].$$

Applying Minkowski's inequality we obtain

$$||\prod_{q}^{p}{}' \left(1 - \frac{D}{A_k}\right) f(x) ||_r \leqq \sum A(\cdots) || \left[\prod_{q_1}^{p_1}{}' \left(1 - \frac{D}{A_{1j}}\right) f_1(x)\right] \cdots \left[\prod_{q_n}^{p_n}{}' \left(1 - \frac{D}{A_{nj}}\right) f(x)\right] ||_r.$$

The following inequality is well known

$$||g_1(x) \cdots g_n(x)||_r \leqq ||g_1||_{r_1} ||g_2||_{r_2} \cdots ||g_n||_{r_n},$$

see Titchmarsh [1939; 394]. Thus

$$||\prod_{q}^{p}{}' \left(1 - \frac{D}{A_k}\right) f(x) ||_r \leqq \sum A(\cdots) ||\prod_{q_1}^{p_1}{}' \left(1 - \frac{D}{A_{1j}}\right) f_1 ||_{r_1} \cdots ||\prod_{q_n}^{p_n}{}' \left(1 - \frac{D}{A_{nj}}\right) f_n ||_{r_n}.$$

Making use of (3) we find that

$$||\prod_{q}^{p}{}' \left(1 - \frac{D}{A_k}\right) f(x) ||_r \leqq ||\varphi_1||_{r_1} ||\varphi_2||_{r_2} \cdots ||\varphi_n||_{r_n} \sum A(\cdots),$$

$$(4) \qquad ||\prod_{q}^{p}{}' \left(1 - \frac{D}{A_k}\right) f(x) ||_r \leqq ||\varphi_1||_{r_1} ||\varphi_2||_{r_2} \cdots ||\varphi_n||_{r_n}.$$

Here we have used Lemma 7.1 again. The inequality (4) together with Theorem 4.2a yields our desired result.

7.2.

THEOREM 7.2a. *Let* $G(t) \in$ class I *and let* $G(t) \sim [H_1(t), \cdots, H_n(t)]$. *If*

$$f_j(x) = \int_{-\infty}^{\infty} H_j(x - t) \, d\beta_j(t) \qquad (-\infty < x < \infty)$$

where $\beta_j(t) \in \uparrow$ $(j = 1, \cdots, n)$ *then*

$$f_1(x)f_2(x) \cdots f_n(x) = \int_{-\infty}^{\infty} G(x-t)\, d\beta(t)$$

where $\beta(t) \in \uparrow$.

It is easily verified that $G(t) \in$ class I implies $H_j(t) \in$ class I, $j = 1, \cdots, n$. By Theorem 6.1 we have

$$f_j(x) = o(e^{\alpha_{j2}x}) \qquad (x \to +\infty),$$

$$f_j(x) = o(e^{\alpha_{j1}x}) \qquad (x \to -\infty),$$

for $j = 1, \cdots, n$ where

$$\alpha_{j1} = \underset{a_{jk}<0}{\text{l.u.b.}}\, (a_{jk}, -\infty), \qquad \alpha_{j2} = \underset{a_{jk}>0}{\text{g.l.b.}}\, (a_{jk}, +\infty).$$

Since $\alpha_{j1} = A_{j,-1}$, $\alpha_{j2} = A_{j,+1}$, and since $\alpha_1 = A_{-1}$, $\alpha_2 = A_{+1}$ our assumptions imply that

$$\sum_{j=1}^{n} \alpha_{j1} \geqq \alpha_1, \qquad \sum_{j=1}^{n} \alpha_{j2} \leqq \alpha_2.$$

It follows that if $f(x) = f_1(x)f_2(x) \cdots f_n(x)$ then

$$f(x) = o(e^{\alpha_2 x}) \qquad (x \to +\infty)$$

$$f(x) = o(e^{\alpha_1 x}) \qquad (x \to -\infty).$$

For every $p_j \geqq 0$, $q_j \leqq 0$ we have

$$\prod_{q_j}^{p_j}{}' \left(1 - \frac{D}{A_{jk}}\right) f_j(x) \geqq 0 \qquad (-\infty < x < \infty).$$

Using Lemma 7.1 we see that

$$\prod_{q}^{p}{}' \left(1 - \frac{D}{A_k}\right) f(x) \geqq 0 \qquad (-\infty < x < \infty).$$

Appealing again to Theorem 6.1 we obtain our desired result.

THEOREM 7.2b. *Let* $G(t) \in$ class II *or* III *and let* $G(t) \sim [H(t), \cdots, H_n(t)]$. *If*

$$f_j(x) = \int_{-\infty}^{\infty} H_j(x-t)\, d\beta_j(t) \qquad (T_j < x < \infty)$$

where $\beta_j(t) \in \uparrow$ $(j = 1, \cdots, n)$ *then*

$$f_1(x)f_2(x) \cdots f_n(x) = \int_{-\infty}^{\infty} G(x-t)\, d\beta(t) \qquad (T < x < \infty)$$

where $\beta(t) \in \uparrow$. *Here* $T = \max(T_1, T_2, \cdots, T_n)$.

The demonstration of this result is exactly like that of the preceding theorem except that we use Theorems 6.2 and 6.3 in place of Theorem 6.1.

7.3. We note that if $0 < \lambda_j < 1$, $j = 1, \cdots, n$, and if $\lambda_1 + \lambda_2 + \cdots + \lambda_n \leqq 1$ then our assumptions will be satisfied if (for any $G(t)$) we set $H_j(t) = \lambda_j G(\lambda_j t)$. Since if

$$f(x) = \int_{-\infty}^{\infty} G(x - t)\varphi(t)\, dt$$

then

$$f(\lambda_j x) = \int_{-\infty}^{\infty} \lambda_j G[\lambda_j(x - t)]\varphi(\lambda_j t)\, dt,$$

we obtain as corollaries the following results.

THEOREM 7.3a. *Let $\lambda_j > 0, j = 1, \cdots, n, \lambda_1 + \lambda_2 + \cdots + \lambda_n \leqq 1$. If*

$$f_j(x) = \int_{-\infty}^{\infty} G(x - t)\varphi_j(t)dt \qquad \varphi_j(t) \in L^{r_j}(-\infty, \infty)$$

where $1 < r_j \leqq \infty$ for $j = 1, \cdots, n$, and if $1 < r \leqq \infty$ where

$$1/r = (1/r_1) + \cdots + (1/r_n)$$

then

$$f_1(\lambda_1 x)f_2(\lambda_2 x) \cdots f_n(\lambda_n x) = \int_{-\infty}^{\infty} G(x - t)\varphi(t)\, dt$$

where $\varphi(t) \in L^r(-\infty, \infty)$. More precisely we have

$$\|\varphi\|_r \leqq \lambda_1^{-1/r_1} \lambda_2^{-1/r_2} \cdots \lambda_n^{-1/r_n} \|\varphi_1\|_{r_1} \|\varphi_2\|_{r_2} \cdots \|\varphi_n\|_{r_n}.$$

THEOREM 7.3b. *Let $G(t) \in$ class I and let $\lambda_j > 0$, $j = 1, \cdots, n$, $\lambda_1 + \lambda_2 + \cdots + \lambda_n \leqq 1$. If*

$$f_j(x) = \int_{-\infty}^{\infty} G(x - t)\, d\beta_j(t) \qquad (-\infty < x < \infty)$$

where $\beta_j(t) \in \uparrow$ $j = 1, \cdots, n$ then

$$f_1(\lambda_1 x)f_2(\lambda_2 x) \cdots f_n(\lambda_n x) = \int_{-\infty}^{\infty} G(x - t)\, d\beta(t) \qquad (-\infty < x < \infty)$$

where $\beta(t) \in \uparrow$.

THEOREM 7.3c. *Let $G(t) \in$ class II or III and let $\lambda_j > 0, j = 1, \cdots, n$, $\lambda_1 + \lambda_9 + \cdots + \lambda_n \leqq 1$. If*

$$f_j(x) = \int_{-\infty}^{\infty} G(x - t)\, d\beta_j(t) \qquad (T_j < x < \infty)$$

where $\beta_j(t) \in \uparrow$ $j = 1, \cdots, n$, then

$$f(\lambda_1 x)f(\lambda_2 x) \cdots f_n(\lambda_n x) = \int_{-\infty}^{\infty} G(x - t)\, d\beta(t) \qquad (T < x < \infty)$$

where $\beta(t) \in \uparrow$, and $T = \max\,[T_1/\lambda_1, T_2/\lambda_2, \cdots, T_n/\lambda_n]$.

7.4. As an example let

$$G(t) = \frac{\Gamma(\mu)2^{-\mu}}{\Gamma\left(\frac{\mu}{2}\right)^2}\left[\operatorname{sech}\frac{t}{2}\right]^{\mu}$$

where $\mu > 0$ and let

$$H_j(t) = \frac{\Gamma(\mu_j)2^{-\mu_j}}{\Gamma\left(\frac{\mu_j}{2}\right)^2}\left[\operatorname{sech}\frac{t}{2}\right]^{\mu_j}$$

where $\mu_j > 0, j = 1, \cdots, n$. See § 9 of III. Here we have

$$A_k = \frac{\mu}{2} + k - 1 \ (k > 0), \qquad A_k = -\frac{\mu}{2} + k + 1 \quad (k < 0),$$

$$A_{jk} = \frac{\mu_j}{2} + k - 1 \ (k > 0), \qquad A_{jk} = -\frac{\mu_j}{2} + k + 1 \quad (k < 0).$$

The conditions of Definition 7.1 are satisfied if $\mu = \mu_1 + \mu_2 + \cdots + \mu_n$. After a logarithmic change of variables we find that if

$$f_j(x) = \int_{0+}^{\infty} \frac{d\beta_j(t)}{(x+t)^{\mu_j}}, \qquad \beta_j(t) \in \uparrow, \qquad (j = 1, \cdots, n),$$

then

$$f_1(x)f_2(x)\cdots f_n(x) = \int_{0+}^{\infty} \frac{d\beta(t)}{(x+t)^{\mu}}, \qquad \beta(t) \in \uparrow.$$

As a second example we set

$$G(t) = \frac{1}{\Gamma(n)} e^{nt}e^{-e^t},$$

$$H_j(t) = e^t e^{-e^t}$$

for $j = 1, \cdots, n$. See § 9 of III. Here we have

$$A_k = n - 1 + k \quad (k > 0), \qquad A_k = -\infty \quad (k < 0),$$

$$A_{jk} = k \quad (k > 0), \qquad A_{jk} = -\infty \quad (k < 0).$$

The conditions of Definition 7.1 are satisfied. After a logarithmic change of variables we find that if

$$f_j(x) = \int_{0+}^{\infty} e^{-xt}\, d\beta_j(t) \qquad \beta_j(t) \in \uparrow \qquad (j = 1, \cdots, n)$$

then

$$f_1(x)f_2(x)\cdots f_n(x) = \int_{0+}^{\infty} e^{-xt}\, d\beta(t). \qquad \beta(t) \in \uparrow.$$

This is a familiar result and may be proved directly. The present methods although not direct are quite general. See also H. Pollard [1946*a*].

8. SUMMARY

8.1. We have developed in this chapter a representation theory corresponding to our previously obtained inversion theory. Roughly speaking, necessary and sufficient conditions for $f(x)$ to be of the form

$$f(x) = \int_{-\infty}^{\infty} G(x - t)\varphi(t)\, dt$$

with $\varphi(t)$ of some prescribed character are that the functions $P_n(D)_y(x)$ possess this character uniformly in n.

CHAPTER VIII

The Weierstrass Transform

1. INTRODUCTION

1.1. In Chapter III we were led in a natural way to see that the most general inversion function $E(s)$ commensurate with our methods should be a function of class II (Pólya-Laguerre):

$$E(s) = e^{-cs^2+bs} \prod_{k=1}^{\infty} \left(1 - \frac{s}{a_k}\right) e^{s/a_k} \tag{1}$$

$$\prod_{k=1}^{\infty} \frac{1}{a_k^2} < \infty.$$

We have thus far treated only such functions for which $c = 0$ and have proved that $E(D)$ is indeed an inversion operator for a suitable convolution transform, namely that for which the kernel has a bilateral Laplace transform equal to $1/E(s)$. In the present chapter we shall treat the complementary case $E(s) = e^{-cs^2}$, $c > 0$, the one factor of (1) hitherto neglected. Since this function has no roots an interpretation of $E(D)$ analogous to that used in Chapter III is denied us, and as a consequence the earlier methods must be modified here.

As a basis of attack consider the known Laplace transform

$$e^{cs^2} = \frac{1}{\sqrt{4\pi c}} \int_{-\infty}^{\infty} e^{-sy} e^{-y^2/4c}\, dy \qquad c > 0, \ -\infty < s < \infty. \tag{2}$$

This shows that if $E(s)$ is the special function e^{-cs^2}, then its reciprocal is still the bilateral Laplace transform of a frequency function

$$G(x) = (4\pi c)^{-1/2} e^{-x^2/4c}. \tag{3}$$

If we adopt this function as the kernel of a convolution transform,

$$f(x) = G * \varphi = \int_{-\infty}^{\infty} G(x - y)\varphi(y)\, dy, \tag{4}$$

we may hope, from previous experience, that the transform will be inverted by the operator e^{-cD^2} if properly interpreted. This hope may be further strengthened by the following formal considerations.

In view of equation (2) it is natural to interpret $e^{cD^2}\varphi(x)$ as

$$e^{cD^2}\varphi(x) = \frac{1}{\sqrt{4\pi c}}\int_{-\infty}^{\infty} e^{yD}\varphi(x)e^{-y^2/4c}\,dy$$

$$= \frac{1}{\sqrt{4\pi c}}\int_{-\infty}^{\infty} \varphi(x-y)e^{-y^2/4c}\,dy,$$

or by equation (4)

$$e^{cD^2}\varphi(x) = f(x).$$

Then if D is treated as a number we obtain the symbolic equation

$$\varphi(x) = e^{-cD^2}f(x), \tag{5}$$

the predicted inversion formula.

Observe that if x is replaced by $\sqrt{c}\,x$ and y by $\sqrt{c}\,y$ in equation (4) it becomes

$$f(\sqrt{c}\,x) = \frac{1}{\sqrt{4\pi}}\int_{-\infty}^{\infty} e^{-(x-y)^2/4}\varphi(\sqrt{c}\,y)\,dy.$$

That is, $f(\sqrt{c}\,x)$ is the convolution transform of $\varphi(\sqrt{c}\,y)$ having kernel equal to the function (3) with $c = 1$. Hence there is no loss in generality in supposing $c = 1$. This we do throughout the remainder of the chapter. Following an accepted terminology, E. Hille [1948; 371], we refer to the resulting transform as the Weierstrass transform.

The chief purpose of the present chapter will be to interpret and prove the inversion formula (5) and then to use it for the purpose of developing representation theory. That is, we seek necessary and sufficient conditions upon a function $f(x)$ in order that it may be the Weierstrass transform of a function φ of prescribed class (such as $\varphi \in L^p$). It will develop that $e^{tD^2}\varphi(x)$ is a solution of the heat equation

$$\frac{\partial^2 u}{\partial x^2} = \frac{\partial u}{\partial t},$$

and much use will be made of this fact. Many results, of interest in themselves, will be proved about such solutions. In particular, we will obtain a characterization of solutions which are positive in a half-plane, $t > 0$. This will be analogous to a classical theorem of Herglotz concerning positive harmonic functions.

1.2. Since the Laplace transform 1.1 (2) is no longer valid when $c < 0$ it does not provide us with an interpretation of e^{-D^2}. However, the complex inversion of that transform, D. V. Widder [1946; 241], suggests an alternative procedure. Replace c by $(4t)^{-1}$, $0 < t \leqq 1$ in 1.1 (2) and invert:

$$e^{-y^2t} = \frac{1}{2\pi i}\int_{a-i\infty}^{a+i\infty} K(s, t)e^{sy}\, ds \qquad -\infty < a < \infty$$

$$K(s, t) = (\pi/t)^{1/2}e^{s^2/4t}.$$

We are thus led to write

$$e^{-tD^2}f(x) = \frac{1}{2\pi i}\int_{a-i\infty}^{a+i\infty} K(s, t)e^{sD}f(x)\, ds$$

$$= \frac{1}{2\pi i}\int_{a-i\infty}^{a+i\infty} K(s, t)f(x + s)\, ds.$$

By a change of variable this becomes

$$e^{-tD^2}f(x) = \frac{1}{2\pi i}\int_{d-i\infty}^{d+i\infty} K(s - x, t)f(s)\, ds, \tag{1}$$

where the constant d is arbitrary and will be chosen so that the vertical line $\sigma = d$ will be one on which $f(s)$ is defined. It will be seen later that it is not appropriate to define e^{-D^2} by setting $t = 1$ in (1), for the integral (1) would then diverge for certain Weierstrass transforms $f(x)$. We choose rather to set

$$e^{-D^2}f(x) = \lim_{t\to 1-} e^{-tD^2}f(x), \tag{2}$$

and this operator will serve to invert every convergent Weierstrass transform. If we set

$$k(x, t) = (4\pi t)^{-1/2}e^{-x^2/4t},$$

the above statement becomes

$$e^{-D^2}\int_{-\infty}^{\infty} k(x - y, 1)\varphi(y)\, dy = \varphi(x).$$

1.3. In order that the reader may see clearly that the methods of the present chapter are in essence the same as those employed earlier, we set forth here in juxtaposition the corresponding functions and operations of Chapters III and VIII. The chief point of contrast is that the inversion operator of the former

$$E(D) = \lim_{n\to\infty} P_n(D)$$

involves the discrete parameter n, whereas 1.2 (2) involves the continuous parameter t. The following table will bring out the analogies in detail.

$E(s) \in E, \quad c = 0$	$E(s) = e^{-s^2}$
1. $E(s) = e^{bs} \prod_{k=1}^{\infty} \left(1 - \frac{s}{a_k}\right) e^{s/a_k}$	1. $E(s) = e^{-s^2}$
2. $G(x) = \frac{1}{2\pi i} \int_{-i\infty}^{i\infty} \frac{e^{sx}}{E(s)} ds$	2. $G(x) = \frac{1}{2\pi i} \int_{-i\infty}^{i\infty} \frac{e^{sx}}{e^{-s^2}} ds = k(x, 1)$
3. $P_n(s) = e^{bs} \prod_{k=1}^{n} \left(1 - \frac{s}{a_k}\right) e^{s/a_k}$	3. $P_t(s) = e^{-ts^2}$
$\lim_{n \to \infty} P_n(s) = E(s)$	$\lim_{t \to 1-} P_t(s) = E(s)$
4. $E_n(s) = E(t)/P_n(s)$	4. $E_t(s) = E(s)/P_t(s)$
5. $G_n(x) = \frac{1}{2\pi i} \int_{-i\infty}^{i\infty} \frac{e^{sx}}{E_n(s)} ds$	5. $G_t(x) = \frac{1}{2\pi i} \int_{-i\infty}^{i\infty} \frac{e^{sx}}{E_t(s)} ds = k(x, 1-t)$
6. $P_n(D)G(x) = G_n(x)$	6. $e^{-tD^2} k(x, 1) = k(x, 1-t)$
7. $\lim_{n \to \infty} G_n(x) = \delta(x)$	7. $\lim_{t \to 1-} k(x, 1-t) = \delta(x)$
8. $\frac{P_{n+1}(D)}{P_n(D)} G_n(x) = G_{n+1}(x)$	8. $\frac{\partial}{\partial t} k(x, t) = \frac{\partial^2}{\partial x^2} k(x, t)$

These analogues will become abundantly clear as the analysis of the chapter progresses. In 7 the Dirac δ-function is intended. To establish 8 from 6 in the continuous case, differentiate with respect to t

$$\frac{\partial}{\partial t} e^{-tD^2} k(x, 1) = -e^{-tD^2} \; D^2 k(x, 1)$$

$$= -\frac{\partial^2}{\partial x^2} e^{-tD^2} k(x, 1)$$

$$\frac{\partial}{\partial t} k(x, 1-t) = -\frac{\partial^2}{\partial x^2} k(x, 1-t).$$

2. THE WEIERSTRASS TRANSFORM

2.1. We now make our formal definition of the Weierstrass transform.
DEFINITION 2.1a.

$$k(x, t) = (4\pi t)^{-1/2} e^{-x^2/4t} \qquad 0 < t < \infty, \qquad -\infty < x < \infty.$$

This is the familiar "source solution" of the heat equation, and $k(x, 1)$ will be the kernel of our transform.

DEFINITION 2.1b. The Weierstrass transform of a function $\varphi(x)$ is the function

$$f(x) = \int_{-\infty}^{\infty} k(x-y, 1)\varphi(y)\, dy \tag{1}$$

whenever the integral converges.

DEFINITION 2.1c. The Weierstrass-Stieltjes transform of a function $\alpha(x)$ is the function

$$f(x) = \int_{-\infty}^{\infty} k(x-y, 1)\, d\alpha(y)$$

whenever the integral converges.

Here $\alpha(y)$ is of bounded variation in every finite interval. If, for example, $\alpha(y)$ is constant except for a single unit jump at the origin, $f(x) = (4\pi)^{-1/2}e^{-x^2/4}$.

We turn next to a theorem of Abelian character which will prove useful to us later. The proof requires an elementary lemma.

LEMMA 2.1. If $\alpha(x)$ is of bounded variation in $a \leqq x \leqq R$ for every $R > 0$, $\alpha(\infty)$ exists, $\beta(x)$ is positive, continuous and non-increasing in $a \leqq x < \infty$, then

$$\int_a^{\infty} \beta(x)\, d\alpha(x)$$

converges.

For if ϵ is given we can determine R so that

$$|\alpha(S) - \alpha(R)| < \epsilon \qquad R \leqq S.$$

Then

$$\int_R^S \beta(x)\, d\alpha(x) = \int_R^S \beta(x)\, d[\alpha(x) - \alpha(R)]$$

$$= \beta(S)[\alpha(S) - \alpha(R)] - \int_R^S [\alpha(x) - \alpha(R)]\, d\beta(x).$$

Hence

$$\left| \int_R^S \beta(x)\, d\alpha(x) \right| < \epsilon\beta(S) + \epsilon[\beta(R) - \beta(S)]$$

$$\leqq \epsilon\beta(a).$$

This proves the lemma.

THEOREM 2.1. *If the integral*

$$\int_{-\infty}^{\infty} k(x-y, t)\, d\alpha(y)$$

converges to a value A when $x = x_0$, $t = 1$, then it also converges for $-\infty < x < \infty$, $0 < t < 1$, *and*

$$\lim_{t \to 1-} \int_{-\infty}^{\infty} k(x_0 - y, t)\, d\alpha(y) = A.$$

For fixed x, x_0, t set

$$\alpha^*(y) = \int_0^y e^{-(x_0 - v)^2/4}\, d\alpha(v)$$

$$\log \beta(y) = -\frac{(x-y)^2}{4t} + \frac{(x_0 - y)^2}{4}.$$

Then

$$(4\pi t)^{-1/2} \int_0^{\infty} \beta(y)\, d\alpha^*(y) = \int_0^{\infty} k(x-y, t)\, d\alpha(y).$$

This integral converges, by Lemma 2.1, since $\beta(y)$ is decreasing for large y and $\alpha^*(\infty)$ exists by hypothesis. If we replace y by $-y$ we may prove by use of the result just proved that

$$\int_{-\infty}^{0} k(x-y, t)\, d\alpha(y)$$

converges.

Finally, note that

$$\int_{-\infty}^{\infty} k(x_0 - y, t)\, d\alpha(y) = (4\pi t)^{-1/2} \int_0^{\infty} e^{-y/4t}\, d[\alpha(x_0 + \sqrt{y}) - \alpha(x_0 - \sqrt{y})].$$

The integral on the right is a Laplace integral which converges at $t = 1$ by hypothesis. Since it represents a continuous function of t in $0 < t \leqq 1$, D. V. Widder [1946; 56], the theorem is proved.

2.2. Let us investigate next the relation of the transform 2.1 (1) to the Laplace transform. It may easily be put into the following form

$$(1) \qquad f(-2x)e^{x^2} = \frac{1}{\sqrt{4\pi}} \int_{-\infty}^{\infty} e^{-xy} e^{-y^2/4} \varphi(y)\, dy,$$

and this is clearly a bilateral Laplace integral. Hence we may infer many of the properties of the Weierstrass transform from known theory, D. V. Widder [1946; 237]. For example, the region of convergence is an interval, as noted above. Or if x is replaced by the complex variable $s = \sigma + i\tau$, the region of convergence is a vertical strip of the complex plane. Also we may deduce at once a complex inversion of the Weierstrass transform.

THEOREM 2.2. *If $\varphi(x)$ is of bounded variation in a neighborhood of a point x_0, has the transform $f(x)$ defined by 2.1 (1), and if*

$$\int_{-\infty}^{\infty} e^{-x^2/4} \left| \varphi(x) \right| dx < \infty, \tag{2}$$

then

$$\lim_{T \to \infty} \frac{1}{\sqrt{4\pi}} \int_{-T}^{T} e^{(x-iy)^2/4} f(iy)\, dy = \frac{\varphi(x_0+) + \varphi(x_0-)}{2}. \tag{3}$$

This is an immediate consequence of Theorem 5a, D. V. Widder [1946; 241] applied to equation (1). Inequality (2) asserts that the integral 2.1 (1) converges absolutely at the origin and hence on the line $\sigma = 0$. If $f(x)$ is the transform of $\varphi(y)$, then $f(x - c)$ is the transform of $\varphi(y - c)$ for any constant c. Hence if 2.1 (1) is known to be absolutely convergent at some other point than the origin, Theorem 2.2 may still be made to apply by a translation.

COROLLARY 2.2. If $\varphi(x)$ satisfies the conditions of Theorem 2.2 and if for some positive constant c

$$f(x) = \int_{-\infty}^{\infty} k(x - y, c)\varphi(y)\, dy,$$

then

$$\lim_{T \to \infty} \int_{-T}^{T} k(y + ix_0, c) f(iy)\, dy = \frac{\varphi(x_0+) + \varphi(x_0-)}{2}.$$

This corollary follows from the theorem by a simple change of variable.

As an example of the theorem consider the pair $f(x) = \varphi(x) = 1$. Equation (3) becomes

$$1 = \frac{1}{\sqrt{4\pi}} \int_{-\infty}^{\infty} e^{(x-iy)^2/4}\, dy.$$

The Cauchy value is not needed since this integral converges absolutely for all x. It reduces to

$$e^{-x^2/4} = \frac{1}{\sqrt{4\pi}} \int_{-\infty}^{\infty} e^{-ixy/2} e^{-y^2/4}\, dy,$$

or by analytic continuation to

$$e^{s^2} = \frac{1}{\sqrt{4\pi}} \int_{-\infty}^{\infty} e^{-sy} e^{-y^2/4}\, dy \qquad -\infty < \sigma < \infty.$$

We thus have a proof of the formula (2) § 1, there assumed. In particular, this result gives the following familiar integral expression for the source solution

$$k(x, t) = \frac{1}{2\pi} \int_{-\infty}^{\infty} e^{-y^2 t} e^{ixy}\, dy = \frac{1}{\pi} \int_{-\infty}^{\infty} e^{-y^2 t} \cos xy\, dy. \tag{4}$$

2.3. We shall need a few simple properties of the source solution. We collect them in

THEOREM 2.3. *If $k(x, t)$ is the function of Definition 2.1a, then*

A. $k_{xx}(x, t) = k_t(x, t) = k(x, t)\,(x^2 - 2t)/4t^2$

B. $|\,k(s, t)\,| \leq At^{-1/2}e^{(\tau^2-\sigma^2)/4t}$

C. $|\,k_{xx}(s, t)\,| \leq At^{-5/2}(\,|\,s\,|^2 + 2t)e^{(\tau^2-\sigma^2)/4t}.$

Here A is a suitable constant and $s = \sigma + i\tau$. Conclusion A follows by differentiation of the equation

$$\log k(x,t) = -\frac{1}{2}\log(4\pi t) - \frac{x^2}{4t}.$$

Conclusions B and C follow directly from the definition of $k(x, t)$.

2.4. Another property of $k(x, t)$ that will be essential to us is contained in

THEOREM 2.4. *If $k(x, t)$ is the function of Definition 2.1a, and if $0 < t_1$, $0 < t_2$, $-\infty < x < \infty$, then*

$$\int_{-\infty}^{\infty} k(x - y, t_1)k(y, t_2)\,dy = k(x, t_1 + t_2).$$

For, the integral is equal to

$$\frac{(t_1t_2)^{-1/2}}{4\pi}\, e^{(4t_1B^2-x^2)/4t_1} \int_{-\infty}^{\infty} e^{-(Ay+B)^2}\,dy, \tag{1}$$

where

$$A^2 = \frac{t_1 + t_2}{4t_1t_2}, \qquad B^2 = \frac{t_2x^2}{4t_1(t_1 + t_2)}.$$

Since the integral appearing in (1) is equal to $\sqrt{\pi}/A$, the desired result follows.

2.5. A companion result, useful when the variable x in $k(x, t)$ is complex is the following.

THEOREM 2.5. *If $k(x, t)$ is the function of Definition 2.1a, and if $0 < t_1 < t_2$, $-\infty < x < \infty$, $-\infty < v < \infty$, then*

$$\int_{-\infty}^{\infty} k(y + ix, t_1)k(iy - v, t_2)\,dy = k(x - v, t_2 - t_1). \tag{1}$$

To prove this note first that by Theorem 2.4

$$\int_{-\infty}^{\infty} k(x - y, t_1)k(y - v, t_2 - t_1)\,dy = k(x - v, t_2).$$

Now apply Corollary 2.2 to this equation with

$$\varphi(y) = k(y - v, t_2 - t_1).$$

The result is (1). Hypothesis 2.2 (2) is evidently satisfied trivially here since for fixed t_1, t_2 and v

$$k(y - v, t_2 - t_1) = O(1) \qquad |y| \to \infty.$$

2.6. We conclude this section by a brief table of Weierstrass transforms. It will serve the double purpose of illustrating the types of functions which can be Weierstrass transforms and of providing concrete examples as checks for our later theory.

$$f(x) = \int_{-\infty}^{\infty} k(x - y, 1)\varphi(y)\, dy \tag{1}$$

	$f(x)$		$\varphi(y)$
1.	1		1
2.	x		y
3.	x^2		$y^2 - 2$
4.	x^n		$H_n(y/2)$
5.	e^{ax}	(all a)	e^{ay-a^2}
6.	e^{x^2}		$5^{-1/2}e^{y^2/5}$
7.	$e^{-x^2/5}$		$5^{1/2}e^{-y^2}$
8.	$e^{ax^2/(1-4a)}$	$(-\infty < a < 1/4)$	$(1 - 4a)^{1/2}e^{ay^2}$
9.	$2\pi^{-1/2}\dfrac{e^{-x^2/4}}{1 - x^2}$		$e^{y^2/4}e^{-\lvert y\rvert/2}$

For all of these pairs except 9 the integral (1) converges in $-\infty < x < \infty$. For 9 the interval of convergence is $-1 < x < 1$. Pairs 1, 2, 3 are special cases of 4, and 6, 7 are included in 8 (easily proved by (2) § 1). The pair 9 results at once from the equation

$$\frac{2a}{a^2 - x^2} = \int_{-\infty}^{\infty} e^{-xy}e^{-a|y|}\, dy \qquad |x| < a,$$

which may be verified by direct integration or by use of the function $g(t)$, § 6.1, Chapter II.

If the formula

$$1 = \int_{-\infty}^{\infty} k(x + 2a - y, 1)\, dy$$

is expanded, the pair 5 is proved. Finally, to establish 4 we must show that

$$x^n = \int_{-\infty}^{\infty} k(x - y, 1) H_n(y/2)\, dy,$$

or

$$(2) \qquad (2x)^n = \frac{1}{\sqrt{\pi}} \int_{-\infty}^{\infty} e^{-(x-y)^2} H_n(y)\, dy,$$

where $H_n(x)$ is the Hermite polynomial, defined by the equation

$$(3) \qquad H_n(x) = (-1)^n e^{x^2} D^n e^{-x^2}.$$

Using (3) in the integral (2), the latter equation reduces, after integration by parts, to

$$1 = \frac{1}{\sqrt{\pi}} \int_{-\infty}^{\infty} e^{-(x-y)^2}\, dy.$$

This proves the desired result.

3. THE INVERSION OPERATOR

3.1. We have already indicated in § 1.2 the heuristic considerations which lead us to the following definition.

DEFINITION 3.1a.

$$(1) \qquad K(s, t) = (\pi/t)^{1/2} e^{s^2/4t} = 2\pi k(is, t).$$

DEFINITION 3.1b. The operator $e^{-D^2}f(x)$ is defined to be

$$(2) \qquad e^{-D^2}f(x) = \lim_{t \to 1-} \frac{1}{2\pi i} \int_{d-i\infty}^{d+i\infty} K(s - x, t) f(s)\, ds$$

whenever the integral converges for $0 < t < 1$ and the limit exists.

At first sight this operator seems to depend on d, but for all functions $f(x)$ to which we shall apply it the result will be independent of d by virtue of Cauchy's integral theorem.

As an example, take $f(x) = e^{-x^2/4}$. The integral (2) diverges for all x when $t = 1$ but converges for all x when $t < 1$. Simple computations show that the limit (2) is zero when $x \neq 0$, fails to exist when $x = 0$.

Note that if $f(s)$ is defined on the imaginary axis, $\sigma = 0$, the constant d may be chosen equal to zero and $e^{-tD^2}f(x)$ takes the form

$$(3) \qquad e^{-tD^2}f(x) = \int_{-\infty}^{\infty} k(y + ix, t) f(iy)\, dy.$$

3.2. Let us introduce a notation for a class of functions to which e^{-D^2} will be applicable and to which all Weierstrass transforms will belong.

DEFINITION 3.2. A function $f(x)$ belongs to class A in an interval $a < x < b$ if and only if it can be extended analytically into the complex plane in such a way that

1. $f(x + iy)$ is analytic in the strip $a < x < b$
2. $f(x + iy) = o(|y| e^{y^2/4})$, $|y| \to \infty$, uniformly in every closed subinterval of $a < x < b$.

For example, $e^{-x^2/4} \in A$ in $-\infty < x < \infty$. That every Weierstrass transform convergent in $a < x < b$ belongs to A in that interval follows from equation 2.2 (1) and the known order of Laplace transforms on vertical lines, D. V. Widder [1946; 92]. If $f(x) \in A$ in $-\infty < x < \infty$ another useful form of $e^{-D^2}f(x)$ is available.

THEOREM 3.2. *If $f(x) \in A$ in $-\infty < x < \infty$, then*

$$e^{-tD^2} f(x) = \int_{-\infty}^{\infty} k(y, t) f(x + iy)\, dy, \tag{1}$$

the integral converging absolutely for $-\infty < x < \infty$, $0 < t < 1$.

Since the function

$$e^{(z-\lambda)^2/4t} f(z) \qquad -\infty < \lambda < \infty, \qquad 0 < t < 1,$$

is an entire function of z, we see by Cauchy's integral theorem that the integrals of this function along the two vertical lines $x = \lambda$ and $x = d$ (any d) are equal if

$$\lim_{|y| \to \infty} \int_{\lambda}^{d} e^{(x-\lambda+iy)^2/4t} f(x + iy)\, dx = 0.$$

But this is evident by assumption 2 of Definition 3.2. Hence

$$\frac{i}{\sqrt{4\pi t}} \int_{-\infty}^{\infty} e^{-y^2/4t} f(\lambda + iy)\, dy = \frac{1}{\sqrt{4\pi t}} \int_{d-i\infty}^{d+i\infty} e^{(z-\lambda)^2/4t} f(z)\, dz,$$

or

$$\int_{-\infty}^{\infty} k(y, t) f(\lambda + iy)\, dy = \frac{1}{2\pi i} \int_{d-i\infty}^{d+i\infty} K(z - \lambda, t) f(z)\, dz.$$

Replacing λ by x and referring to 3.1 (2) we obtain our result. Note that this theorem confirms our earlier remark that $e^{-D^2} f(x)$ will generally be independent of the constant d appearing in Definition 3.1b.

COROLLARY 3.2. If $f(x) \in A$ in $-\infty < x < \infty$, then

$$e^{-tD^2} f(x) = \int_{-\infty}^{\infty} k(y, t) \cos yD\, f(x)\, dy, \tag{2}$$

where

$$\cos yD\, f(x) = \sum_{k=0}^{\infty} \frac{(-1)^k}{(2k)!} f^{(2k)}(x) y^{2k}.$$

For, by Taylor's theorem

$$f(x + iy) = e^{iyD} f(x) = \cos yD\, f(x) + i \sin yD\, f(x).$$

The imaginary part being an odd function of y disappears when substituted in the integral (1) so that the corollary is proved. This latter form of the operator puts into evidence the fact that $e^{-tD^2}f(x)$ is real when $f(x)$ is real.

3.3. We show next that $e^{tD^2}\varphi(x)$ is a solution of the heat equation when it is defined. Since we shall have frequent use for such solutions let us introduce an abbreviation.

DEFINITION 3.3. A function $u(x, t) \in H$ in a domain D if and only if $u(x, t) \in C^2$ and $u_{xx}(x, t) = u_t(x, t)$ there; $u(x, t) \in H$ in a region R (perhaps closed) if R can be enclosed in a domain in which $u(x, t) \in H$.

For example, $u(x, t) \in H$ in $x^2 + t^2 \leqq 1$ implies $u(x, t) \in H$ in $x^2 + t^2 \leqq \rho^2$ for some $\rho > 1$.

THEOREM 3.3. *If*

$$u(x, t) = \int_{-\infty}^{\infty} k(x - y, t)\, d\alpha(y),$$

the integral converging in the strip $0 < t < c$, *then* $u(x, t) \in H$ *there.*

For fixed t this integral is the product of the entire function $e^{-x^2/4t}$ by a Laplace transform which converges for all x. Hence $u(x, t)$ is an entire function of x, and differentiation under the integral sign is permitted, D. V. Widder [1946; 57]. Hence

$$u_{xx} = \int_{-\infty}^{\infty} k_{xx}(x - y, t)\, d\alpha(y). \tag{1}$$

To prove that

$$u_t = \int_{-\infty}^{\infty} k_t(x - y, t)\, d\alpha(y) = \int_{-\infty}^{\infty} k_t(y, t)\, d_y\alpha(x - y), \tag{2}$$

it will be sufficient to show that for any fixed x the integral (2) is uniformly convergent for $\delta \leqq t \leqq c - \delta$, where δ is an arbitrary positive number less than c. The integral (2) is the sum of two others corresponding to the intervals of integration $(0, \infty)$ and $(-\infty, 0)$. By Theorem 2.3 the first is

$$\frac{1}{4t^2}\int_0^{\infty} k(y, t)\,(y^2 - 2t)\, d_y\alpha(x - y) =$$

$$\frac{1}{4t^2\sqrt{\pi t}}\int_0^{\infty} e^{-y/t}(2y - t)\, d_y\alpha(x - 2\sqrt{y}).$$

Since the latter integral is the sum of two Laplace integrals (in $1/t$) we may appeal to known theory, D. V. Widder [1946; 54] to verify the desired uniform convergence. The integral over the range $(-\infty, 0)$ is treated similarly. Another appeal to Theorem 2.3 shows that the integrals (1) and (2) are equal, and the proof is complete.

COROLLARY 3.3. If the Weierstrass transform of $\varphi(x)$ converges for some x, then $e^{tD^2}\varphi(x) \in H$ in the strip $0 < t < 1$.

By hypothesis the integral $e^{D^2}\varphi(x)$ will converge for some x_0. By Theorem 2.1 the integral $e^{tD^2}\varphi(x)$ will converge for all x, $0 < t < 1$ and Theorem 3.3 is applicable.

3.4. We prove the following companion result.

THEOREM 3.4. *If $f(d + iy) = o(|y| e^{y^2/4})$, $|y| \to \infty$, for some number d, then the function*

$$u(x, t) = e^{-(1-t)D^2}f(x) = \frac{1}{2\pi i}\int_{d-i\infty}^{d+i\infty} K(s - x, 1 - t)f(s)\, ds$$

will belong to H in the strip $0 < t < 1$.

By the definition of $K(s, t)$ and by Theorem 2.3 we have

$$u = \int_{-\infty}^{\infty} k(y - id + ix, 1 - t)f(d + iy)\, dy$$

$$u_{xx} = -\int_{-\infty}^{\infty} k_{xx}(y - id + ix, 1 - t)f(d + iy)\, dy = u_t \tag{1}$$

provided that differentiation under the integral sign is valid. We show that for every $R > 0$ the integral (1) is uniformly convergent in $|x| \leqq R$, $\delta \leqq t \leqq 1 - \delta$, $0 < \delta < 1$. By C of Theorem 2.3 the following will be a dominant integral independent of x and t in that region

$$A\delta^{-5/2}\int_{-\infty}^{\infty} (4R^2 + y^2 + 2)e^{R^2/\delta}e^{-y^2/4(1-\delta)}O(|y| e^{y^2/4})\, dy.$$

This integral converges since $0 < \delta < 1$, and the proof is complete.

4. INVERSION

4.1. Let us show in this section how the operator e^{-D^2} inverts the Weierstrass transform in the special case in which $\varphi(x)$ is bounded and continuous. The essentials of the method will thus be put clearly into evidence in the absence of the technical difficulties which more general hypotheses will introduce into later work.

THEOREM 4.1. *If $\varphi(y)$ is bounded and continuous in $-\infty < y < \infty$, and if*

$$f(x) = \int_{-\infty}^{\infty} k(x - y, 1)\varphi(y)\, dy,$$

then

$$e^{-D^2}f(x) = \varphi(x) \qquad -\infty < x < \infty.$$

By Definition 3.1b, with $d = 0$,

$$e^{-D^2}f(x) \doteq \lim_{t\to 1-} \int_{-\infty}^{\infty} k(y + ix, t) f(iy)\, dy$$

$$= \lim_{t\to 1-} \int_{-\infty}^{\infty} k(y + ix, t)\, dy \int_{-\infty}^{\infty} k(iy - z, 1)\varphi(z)\, dz.$$

By B of Theorem 2.3 and the boundedness of $\varphi(z)$ Fubini's theorem is applicable, so that

$$e^{-D^2}f(x) = \lim_{t\to 1-} \int_{-\infty}^{\infty} \varphi(z)dz \int_{-\infty}^{\infty} k(y + ix, t)k(iy - z, 1)\, dy$$

$$= \lim_{t\to 1-} \int_{-\infty}^{\infty} k(x - z, 1 - t)\varphi(z)\, dz.$$

We have here used Theorem 2.5. Finally, by a change of variable

$$(1) \qquad e^{-D^2}f(x) = \lim_{t\to 1-} \frac{1}{\sqrt{\pi}} \int_{-\infty}^{\infty} e^{-z^2}\varphi(x + 2z\sqrt{1 - t})\, dz.$$

Now we may use Lebesgue's limit theorem, since the integrand is dominated by a constant multiple of the integrable function e^{-z^2}. Hence the limit (1) is $\varphi(x)$, and the theorem is proved.

5. TYCHONOFF'S UNIQUENESS THEOREM

5.1. We have seen that $e^{tD^2}\varphi(x)$ and $e^{-(1-t)D^2}f(x)$ satisfy the heat equation. To capitalize on this fact we need to know a few elementary facts about solutions of that equation. In particular we shall need to know to what extent a solution $u(x, t)$ is uniquely determined at later times by its values at a given time t. One form of this uniqueness theorem is due to A. Tychonoff [1935; 199] and is the form presented here.

5.2. We show first that a function of class H, like a harmonic function, cannot take on its minimum (or maximum) value in the interior of a region where it belongs to H. We need in fact a slightly more general result. We shall not assume that $u(x, t)$ belongs to H on the boundary of our region nor that it approaches a limit as we approach the boundary. We shall thus be able to apply the theorem to functions like the source solution $k(x, t)$ which approaches no limit as (x, t) approaches the origin. We will need to consider only rectangular regions.

To save writing let us introduce the following notation. Denote by D the set of points (x, t) for which $|x| < R$, $0 < t \leq c$. That is, it is the interior of a rectangle plus its upper boundary. $\bar{D}$ shall be the closure of D, the interior of the rectangle plus all of its boundary. And B shall be $\bar{D} - D$, the lower side and the two vertical sides. The top points of the vertical sides form a part of B.

Theorem 5.2. *If*

1. $u(x, t) \in H$ *in* D
2. $\underline{\lim}_{\substack{x \to x_0 \\ t \to t_0}} u(x, t) \geqq 0$ *for all* (x_0, t_0) *on* B,

then

$$u(x, t) \geqq 0 \text{ in } D. \tag{1}$$

In hypothesis 2 it is assumed that $(x, t) \to (x_0, t_0)$ through points of D. As mentioned above, the theorem avoids an assumption about the existence of a limit on B. By the definition of "$\underline{\lim}$," for each $\epsilon > 0$ and each (x_0, t_0) of B there is a δ_0 such that $u(x, t) > -\epsilon$ at all points (x, t) of D within a distance δ_0 of (x_0, t_0). Since B is a closed set we may use the Heine-Borel theorem to show that there is rim of points B_δ in D all a distance $<\delta$ from B where $u(x, t) > -\varepsilon$.

Let us now make an assumption contrary to (1),

$$u(x_1, t_1) = -l < 0 \qquad (x_1, t_1) \in D,$$

and deduce a contradiction. Form the equation

$$v(x, t) = u(x, t) + r(t - t_1) \tag{2}$$

where r is a positive number to be determined so that $v(x, t)$ will take on a minimum in D. Whatever it is, $v(x_1, t_1) = -l$. Hence we need only determine r so that $v(x, t) > -l$ in B_δ. But there

$$v(x, t) > -\epsilon + r(t - t_1) > -\epsilon - rt_1.$$

Hence if we choose r so that $rt_1 < l$ and then $\epsilon < l - rt_1$, we have $v(x, t) > -l$ in B_δ, as desired. Hence $v(x, t)$ must take on a minimum in D, in fact in $D - B_\delta$, say at (x_2, t_2). But at such an *interior* minimum $v_x(x_2, t_2) = 0$, $v_t(x_2, t_2) \leqq 0$ (the inequality being possible if $t_2 = c$) and $v_{xx}(x_2, t_2) \geqq 0$. That is, $v_{xx}(x_2, t_2) - v_t(x_2, t_2) \geqq 0$. But since $u(x, t) \in H$, equation (2) gives $v_{xx} - v_t = -r < 0$ for all (x, t) in D, and the desired contradiction is at hand.

Of course it follows from this theorem that if $u(x, t) \in H$ in $\bar{D}$, then it has a minimum value for all points of $\bar{D}$ which is taken on at a point of B.

5.3. We can now prove the useful theorem of Tychonoff.

Theorem 5.3. *If*

1. $u(x, t) \in H$ *in the strip* $0 < t \leqq c$
2. $\lim_{\substack{x \to x_0 \\ t \to 0+}} u(x, t) = 0$ *for all* x_0, $-\infty < x_0 < \infty$
3. $f(x) = \max_{0<t\leqq c} |u(x, t)|$
4. $f(x) = O(e^{ax^2})$, $|x| \to \infty$ *for some* a,

then $u(x, t) = 0$ *throughout the strip* $0 < t \leqq c$.

Define

$$U_R(x, t) = f(-R)k(x + R, t) + f(R)k(x - R, t),$$

where $k(x, t)$ is the source solution and R is an arbitrary constant >0. Using the notation of § 5.2, let us show that the two functions $(4\pi c)^{1/2}U_R(x, t) \pm u(x, t)$ satisfy the hypotheses of the previous theorem. They belong to H in D and as (x, t) approaches the points of B on the x-axis

$$\varliminf_{\substack{x\to x_0 \\ t\to 0+}} (4\pi c)^{1/2}U_R(x, t) \pm u(x, t) \geqq \varliminf_{\substack{x\to x_0 \\ t\to 0+}} u(x, t) = 0,$$

$-R \leqq x_0 \leqq R$. Next consider the vertical sides of B. We have

$$U_R(\pm R, t) \geqq f(\pm R)k(0, t) \geqq f(\pm R)\,(4\pi c)^{-1/2}$$

for $0 < t \leqq c$. By the definition of $f(\pm R)$,

$$|u(\pm R, t)| \leqq f(\pm R) \leqq (4\pi c)^{1/2}U_R(\pm R, t).$$

This shows that the two functions in question are $\geqq 0$ on the vertical sides of B. By Theorem 5.2 they must be $\geqq 0$ throughout D,

$$-(4\pi c)^{1/2}U_R(x, t) \leqq u(x, t) \leqq (4\pi c)^{1/2}U_R(x, t).$$

Now hold (x, t) fixed and allow R to become infinite. By hypothesis 4 $U_R(x, t)$ will tend to zero if $t < 1/(4a)$. If $4ac < 1$ the proof is complete. Otherwise, repeat the above argument with $u(x, t)$ replaced by $u(x, t + 1/(4a))$, and so forth.

It is important to note that the approach of (x, t) to $(x_0, 0)$ in hypothesis 2 is a two-dimensional one. The theorem would be false if that hypothesis read $u(x, 0+) = 0$, as the example $k(x, t)x/t$ shows. This function belongs to H for $t > 0$ and approaches zero along *every* vertical line as $t \to 0+$. But it is not identically zero!

6. THE WEIERSTRASS TRANSFORM OF BOUNDED FUNCTIONS

6.1. We can now already characterize those functions which are Weierstrass transforms of bounded functions. Here again, as in § 4, the technical difficulties which will confront us for other classes are lacking, so that the essence of the method will not be clouded by detail.

6.2. We need next a result concerning solutions of the heat equation.

LEMMA 6.2. If

1. $u(x, t) \in H$ $\qquad -\infty < x < \infty, \quad 0 < t < c$

2. $|u(x, t)| < M$ $\qquad -\infty < x < \infty, \quad 0 < t < c,$

then for $0 < \delta < c$, $0 < t < c - \delta$, $-\infty < x < \infty$

$$u(x, t + \delta) = \int_{-\infty}^{\infty} k(x - y, t)u(y, \delta)\, dy.$$

Denote this integral, which converges for all x and all positive t, by $v(x, t + \delta)$. It will be enough to show that $v - u$ satisfies the hypotheses of Theorem 5.3. As in § 4.1

$$v(x, t + \delta) = \frac{1}{\sqrt{\pi}} \int_{-\infty}^{\infty} e^{-y^2}\, u(x + 2y\sqrt{t}, \delta)\, dy,$$

and since $u(x, t)$ is bounded, Lebesgue's limit theorem enables us to show that $v(x, t + \delta) \to u(x_0, \delta)$ as $(x, t) \to (x_0, 0+)$. By continuity $u(x, t + \delta)$ approaches the same value. Both functions belong to H and both satisfy condition 4 of Theorem 5.3 ($a = 0$) since they are bounded. Hence $v - u$ is identically zero, as stated.

THEOREM 6.2. *Conditions* 1 *and* 2 *of Lemma* 6.2 *are necessary and sufficient that*

$$(1) \qquad u(x, t) = \int_{-\infty}^{\infty} k(x - y, t)\varphi(y)\, dy \qquad -\infty < x < \infty, \quad 0 < t < c,$$

where $|\varphi(y)| < M$, $-\infty < y < \infty$.

We prove first the necessity, assuming the representation (1). That condition 1 is satisfied we see by Theorem 3.3. Condition 2 follows from equation (2), § 1.1. In fact, both conditions are satisfied in the half-plane $0 < t < \infty$.

In proving the converse we begin by use of Lemma 6.2. Then allowing δ to approach zero, we have

$$(2) \qquad u(x, t) = \lim_{\delta \to 0+} \int_{-\infty}^{\infty} k(x - y, t)u(y, \delta)\, dy.$$

In the first instance $t < c - \delta$, but since $\delta \to 0$, t may be taken as any number $< c$. We may now use a weak compactness theorem, D. V. Widder [1946; 33], to conclude the proof. By that theorem there exists a function $\varphi(y)$, $|\varphi(y)| < M$, and a sequence δ_n, $n = 1, 2, \cdots$, tending to 0 such that

$$\lim_{n \to \infty} \int_{-\infty}^{\infty} k(x - y, t)u(y, \delta_n)\, dy = \int_{-\infty}^{\infty} k(x - y, t)\varphi(y)\, dy.$$

This limit, through a subset of the values of δ used in (2), must have the same value, $u(x, t)$, as given by (2). This completes the proof.

6.3. We turn next to the principal result of this section.

THEOREM 6.3. *The conditions*

1. $f(x) \in A \qquad -\infty < x < \infty,$
2. $|e^{-tD^2} f(x)| < M \qquad -\infty < x < \infty, \qquad 0 < t < 1$

are necessary and sufficient that

$$f(x) = \int_{-\infty}^{\infty} k(x - y, 1)\varphi(y)\, dy \qquad -\infty < x < \infty, \tag{1}$$

where $|\varphi(y)| < M$, $-\infty < y < \infty$.

The conditions are necessary. For if (1) holds

$$e^{-tD^2} f(x) = \int_{-\infty}^{\infty} k(x - y, 1 - t)\varphi(y)\, dy, \tag{2}$$

as was proved in § 4.1. Since the integral (1) converges absolutely for all x it is clear that $f(x) \in A$ in $-\infty < x < \infty$. Condition 2 follows from (2) and from (2) § 1.1.

Conversely, condition 1 insures that the function

$$u(x, t) = e^{-(1-t)D^2} f(x)$$

$\in H$ in $-\infty < x < \infty$, $0 < t < 1$, by Theorem 3.4. Moreover, $u(x, t)$ is bounded there so that Theorem 6.2 is applicable to conclude that

$$u(x, t) = \int_{-\infty}^{\infty} k(x - y, t)\varphi(y)\, dy \qquad -\infty < x < \infty,\ 0 < t < 1,$$

where $|\varphi(y)| < M$, $-\infty < y < \infty$. But this integral converges for all $t > 0$, so by continuity

$$u(x, 1-) = \int_{-\infty}^{\infty} k(x - y, 1)\varphi(y)\, dy.$$

Hence our result will be proved if we show $u(x, 1-) = f(x)$. By equation (1) § 3.2 and by a change of variable

$$u(x, t) = \int_{-\infty}^{\infty} k(y, 1 - t) f(x + iy)\, dy$$

$$= \int_{-\infty}^{\infty} k(y, 1) f(x + iy\sqrt{1 - t})\, dy.$$

Since $f(x) \in A$ the last integral is dominated by

$$N \int_{-\infty}^{\infty} k(y, 1) e^{y^2/8} |y|\, dy < \infty \qquad 1/2 < t < 1$$

for a suitable constant N. Hence another application of Lebesgue's limit theorem shows that

$$\lim_{t \to 1-} u(x, t) = \int_{-\infty}^{\infty} k(y, 1) f(x)\, dy = f(x).$$

This completes the proof of the theorem.

7. INVERSION, GENERAL CASE

7.1. In § 4 we showed how to invert the Weierstrass transform of a function $\varphi(x)$ under simplified hypotheses. We turn next to the general case. We observe that the inversion provided by Theorem 2.2 required $\varphi(x)$ to be of bounded variation in a neighborhood of every point where the inversion was to be effective. The present method, with no local condition imposed upon $\varphi(x)$, will provide inversion for almost all x. We assume, as always, that $\varphi(x)$ is absolutely integrable in every finite interval. Note that formula (3) § 2.2 is closely related to our definition of $e^{-D^2}f(x)$. It adopts a different method of summability.

7.2. Let us prove first the following preliminary result.

LEMMA 7.2. *If*

1. $\int_{-\infty}^{\infty} k(x_0 - y, 1)\varphi(y)\, dy$ *converges for some* x_0

2. $\int_a^y [\varphi(v) - \varphi(a)]\, dv = o(y - a) \qquad y \to a+,$

then

$$\lim_{t\to 0+} \int_a^{\infty} k(a - y, t)\varphi(y)\, dy = \frac{\varphi(a)}{2}. \tag{1}$$

Choose an arbitrary positive δ and write the integral (1) as the sum of two others, $I_1(t)$, $I_2(t)$, corresponding respectively to the intervals of integration $(a, a + \delta)$, $(a + \delta, \infty)$. Set

$$\beta(v) = e^{-(v-a)^2/4t} e^{(v-x_0)^2/4}$$

$$\alpha(v) = \int_{a+\delta}^{v} e^{-(y-x_0)^2/4}\, \varphi(y)\, dy.$$

Then $\alpha(a + \delta) = 0$, and $\alpha(+\infty)$ exists by hypothesis. Moreover, $\beta'(v) < 0$, $a + \delta \leqq v < \infty$ for all t sufficiently small, so that $\beta(v)$ is decreasing and $\beta(+\infty) = 0$. Hence for small t

$$I_2(t) = (4\pi t)^{-1/2} \int_{a+\delta}^{\infty} \beta(v)\, d\alpha(v)$$

$$= -(4\pi t)^{-1/2} \int_{a+\delta}^{\infty} \alpha(v)\, d\beta(v).$$

If M is a constant not larger than $|\alpha(t)|$,

$$|I_2(t)| \leqq (4\pi t)^{-1/2} M \int_{a+\delta}^{\infty} d[-\beta(v)] = \frac{M\beta(a+\delta)}{(4\pi t)^{1/2}}$$

$$\leqq \frac{Me^{-\delta^2/4t} e^{(a+\delta-x_0)^2/4}}{(4\pi t)^{1/2}} = o(1) \qquad t \to 0+.$$

To $I_1(t)$ we apply Theorem 2b, D. V. Widder [1946; 278], with $k = 1/t$, $h(v) = -(v - a)^2/4$, $h'(a) = 0$, $h''(a) = -1/2$.
We conclude that

$$I_1(t) \sim \varphi(a)/2 \qquad t \to 0+.$$

This completes the proof.

By use of this lemma we can now prove

THEOREM 7.2. *If*

1. $\int_{-\infty}^{\infty} k(x_0 - y, 1)\varphi(y)\, dy$ *converges for some* x_0

2. $\int_a^y [\varphi(v) - \varphi(a)]\, dv = o(|y - a|) \qquad y \to a,$

then

$$\lim_{t\to 0+} \int_{-\infty}^{\infty} k(a - y, t)\varphi(y)\, dy = \varphi(a). \tag{2}$$

For, write the integral (2) as the sum of two others corresponding to the intervals of integration $(-\infty, a)$, (a, ∞) The second of these approaches $\varphi(a)/2$ by the lemma. The first does also since it can be reduced to the integral

$$\int_{-a}^{\infty} k(a + y, t)\varphi(-y)\, dy,$$

to which the lemma may again be applied.

COROLLARY 7.2a. If hypothesis 2 is replaced by: $\varphi(a+)$ and $\varphi(a-)$ exist, then

$$\lim_{t\to 0+} \int_{-\infty}^{\infty} k(a - y, t)\varphi(y)\, dy = \frac{\varphi(a+) + \varphi(a-)}{2}.$$

COROLLARY 7.2b. If hypothesis 2 is omitted, (2) holds for almost all a.

For, it is known, E. C. Titchmarsh [1939; 362], that hypothesis 2 holds automatically at the Lebesgue set for $\varphi(x)$, and hence almost everywhere.

7.3. The previous theorem is the basis for the inversion of the Weierstrass transform in the most general case. However, to apply it we need the following preliminary result, symbolically equivalent to the equation

$$e^{-tD^2}e^{D^2}\varphi(x) = e^{(1-t)D^2}\varphi(x).$$

THEOREM 7.3. *If the Weierstrass transform*

$$f(x) = \int_{-\infty}^{\infty} k(x - u, 1)\, d\alpha(u) \tag{1}$$

converges in the interval $a < x < b$, *then for* $a < d < b$, $0 < t < 1$, $-\infty < x < \infty$,

$$e^{-tD^2}f(x) = \frac{1}{2\pi i}\int_{d-i\infty}^{d+i\infty} K(s-x, t)f(s)\,ds = \int_{-\infty}^{\infty} k(x-u, 1-t)\,d\alpha(u).$$

Choose two constants ξ, η such that $a < \xi < d < \eta < b$. By Lemma 2.1c of Chapter VI

$$\begin{aligned} \alpha(u) &= o(e^{(u-\eta)^2/4}) && u \to +\infty \\ &= o(e^{(u-\xi)^2/4}) && u \to -\infty. \end{aligned} \tag{2}$$

Now integrate (1) by parts. The integrated part vanishes by (2) when $\xi < x < \eta$, so that

$$f(x) = \int_{-\infty}^{\infty} k_1(x-y, 1)\alpha(y)\,dy,$$

where

$$k_1(x, t) = \frac{\partial}{\partial x} k(x, t) = \frac{-x}{2t} k(x, 1).$$

Hence

$$e^{-tD^2}f(x) = \frac{1}{2\pi i}\int_{d-i\infty}^{d+i\infty} K(s-x, t)\,ds \int_{-\infty}^{\infty} k_1(s-u, 1)\alpha(u)\,du. \tag{3}$$

Set $s = d + iy$ and note that the resulting integral is dominated by

$$A\int_{-\infty}^{\infty} e^{[(x-d)^2-y^2]/4t}\,dy \int_{-\infty}^{\infty} \sqrt{y^2 + (d-u)^2}\, e^{[y^2-(d-u)^2]/4}\,|\,\alpha\,(u)\,|\,du$$

by virtue of Theorem 2.3. But this converges for $0 < t < 1$ by (2), so that we may apply Fubini's theorem to (3) and obtain

$$e^{-tD^2}f(x) = \frac{1}{2\pi i}\int_{-\infty}^{\infty} \alpha(u)\,du \int_{d-i\infty}^{d+i\infty} K(s-x, t)k_1(s-u, 1)\,ds.$$

By Theorem 2.5 we have for $-\infty < x < \infty$, $-\infty < u < \infty$,

$$\frac{1}{2\pi i}\int_{d-i\infty}^{d+i\infty} K(s-x, t)k_1(s-u, 1)\,ds = k_1(x-u, 1-t),$$

so that

$$e^{-tD^2}f(x) = \int_{-\infty}^{\infty} k_1(x-u, 1-t)\alpha(u)\,du = \int_{-\infty}^{\infty} k(x-u, 1-t)\,d\alpha(u).$$

The final integration by parts is again justified for $-\infty < x < \infty$ by (2)

7.4. The inversion of the Weierstrass-Lebesgue transform is now an immediate consequence of the foregoing results.

THEOREM 7.4. *If*

1. $f(x) = \int_{-\infty}^{\infty} k(x-u, 1)\varphi(u)\, du$, *the integral converging in* $a < x < b$,

2. $\int_x^y [\varphi(u) - \varphi(x)]\, du = o(|y-x|)$, $y \to x$, *for some* x, $-\infty < x < \infty$,

then

$$e^{-D^2}f(x) = \lim_{t\to 1-} \frac{1}{2\pi i}\int_{d-i\infty}^{d+i\infty} K(s-x, t)f(s)\, ds = \varphi(x). \tag{1}$$

For, by Theorem 7.3

$$e^{-tD^2}f(x) = \int_{-\infty}^{\infty} k(x-u, 1-t)\varphi(u)\, du.$$

Theorem 7.2 is applicable to this integral (replacing t by $1-t$), and the proof is complete. As in Corollary 7.2b, hypothesis 2 may be omitted, in which case the conclusion (1) is valid for almost all x.

7.5. Our next conclusion enables us to invert the Weierstrass-Stieltjes integral.

THEOREM 7.5. *If* $\alpha(u)$ *is a normalized function of bounded variation in every finite interval and if*

$$f(x) = \int_{-\infty}^{\infty} k(x-u, 1)\, d\alpha(u),$$

the integral converging for $a < x < b$, *then for* $a < d < b$ *and any two real numbers* x_1, x_2

$$\int_{x_1}^{x_2} e^{-D^2}f(x)\, dx = \lim_{t\to 1-} \frac{1}{2\pi i}\int_{x_1}^{x_2} dx \int_{d-i\infty}^{d+i\infty} K(s-x, t)f(s)\, ds = \alpha(x_2) - \alpha(x_1).$$

By Theorem 7.3

$$e^{-tD^2}f(x) = \int_{-\infty}^{\infty} k(x-u, 1-t)\, d\alpha(u)$$

$$= \int_{-\infty}^{\infty} k_1(x-u, 1-t)\alpha(u)\, du \qquad 0 < t < 1, \quad -\infty < x < \infty. \tag{1}$$

The relations 7.3 (2) not only justify this integration by parts, but show, in conjunction with Theorem 2.3, that the integral (1) converges uniformly for $x_1 \leqq x \leqq x_2$. Hence we have

$$\int_{x_1}^{x_2} e^{-tD^2}f(x)\, dx = \int_{x_1}^{x_2} dx \int_{-\infty}^{\infty} k_1(x-u, 1-t)\, \alpha(u)\, du$$

$$= \int_{-\infty}^{\infty} k(x_2-u, 1-t)\alpha(u)\, du$$

$$- \int_{-\infty}^{\infty} k(x_1-u, 1-t)\alpha(u)\, du.$$

An application of Theorem 7.2 to each of these integrals now yields the desired result.

7.6. In § 7.2 we discussed the behaviour of $e^{-tD^2}f(x)$ as (x, t) approached $(x_0, 0)$, the approach being along a normal to the x-axis. For certain purposes (compare § 5.3 and the example at the end) it is important to let (x, t) approach the boundary in an arbitrary two-dimensional way. We shall need the following result.

THEOREM 7.6. *If $\varphi(x) \in C$ at $x = a$ and if*

$$u(x, t) = \int_{-\infty}^{\infty} k(x - y, t)\varphi(y)\, dy, \tag{1}$$

the integral converging at $(x_0, 1)$, then

$$\lim_{\substack{x \to a \\ t \to 0+}} u(x, t) = \varphi(a).$$

Choose an arbitrary $\delta > 0$ and make the usual decomposition of (1) into $I_1(x, t)$, $I_2(x, t)$, $I_3(x, t)$, these functions corresponding respectively to the intervals $(-\infty, a - \delta)$, $(a - \delta, a + \delta)$, $(a + \delta, \infty)$. Define $\alpha(v)$ as in § 7.2, but write

$$\beta(v) = e^{-(v-x)^2/4t}e^{(v-x_0)^2/4},$$

so that

$$I_3(x, t) = (4\pi t)^{-1/2} \int_{a+\delta}^{\infty} \beta(v)\, d\alpha(v).$$

If $|x - a| < \delta/2$ and t is so small that $t(a - x_0 + \delta) < \delta/2$, then $\beta'(v) < 0$ for $a + \delta \leqq v < \infty$, and we have, as in § 7.2, for a suitable constant M

$$\begin{aligned} |I_3(x, t)| &\leqq M(4\pi t)^{-1/2}\beta(a + \delta) \\ &\leqq M(4\pi t)^{-1/2}e^{-(a+\delta-x)^2/4t}e^{(a+\delta-x_0)^2/4} \\ &\leqq M(4\pi t)^{-1/2}e^{-\delta^2/16t}e^{(a+\delta-x_0)^2/4} \qquad |x - a| < \delta/2. \end{aligned}$$

Hence

$$\lim_{\substack{x \to a \\ t \to 0+}} I_3(x, t) = 0.$$

A similar argument applies to $I_1(x, t)$. Also

$$|I_2(x, t) - \varphi(a)| \leqq \underset{|y-a| \leqq \delta}{\text{l.u.b.}} |\varphi(y) - \varphi(a)| + \frac{|\varphi(a)|}{\sqrt{4\pi}} \int_{|y| \geqq \delta/2t} e^{-y^2/4}\, dy,$$

when $|x - a| < \delta/2$. That is,

$$\overline{\lim_{\substack{x \to a \\ t \to 0+}}} |u(x, t) - \varphi(a)| \leqq \underset{|y-a| \leqq \delta}{\text{l.u.b.}} |\varphi(y) - \varphi(a)|.$$

By our hypothesis of continuity at a, the right-hand side approaches zero with δ, so that the proof is complete.

COROLLARY 7.6. If $|\varphi(x)| \leqq M$, $|x-a| \leqq \delta$ and if (1) converges at $(x_0, 1)$, then

$$\overline{\lim_{\substack{x\to a\\ t\to 0+}}} |u(x,t)| \leqq M.$$

For, under the modified hypotheses

$$|I_2(x,t)| \leqq M \int_{-\delta}^{\delta} k(y,t)\,dy \leqq M.$$

8. FUNCTIONS OF L^p

8.1. This section is devoted to known results from real-variable theory (compare S. Bochner and K. Chandrasekharan [1949; 98]). Proofs are included for the reader's convenience. We recall that $f(x) \in L^p$ (in $-\infty < x < \infty$) means that

$$(1) \qquad I = \int_{-\infty}^{\infty} |f(x)|^p\,dx < \infty$$

and that the *norm* of f is

$$||f(x)||_p = I^{1/p}.$$

We are concerned here with real $p \geqq 1$. We shall frequently omit the subscript p on the notation for the norm, when no ambiguity results.

8.2. The first result needed is

THEOREM 8.2. *If $f(x) \in L^p$, $p \geqq 1$, and if*

$$\tau(h) = ||f(x+h) - f(x)||_p,$$

then

A. *$\tau(h)$ is bounded* $\qquad -\infty < h < \infty$

B. $\tau(0+) = 0$.

Conclusion A is a result of Minkowski's inequality,

$$\tau(h) \leqq ||f(x+h)|| + ||f(x)|| = 2\,||f(x)||.$$

Let us first prove B under the additional assumption that $f(x) \in C$. Then for arbitrary positive R

$$\tau(h)^p = \left\{ \int_{|x|\leqq R} + \int_{|x|\geqq R} \right\} |f(x+h) - f(x)|^p\,dx = I_1 + I_2,$$

and by Minkowski's inequality

$$I_2^{1/p} \leqq \left[\int_{|x-h|\geqq R} |f(x)|^p\,dx \right]^{1/p} + \left[\int_{|x|\geqq R} |f(x)|^p \right]^{1/p}$$

If $|x| \geqq R$ and $0 \leqq h \leqq \delta$, then $|x - h| \geqq R - \delta$. Given $\epsilon > 0$, we can determine R, using 8.1 (1), so that $I_2 < \epsilon$ for $0 \leqq h \leqq \delta$. But with this fixed R, I_1 tends to zero with h by Lebesgue's limit theorem. The integrand has a bound independent of h in $0 \leqq h \leqq \delta$ by the continuity of $f(x)$ in $-R \leqq x \leqq R + \delta$. Hence

$$\overline{\lim_{h \to 0+}}\, \tau(h)^p \leqq \epsilon$$

and $\tau(0+) = 0$. Since the class of continuous functions is dense in L^p, E. W. Hobson [1926; 250], the general case is easily reduced to the special case just considered as indicated briefly by the following inequalities:

$$\|f(x) - g(x)\| < \epsilon \qquad f \in L^p, \quad g \in C$$

$$\|f(x + h) - g(x + h)\| < \epsilon$$

$$\|g(x + h) - g(x)\| < \epsilon \qquad 0 < h < h_0.$$

These imply

$$\|f(x + h) - f(x)\| < 3\epsilon \qquad 0 < h < h_0,$$

as required. This completes the proof.

8.3. Our principal result, the one needed in later sections, is

THEOREM 8.3. *If $f(x) \in L^p$, $p \geqq 1$, and if*

$$f_h(x) = \frac{1}{2h} \int_{x-h}^{x+h} f(y)\, dy,$$

then

$$\lim_{h \to 0+} \|f - f_h\| = 0.$$

Denote by $R(y)$ a function which is 1/2 for $|y| < 1$ and 0 for $|y| \geqq 1$. The integral of $R(y)$ over $(-\infty, \infty)$ is 1, so that

$$f_h(x) - f(x) = \int_{-\infty}^{\infty} R(y)\, [f(x + yh) - f(x)]\, dy.$$

By Hölder's inequality ($p > 1$) or directly ($p = 1$),

$$|f_h(x) - f(x)|^p \leqq \int_{-\infty}^{\infty} R(y)\, |f(x + yh) - f(x)|^p\, dy$$

$$\int_{-\infty}^{\infty} |f_h(x) - f(x)|^p\, dx \leqq \int_{-\infty}^{\infty} dx \int_{-\infty}^{\infty} R(y)\, |f(x + yh) - f(x)|^p\, dy$$

$$\leqq \int_{-\infty}^{\infty} R(y) \tau(yh)^p\, dy \tag{1}$$

provided that Fubini's theorem is applicable. Since the integrand of the double integral is $\geqq 0$ we have only to check that (1) converges. It does so since $\tau(yh)$ is bounded and $R(y)$ is integrable.

Finally let $h \to 0+$ in (1). We may apply Lebesgue's limit theorem by virtue of A, Theorem 8.2. Hence the limit is zero, and the theorem is proved.

9. WEIERSTRASS TRANSFORMS OF FUNCTIONS IN L^p

9.1. This section is devoted to the derivation of necessary and sufficient conditions on a function $f(x)$ in order that it may be the Weierstrass transform of a function $\varphi \in L^p$, $p > 1$. The result is derived from another, of interest in itself, concerning temperature functions which arise from initial temperature functions belonging to L^p.

9.2. The theorem needed about solutions of the heat equation is

THEOREM 9.2. *The conditions*

1. $u(x, t) \in H \qquad -\infty < x < \infty, \quad 0 < t < c$
2. $||u(x, t)||_p < M \qquad p > 1, \quad 0 < t < c$

are necessary and sufficient that

$$u(x, t) = \int_{-\infty}^{\infty} k(x - y, t)\varphi(y)\, dy \qquad -\infty < x < \infty,\ 0 < t < c, \tag{1}$$

where

$$||\varphi(y)||_p < M. \tag{2}$$

Assuming (1) we see by Hölder's inequality that for fixed x and t

$$\int_{-\infty}^{\infty} k(x - y, t)\,|\varphi(y)|\, dy \leqq ||k(x - y, t)||_q\ ||\varphi(y)||_p.$$

Since the right-hand side is finite, (1) converges absolutely for $-\infty < x < \infty$, $0 < t < \infty$. By Theorem 3.3, $u(x, t) \in H$ there.

To prove condition 2 we have

$$|u(x, t)| \leqq ||k(x - y, t)^{1/p}\varphi(y)||_p\, ||k(x - y, t)^{1/q}||_q$$

$$\int_{-\infty}^{\infty} |u(x, t)|^p\, dx \leqq \int_{-\infty}^{\infty} dx \int_{-\infty}^{\infty} k(x - y, t)\,|\varphi(y)|^p\, dy$$

$$= \int_{-\infty}^{\infty} |\varphi(y)|^p\, dy < M \qquad 0 < t < \infty.$$

Fubini's theorem is applicable since $k \geqq 0$ and $\varphi \in L^p$ by (2). The necessity of the conditions is established.

To prove the sufficiency consider the function

$$u_h(x, t) = \frac{1}{2h}\int_{x-h}^{x+h} u(y, t)\, dy \qquad 0 < h,$$

which belongs to class H where $u(x, t)$ does. This follows by direct computation of the partial derivatives of u_h:

$$\frac{\partial^2}{\partial x^2} u_h(x, t) = \frac{\partial}{\partial t} u_h(x, t) = \frac{1}{2h} [u_x(x + h, t) - u_x(x - h, t)].$$

Moreover, by Hölder's inequality

$$|u_h(x, t)| \leqq \frac{1}{2h} ||u(x, t)||_p (2h)^{1/q} \leqq \frac{M}{(2h)^{1/p}}$$

for $-\infty < x < \infty$, $0 < t < c$. Hence by Lemma 6.2

$$u_h(x, t + \delta) = \int_{-\infty}^{\infty} k(x - y, t) u_h(y, \delta)\, dy \tag{3}$$

for $0 < \delta < c$, $0 < t < c - \delta$, $-\infty < x < \infty$. Let $h \to 0+$. By the law of the mean $u_h(x, t + \delta) \to u(x, t + \delta)$. By weak convergence, E. C. Titchmarsh [1939; 389], the integral (3) approaches the same integral with u_h replaced by u. The weak convergence theorem is applicable since $k \in L^q$ and since $u_h \to u$ "in mean of index p" by Theorem 8.3. Hence equation (3) also holds if the subscript h is omitted throughout.

Now let $\delta \to 0$,

$$u(x, t) = \lim_{\delta \to 0} \int_{-\infty}^{\infty} k(x - y, t) u(y, \delta)\, dy.$$

Using weak compactness, D. V. Widder [1946; 33], we complete the proof as in §6.2, obtaining equation (1) as desired.

9.3. We can now relate the previous result to the operator e^{-tD^2} as follows.

THEOREM 9.3. *The conditions*

1. $f(x) \in A$ $\qquad -\infty < x < \infty$
2. $||e^{-tD^2} f(x)||_p < M$ $\qquad p > 1, \quad 0 < t < 1$

are necessary and sufficient that

$$f(x) = \int_{-\infty}^{\infty} k(x - y, 1) \varphi(y)\, dy \qquad -\infty < x < \infty, \tag{1}$$

where

$$||\varphi(y)||_p < M. \tag{2}$$

If (2) holds, the integral (1) converges absolutely for all x since (1) is 9.2 (1) with $t = 1$. Hence $f(x) \in A$, $-\infty < x < \infty$.

Also

$$e^{-tD^2} f(x) = \int_{-\infty}^{\infty} k(x - y, 1 - t) \varphi(y)\, dy \tag{3}$$

by Theorem 7.3. Now apply Theorem 9.2 to the integral (3) (replacing t by $1 - t$) to obtain condition 2.

Conversely, condition 1 guarantees, by Theorem 3.4, that

$$u(x, t) = e^{-(1-t)D^2} f(x) \tag{4}$$

belongs to H in $-\infty < x < \infty$, $0 < t < 1$. Condition 2 enables us to apply Theorem 9.2 to the function (4), so that

$$u(x, t) = \int_{-\infty}^{\infty} k(x - y, t)\varphi(y)\, dy,$$

where $\varphi(y)$ satisfies (2). As we have seen, this integral converges absolutely for all $t > 0$, so by continuity

$$u(x, 1-) = \int_{-\infty}^{\infty} k(x - y, 1)\varphi(y)\, dy.$$

On the other hand, we showed in § 6.3 that when $f(x) \in A$, $-\infty < x < \infty$, then the function (4) $\to f(x)$ as $t \to 1-$. Hence (1) holds, and the proof is complete.

10. WEIERSTRASS-STIELTJES TRANSFORMS

10.1. Theorems 9.2 and 9.3 are no longer true if $p = 1$ throughout. To show this for Theorem 9.2 take $u(x, t) = k(x, t)$, the source solution itself. Then

$$\|u(x, t)\| = \int_{-\infty}^{\infty} k(x, t)\, dx = 1,$$

so that conditions 1 and 2 are satisfied in $-\infty < x < \infty$, $0 < t < \infty$. But the equation

$$k(x, t) = \int_{-\infty}^{\infty} k(x - y, t)\varphi(y)\, dy, \tag{1}$$

with $\varphi(y) \in L$, is impossible. For, by Corollary 7.2b, we should have

$$\lim_{t\to 0+} k(x, t) = 0 = \varphi(x)$$

for almost all $x \neq 0$. That is, the integral (1) would be identically zero, contradicting equation (1). On the other hand (1) is true if $\varphi(y)\, dy$ is replaced by $d\alpha(y)$ where $\alpha(y)$ is constant except for a single jump at $y = 0$. That is, $k(x, 1)$ may be a Weierstrass-Stieltjes transform.

10.2. The correct conclusion when $p = 1$ is the following.

THEOREM 10.2. *The conditions of Theorem* 9.2 *with* $p = 1$ *are necessary and sufficient that*

$$u(x, t) = \int_{-\infty}^{\infty} k(x - y, t)\, d\alpha(y) \qquad -\infty < x < \infty, \quad 0 < t < c, \tag{1}$$

where

$$\int_{-\infty}^{\infty} |d\alpha(y)| < M. \tag{2}$$

Under assumption (2) the integral (1) converges absolutely in the half-plane $t > 0$. Indeed

$$|u(x, t)| \leqq \frac{1}{\sqrt{4\pi t}} \int_{-\infty}^{\infty} |d\alpha(y)| < \frac{M}{\sqrt{4\pi t}}.$$

Hence by Theorem 3.3 $u \in H$ there. Condition 2 is also satisfied since

$$\int_{-\infty}^{\infty} |u(x, t)|\, dx \leqq \int_{-\infty}^{\infty} dx \int_{-\infty}^{\infty} k(x-y, t)\, |d\alpha(y)|$$

(3)
$$\leqq \int_{-\infty}^{\infty} |d\alpha(y)| < M \qquad 0 < t < \infty.$$

Thus both conditions are necessary.

To prove the converse we again introduce the function $u_h(x, t)$ as in the proof of Theorem 9.2. It is again bounded since

$$|u_h(x, t)| \leqq \frac{1}{2h} \int_{-\infty}^{\infty} |u(y, t)|\, dy < \frac{M}{2h}.$$

The rest of the proof is the same as for Theorem 9.2 except that the Helly and Helly-Bray theorems are used instead of Theorems 17a or 17b, D. V. Widder [1946; 29–33]. The procedure is standard (compare D. V. Widder [1946; 307]) and is omitted here. The conclusion is (1) (2), as desired.

10.3. The previous theorem leads to the following representation theorem for the Weierstrass-Stieltjes transform.

THEOREM 10.3. *The conditions of Theorem* 9.3 *with* $p = 1$ *are necessary and sufficient that*

(1)
$$f(x) = \int_{-\infty}^{\infty} k(x-y, 1)\, d\alpha(y),$$

where

(2)
$$\int_{-\infty}^{\infty} |d\alpha(y)| < M.$$

We saw in the previous section ($t = 1$) that the integral (1) converges absolutely when (2) holds. Hence $f(x) \in A$ in $-\infty < x < \infty$. Moreover,

$$e^{-tD^2} f(x) = \int_{-\infty}^{\infty} k(x-y, 1-t)\, d\alpha(y).$$

The change in the order of integration needed to prove this is valid if

$$\int_{-\infty}^{\infty} e^{-y^2/4t} e^{y^2/4}\, dy \int_{-\infty}^{\infty} e^{-v^2/4}\, |d\alpha(v)| < \infty.$$

This is true if $0 < t < 1$. Now inequality 10.2 (3), with t replaced by $1 - t$ completes the proof of the necessity.

For the converse, the proof of Theorem 9.3 may be used, appealing to Theorem 10.2 instead of 9.2 and replacing $\varphi(y)\, dy$ by $d\alpha(y)$.

11. POSITIVE TEMPERATURE FUNCTIONS

11.1. In § 5.3 we noted that $u(x, t) = k(x, t)x/t$ is a temperature function not identically zero even though $u(x, 0+) = 0$ for all x. However, this function takes on both positive and negative values in the half-plane $t > 0$. We wish to show now that any temperature function which is known to be non-negative for $t > 0$ and to vanish for $t = 0$ must be identically zero (D. V. Widder [1944; 85]).

11.2. The integral

$$\int_{-\infty}^{\infty} k(x-y, t)u(y, \delta)\, dy \tag{1}$$

generally gives the temperature of an infinite bar at time $t + \delta$ when its temperature at time $t = \delta$ is $u(x, \delta)$. This is not always the case as the example of § 11.1, with $\delta = 0$, shows. We shall show that it is so for non-negative functions $u(x, t)$, but we must first show that the integral (1) always converges for such functions. To do so we need the following preliminary result,

LEMMA 11.2. If

$$u(x, t) = \int_{-A}^{A} k(x-y, t)\varphi(y)\, dy$$

and $f(x)$ is the function of Theorem 5.3, then

$$f(x) = O(1/|x|) \qquad |x| \to \infty.$$

It is of course assumed that $\varphi(y) \in L$ in $(-A, A)$. The inequality

$$|x| e^{-x^2} < 1 \qquad -\infty < x < \infty$$

is trivial. Setting $x = y/\sqrt{4t}$ therein, we have

$$k(y, t) < 1/|y| \qquad 0 < |y| < \infty, \quad 0 < t < \infty.$$

Hence when $|x| > A$

$$|u(x, t)| \leqq \int_{-A}^{A} \frac{|\varphi(y)|}{|y-x|}\, dy \leqq \frac{1}{|x| - A} \int_{-A}^{A} |\varphi(y)|\, dy.$$

This proves the result.

THEOREM 11.2. *If $u(x, t) \geqq 0$ and $\in H$ in the strip $0 < t - c$, then the integral* (1), *with $0 < \delta < c$, converges in the strip $0 < t < c - \delta$ and is $\leqq u(x, t + \delta)$ there.*

We prove later that in this ambiguous conclusion ($\leqq$) only the equality can hold.

Consider the function

$$v(x, t) = u(x, t + \delta) - \int_{-A}^{A} k(x-y, t)u(y, \delta)\, dy \tag{2}$$

for arbitrary positive constants A, δ, the latter $< c$. By use of Theorem 5.2, we show that $v(x, t) \geqq 0$ in the strip $0 < t < c - \delta$. By Theorem 7.6

$$\lim_{\substack{x \to x_0 \\ t \to 0+}} v(x, t) = u(x_0, \delta) \geqq 0 \qquad |x_0| > A$$

$$= 0 \qquad |x_0| < A.$$

However, the integral (2) generally approaches no limit as $(x, t) \to (A, 0)$ or $(-A, 0)$ in the two-dimensional way. But by Corollary 7.6 it has a limit superior $\leqq u(A, \delta)$ or $u(-A, \delta)$, respectively. That is,

$$\varliminf_{\substack{x \to \pm A \\ t \to 0+}} v(x, t) \geqq u(\pm A, \delta) - u(\pm A, \delta) = 0.$$

Hence $v(x, t)$ satisfies hypothesis 2 of Theorem 5.2 at all points $(x_0, 0)$ of the x-axis.

Now to produce a contradiction assume that $v(x_0, t_0) = -l < 0$ at some point (x_0, t_0), $0 < t_0 < c - \delta$. By Lemma 11.2 we can determine R so large that $v(x, t) \geqq -l/2$ on the segments $x = \pm R$, $0 < t < c - \delta$ because $u(x, t) \geqq 0$ and the integral (2) tends uniformly to zero for $0 < t < \infty$ as $|x| \to \infty$. That is, $v(x, t)$ must have a minimum inside a rectangular region D of the type described in Theorem 5.2. This contradicts that theorem, so that the assumption $v < 0$ must have been false. The contrary assumption gives

$$\int_{-A}^{A} k(x - y, t)u(y, \delta)\, dy \leqq u(x, t + \delta)$$

in the strip $0 < t < c - \delta$. Since this integral increases with A the convergence of (1) is guaranteed and the theorem is proved.

11.3. By use of the foregoing result we can now prove the required uniqueness theorem.

THEOREM 11.3. *If*

1. $u(x, t) \in H$ $\qquad 0 \leqq t < c$
2. $u(x, t) \geqq 0$ $\qquad 0 \leqq t < c$
3. $u(x, 0) = 0$ $\qquad -\infty < x < \infty,$

then

$$u(x, t) = 0 \qquad -\infty < x < \infty, \quad 0 \leqq t < c.$$

Observe that we are assuming $u(x, t) \in H$ on the lower boundary of the strip. It is for this reason that we can give the boundary condition 3 without the use of limits. Set

$$v(x, t) = \int_0^t u(x, y)\, dy.$$

By 1 and 3 we have

$$v_{xx}(x, t) = u(x, t) = v_t \qquad 0 \leqq t < c,$$

so that $v \in H$ there. It satisfies all the hypotheses of the theorem but has the additional properties of being convex in x and non-decreasing in t. Moreover, if $v(x, t)$ vanishes identically, the same is true of $u(x, t)$. Hence there is no loss in generality if we include these additional properties as hypotheses on $u(x, t)$.

Let δ be an arbitrary positive number $<c$ and set $x = 0, t = t_0 < c - \delta$ in the integral 11.2 (1). Then

$$\int_{-\infty}^{\infty} e^{-y^2/4t_0} u(y, \delta)\, dy < \infty$$

by Theorem 11.2. Since $u(x, t)$ is non-negative and non-decreasing in t the function $f(x)$ of Tychonoff's theorem is

$$f(x) = \max_{0 \leqq t \leqq \delta} u(x, t) = u(x, \delta).$$

By the convexity of $f(x)$ we have for $x > 0$

$$2x\, f(x) \leqq \int_0^{2x} f(y)\, dy$$

(the area under a convex curve $\geqq$ the area under a tangent). Hence

$$2x\, f(x) e^{-x^2/t_0} \leqq \int_0^{2x} e^{-y^2/4t_0} f(y)\, dy \leqq \int_{-\infty}^{\infty} e^{-y^2/4t_0} u(y, \delta)\, dy.$$

Since the latter integral is known to converge we have

(1) $$f(x) = O(e^{x^2/t_0}) \qquad x \to +\infty.$$

If $x < 0$ we have

$$-2x\, f(x) e^{-x^2/t_0} \leqq \int_{2x}^{0} e^{-y^2/4t_0} f(y)\, dy \leqq \int_{-\infty}^{\infty} e^{-y^2/4t_0} u(y, \delta)\, dy$$

and (1) is also valid as $x \to -\infty$. Hence we may apply Tychonoff's theorem and conclude that $u(x, t)$ is zero in the strip $0 < t \leqq \delta$. Since δ was arbitrary the proof is complete.

Corollary 11.3. *If conditions 1 and 2 of the theorem hold, then*

$$u(x, t + \delta) = \int_{-\infty}^{\infty} k(x - y, t) u(y, \delta)\, dy \qquad 0 < t < c - \delta.$$

For, by Theorem 11.2 the function

$$u(x, t + \delta) - \int_{-\infty}^{\infty} k(x - y, t) u(y, \delta)\, dy$$

is $\geqq 0$ in the strip $0 < t < c - \delta$. Indeed it clearly satisfies all conditions of Theorem 11.3 (with c replaced by $c - \delta$) and is consequently identically zero.

12. WEIERSTRASS-STIELTJES TRANSFORMS OF INCREASING FUNCTIONS

12.1. In § 10 we discussed Weierstrass-Stieltjes transforms of functions of bounded variation. One subclass consists of those transforms for which the function to be transformed is non-decreasing. The latter class is also a subclass of transforms of unbounded non-decreasing functions. Illustrative examples are provided from the table of § 2.6 by the pairs 1, 3 ($\varphi = y^2, f = x^2 + 2$), 5, 6, 7, 8, 9 and by the function $\alpha(y)$ given immediately after Definition 2.1c. Note that the increasing function

$$\alpha(y) = \int_0^y e^{v^2/5}\, dv$$

of the pair 6, for example, is not of bounded variation on $(-\infty, \infty)$. It is important to include functions like this one, if the class of their transforms is to be neatly characterized by use of the operator e^{-tD^2}. The situation is analogous to one in Laplace transform theory. According to Bernstein's theorem the class of functions $f(x)$ "completely monotonic" on $0 < x < \infty$ is equivalent to the class

$$f(x) = \int_0^\infty e^{-xy}\, d\alpha(y) \qquad 0 < x < \infty$$

with $\alpha(y)$ non-decreasing. Here also the variation of $\alpha(y)$ may be infinite, as for example when $\alpha(y) = y$ and $f(x) = 1/x$.

12.2. As in §§ 6, 9, 10 we need a preliminary result about temperature functions. The theorem will be the analogue of a familiar one by A. Herglotz [1911; 501] concerning positive harmonic functions.

THEOREM 12.2. *A necessary and sufficient condition that*

$$u(x, t) = \int_{-\infty}^{\infty} k(x - y, t)\, d\alpha(y), \tag{1}$$

where $\alpha(y)$ is non-decreasing and the integral converges in the strip $0 < t < c$ is that $u(x, t) \in H$, $u(x, t) \geqq 0$ there.

The necessity of the condition follows by inspection of the integral (1).

To prove the sufficiency set

$$\beta_\delta(x) = \int_{-\infty}^{x} k(y, t_0) u(y, \delta)\, dy, \tag{2}$$

where $0 < \delta < c$, $0 < t_0 < c - \delta$. By Theorem 11.2

$$0 \leqq \beta_\delta(x) \leqq u(0, t_0 + \delta). \tag{3}$$

By Corollary 11.3

$$u(x, t + \delta) = \int_{-\infty}^{\infty} k(x - y, t)u(y, \delta)\, dy \qquad 0 < t < c - \delta$$

$$= \int_{-\infty}^{\infty} \frac{k(x - y, t)}{k(y, t_0)}\, d\beta_\delta(y).$$

Hence

$$u(x, t) = \lim_{\delta \to 0+} \int_{-\infty}^{\infty} \frac{k(x - y, t)}{k(y, t_0)}\, d\beta_\delta(y) \qquad 0 < t < c.$$

By virtue of (3) we may now apply the Helly and Helly-Bray theorems, D. V. Widder [1941; 26–32], in the standard way to obtain

$$u(x, t) = \int_{-\infty}^{\infty} \frac{k(x - y, t)}{k(y, t_0)}\, d\beta(y)$$

where $\beta(y)$ is non-decreasing and bounded. We thus obtain (1) where

$$\alpha(y) = \int_0^y \frac{1}{k(x, t_0)}\, d\beta(x) \qquad -\infty < y < \infty.$$

This function is non-decreasing, so that the theorem is proved.

12.3. As a consequence of the previous theorem we can strengthen Theorem 11.3 by dropping the demand that $u(x, t) \in H$ *on* the x-axis and by replacing condition 3 by $u(x, 0+) = 0$.

THEOREM 12.3. *If*

1. $u(x, t) \in H \qquad 0 < t < c$
2. $u(x, t) \geq 0 \qquad 0 < t < c$
3. $u(x, 0+) = 0 \qquad -\infty < x < \infty,$

then

$$u(x, t) = 0 \qquad -\infty < x < \infty,\ 0 < t < c.$$

For, by hypotheses 1 and 2 we have the integral representation 12.2 (1) for $u(x, t)$. Let y_0 be any real number and let δ be any positive number. Then

$$u(y_0, t) \geq \int_{y_0}^{y_0+\delta} k(y_0 - y, t)\, d[\alpha(y) - \alpha(y_0)] \tag{1}$$

$$\geq \frac{1}{2t} \int_0^{\delta} yk(y, t)\, [\alpha(y + y_0) - \alpha(y_0)]\, dy.$$

We have here integrated by parts and expressed the derivative of k in terms of k. Set

$$p(\delta) = \underset{0 \leqq y \leqq \delta}{\text{g.l.b.}} \frac{\alpha(y + y_0) - \alpha(y_0)}{y}.$$

Then $p(\delta) \geqq 0$ by Theorem 12.2. Moreover,

$$u(y_0, t) \geqq \frac{p(\delta)}{2t} \int_0^\delta y^2 k(y, t)\, dy$$

$$\geqq \frac{p(\delta)}{2} \int_0^{\delta/\sqrt{t}} y^2 k(y, 1)\, dy.$$

Now let $t \to 0+$, using hypothesis 3 and pair 3 of § 2.6 to obtain

$$0 \geqq p(\delta)/2,$$

so that $p(\delta) = 0$. Allowing δ to approach zero we have

$$\varliminf_{y \to 0+} \frac{\alpha(y + y_0) - \alpha(y_0)}{y} = 0. \tag{2}$$

That is, the lower derivate on the right for $\alpha(y)$ is zero at y_0. If the range of integration in the integral (1) is changed to $(y_0 - \delta, y_0)$ we easily see that (2) still holds when $y \to 0-$. By (2) it is clear that $\alpha(y_0+) = \alpha(y_0)$. Similarly $\alpha(y_0-) = \alpha(y_0)$, and $\alpha(y)$ is continuous. But a continuous function with any derivate constantly zero is constant (see, for example, C. J. de la Vallée Poussin [1914; 99]). But then $u(x, t)$ is identically zero by equation 12.2 (1), and the theorem is proved.

12.4. We come finally to the representation of functions as Weierstrass-Stieltjes transforms of non-decreasing functions.

THEOREM 12.4. *The conditions*

1. $f(x) \in A$ $\qquad a < x < b$
2. $e^{-tD^2} f(x) \geqq 0$ $\qquad 0 < t < 1, \quad -\infty < x < \infty$

are necessary and sufficient that

$$f(x) = \int_{-\infty}^{\infty} k(x - y, 1)\, d\alpha(y), \tag{1}$$

where $\alpha(y)$ is non-decreasing and the integral converges in $a < x < b$.

As we have seen, it is no restriction to assume that the origin lies in (a, b). We prove the necessity, assuming (1). The absolute convergence of the integral (1) in (a, b) insures condition 1. By definition

$$e^{-tD^2} f(x) = \int_{-\infty}^{\infty} k(y + ix, t)\, dy \int_{-\infty}^{\infty} k(iy - v, 1)\, d\alpha(v)$$

$$\ll \frac{e^{x^2/4t}}{4\pi\sqrt{t}} \int_{-\infty}^{\infty} e^{-y^2/4t} e^{y^2/4}\, dy \int_{-\infty}^{\infty} e^{-v^2/4}\, d\alpha(v) < \infty.$$

By Fubini's theorem and Theorem 2.5

$$e^{-tD^2} f(x) = \int_{-\infty}^{\infty} d\alpha(v) \int_{-\infty}^{\infty} k(y + ix, t) k(iy - v, 1)\, dy$$

$$= \int_{-\infty}^{\infty} k(x - y, 1 - t)\, d\alpha(y) \qquad -\infty < x < \infty,\ 0 < t < 1.$$

Since $\alpha(y) \in \uparrow$ it is clear that condition 2 is satisfied throughout the whole infinite strip $0 < t < 1$.

For the sufficiency define

$$u(x, t) = e^{-(1-t)D^2} f(x).$$

By conditions 1 and 2 it is well defined and $\geqq 0$ in the strip $0 < t < 1$. By Theorem 3.4 it belongs to H there. Hence we may apply Theorem 12.2 to obtain

$$u(x, t) = \int_{-\infty}^{\infty} k(x - y, t)\, d\alpha(y), \tag{2}$$

where $\alpha(y) \in \uparrow$ and the integral converges throughout the strip $0 < t < 1$. Since $f(x) \in A$ in $a < x < b$ we have by Cauchy's integral theorem, applied to the integral 3.1 (3), that

$$u(x, t) = \int_{-\infty}^{\infty} k(y, 1) f(x + iy\sqrt{1 - t})\, dy$$

$$\ll M \int_{-\infty}^{\infty} |y|\, e^{-y^2/4} e^{y^2(1-\delta)/4}\, dy \qquad A \leqq x \leqq B,\quad \delta \leqq t < 1.$$

where M is some constant, $a < A < B < b$, and δ is an arbitrary positive constant less than 1. Since the dominant integral converges and is independent of t we have by Lebesgue's limit theorem that

$$u(x, 1-) = f(x) \qquad a < x < b. \tag{3}$$

We show next that the integral (2) converges when $t = 1$ and $a < x < b$. For every x, t, R $(0 < t < 1,\ R > 0)$

$$u(x, t) \geqq \int_{-R}^{R} k(x - y, t)\, d\alpha(y). \tag{4}$$

The integral is dominated by

$$\int_{-R}^{R} \frac{d\alpha(y)}{\sqrt{4\pi\delta}} \qquad 0 < \delta \leqq t < 1.$$

By allowing t to approach 1 in inequality (4) we obtain

$$f(x) \geqq \int_R^R k(x-y, 1)\, d\alpha(y) \qquad a < x < b,$$

and since $\alpha(y) \in \uparrow$ the convergence of (2) is immediate.

Finally, by Theorem 2.1,

$$u(x, 1-) = \int_{-\infty}^{\infty} k(x-y, 1)\, d\alpha(y),$$

and the proof is completed by equation (3).

13. TRANSFORMS OF FUNCTIONS WITH PRESCRIBED ORDER CONDITIONS

13.1. The behaviour of $\varphi(y)$ for large $|y|$ affects the width of the strip in which the integral

$$(1) \qquad \int_{-\infty}^{\infty} k(x-y, t)\varphi(y)\, dy$$

converges. For example, if $\varphi(y)$ is bounded (1) converges absolutely in the half-plane $t > 0$. More generally $\varphi(y) = O(e^{ay^2})$, $|y| \to \infty$, implies that (1) converges for $0 < t < 1/(4a)$. Let us now obtain conditions on a temperature function that it can be equal to an integral (1) with $\varphi(y)$ satisfying order conditions of the above type. We can then apply the result to the Weierstrass transform.

13.2.

THEOREM 13.2. *The conditions*

1. $u(x, t) \in H \qquad 0 < t < \delta, \quad -\infty < x < \infty$

2. $|u(x, t)| < \dfrac{Me^{ax^2/(1-4at)}}{\sqrt{1-4at}} \qquad 0 < t < \delta, \quad -\infty < x < \infty,$

for some $a > 0$ and some $\delta \leqq 1/(4a)$, are necessary and sufficient that

$$u(x, t) = \int_{-\infty}^{\infty} k(x-y, t)\varphi(y)\, dy \qquad 0 < t < \delta, \quad -\infty < x < \infty,$$

the integral converging absolutely in $-\infty < x < \infty$, $0 < t < 1/(4a)$ and

$$|\varphi(y)| < Me^{ay^2} \qquad -\infty < y < \infty.$$

The proof of the necessity is made by use of Theorem 3.3 and the pair 8 of § 2.6.

Conversely, condition 2 is equivalent to

$$-Mv(x, t) < u(x, t) < Mv(x, t) \qquad 0 < t < \delta,$$

where

$$v(x, t) = \int_{-\infty}^{\infty} k(x - y, t)\, d\beta(y)$$

$$\beta(y) = \int_0^y e^{av^2}\, dv.$$

Hence $Mv - u$ and $Mv + u$ are functions which satisfy the conditions of Theorem 12.2 in the strip $0 < t < \delta$. Hence

$$Mv(x, t) \pm u(x, t) = \int_{-\infty}^{\infty} k(x - y, t)\, d[M\beta(y) \pm \alpha(y)]$$

for a suitable function $\alpha(y)$. By the theorem used, the functions $M\beta(y) \pm \alpha(y)$ are both non-decreasing, so that

$$(1) \qquad -M \int_y^{y+h} e^{av^2}\, dv \leqq \alpha(y + h) - \alpha(y) \leqq M \int_y^{y+h} e^{av^2}\, dv \quad 0 < h.$$

From these inequalities it is clear that $\alpha(y)$ is absolutely continuous in any finite interval. Hence by a familiar theorem, E. C. Titchmarsh [1939; 364], $\alpha(y)$ is an integral of some function $\varphi(y)$,

$$\alpha(y) = \int_0^y \varphi(v)\, dv - \alpha(0),$$

where $\alpha'(y) = \varphi(y)$ almost everywhere. By (1)

$$(2) \qquad |\varphi(y)| < Me^{ay^2}$$

except perhaps at a set of measure zero. Redefine $\varphi(y)$, if necessary, so that (2) is valid for all y.

$$u(x, t) = \int_{-\infty}^{\infty} k(x - y, t)\, d\alpha(y) = \int_{-\infty}^{\infty} k(x - y, t)\varphi(y)\, dy$$

in the strip $0 < t < \delta$. This completes the proof.

13.3.

THEOREM 13.3. *The conditions*

1. $f(x) \in A \qquad -\infty < x < \infty$
2. $\left|e^{-tD^2} f(x)\right| \leqq \dfrac{Me^{ax^2/(1-4a+4at)}}{\sqrt{1 - 4a + 4at}} \qquad 0 < t < 1, \quad -\infty < x < \infty$

for some $a < 1/4$ are necessary and sufficient that

$$(1) \qquad f(x) = \int_{-\infty}^{\infty} k(x - y, 1)\varphi(y)\, dy,$$

where the integral converges for all x and

$$|\varphi(y)| < Me^{ay^2} \qquad -\infty < y < \infty. \tag{2}$$

If (1) (2) hold with $a < 1/4$, then the integral (1) converges absolutely in $-\infty < x < \infty$, and $f(x) \in A$ there. By Theorem 7.3

$$e^{-tD^2} f(x) = \int_{-\infty}^{\infty} k(x - y, 1 - t)\varphi(y)\, dy \qquad -\infty < x < \infty, \quad 0 < t < 1.$$

Hence by the necessary part of Theorem 13.2 with $\delta = 1$

$$|e^{-tD^2} f(x)| < \frac{Me^{ax^2(1-4a+4at)}}{\sqrt{1 - 4a + 4at}} \qquad -\infty < x < \infty, \quad 0 < t < 1.$$

Conversely, set

$$u(x, t) = e^{-(1-t)D^2} f(x).$$

By Theorem 3.4 it belongs to H in $0 < t < 1$, and

$$|u(x, t)| < \frac{Me^{ax^2/(1-4at)}}{\sqrt{1 - 4at}} \qquad 0 < t < 1$$

by hypothesis 2. By the sufficient part of Theorem 13.2 with $\delta = 1$

$$u(x, t) = \int_{-\infty}^{\infty} k(x - y, t)\varphi(y)\, dy \qquad -\infty < x < \infty, \quad 0 < t < 1, \tag{3}$$

where $\varphi(y)$ satisfies (2). But in the presence of (2) the integral (3) converges absolutely in $0 < t < 1/(4a)$ and hence on the line $t = 1$. By Theorem 2.1

$$u(x, 1-) = \int_{-\infty}^{\infty} k(x - y, 1)\varphi(y)\, dy \qquad -\infty < x < \infty.$$

But we saw in § 12.4 that

$$\lim_{t\to 1-} e^{-(1-t)D^2} f(x) = f(x)$$

in any interval in which $f(x) \in A$. Hence we have for all x

$$u(x, 1-) = f(x) = \int_{-\infty}^{\infty} k(x - y, 1)\varphi(y)\, dy,$$

and the theorem is proved.

14. SUMMARY

14.1. The principal results of the present chapter are the following.

A. If $E(s)$ is the special function e^{-s^2} of class E, then as for other functions of this class its reciprocal e^{s^2} is the bilateral Laplace transform of a frequency function $k(x, 1) = (4\pi)^{-1/2}e^{-x^2/4}$,

$$e^{s^2} = \int_{-\infty}^{\infty} e^{-sy}k(y, 1)\,dy.$$

B. If this frequency function is taken as the kernel of a convolution transform, the latter, known as the Weierstrass transform,

$$f(x) = \int_{-\infty}^{\infty} k(x - y, 1)\varphi(y)\,dy, \tag{1}$$

is inverted by e^{-D^2} in the following sense:

$$e^{-D^2}f(x) = \lim_{t\to 1-} \frac{1}{2\pi i}\int_{c-i\infty}^{c+i\infty} K(s - x, t)f(s)\,ds = \varphi(x)$$

$$K(s, t) = (\pi/t)^{1/2}e^{s^2/4t}.$$

C. Necessary and sufficient conditions, couched in terms of $e^{-tD^2}f(x)$, are available to guarantee a representation (1) with $\varphi(y)$ in a prescribed class.

In conclusion we emphasize again that the basic ideas of this chapter are the same as the guiding ones for the rest of the book. Any apparent difference is essentially due to the contrast of a discrete parameter with a continuous one.

CHAPTER IX

Complex Inversion Theory

1. INTRODUCTION

1.1. In this chapter we shall obtain a complex inversion theory for a suitably restricted class of kernels, those of the form

(1) $$G(t) = (1/2\pi i) \int_{-i\infty}^{i\infty} [E(s)]^{-1} e^{st} \, ds$$

where

$$E(s) = \prod_{k=1}^{\infty} \left(1 - \frac{s^2}{a_k^2}\right),$$

the a_k being real and such that

(2) $$\lim_{k \to \infty} k/a_k = \Omega \qquad (0 < \Omega < \infty).$$

In order to see what to expect let us consider the example $a_k = (2k - 1)/2$. Here $E(s) = \cos \pi s$ and $G(t) = (1/2\pi) \operatorname{sech} \frac{1}{2}t$. If in the equation

(3) $$f(z) = (1/2\pi) \int_{-\infty}^{\infty} \operatorname{sech} \tfrac{1}{2}(z - t) \, \varphi(t) \, dt$$

we set $f(z) = F(e^z)e^{z/2}$, $\varphi(t) = \pi\Phi(e^t)e^{t/2}$ equation (3) becomes (see § 5.1 of I) the Stieltjes transform

(3′) $$F(z) = \int_{0+}^{\infty} \frac{\Phi(t)}{z + t} \, dt.$$

It is well known, see D. V. Widder [1941; 326], that $F(z)$ is an analytic function in the sector $|\arg z| < \pi$. Correspondingly $f(z)$ is an analytic function in the strip $|\operatorname{Im} z| < \pi$. The following formulas may be established by direct computation:

(4) $$\frac{z}{z^2 + \pi^2} = \int_0^{\infty} \cos \pi s \, e^{-sz} \, ds;$$

(5) $$\cos \pi s = \frac{1}{2\pi i} \int_C \frac{z}{z^2 + \pi^2} e^{sz} \, dz.$$

Here C is a rectifiable closed curve containing the points $-i\pi$ and $i\pi$ in its interior and proceeding counterclockwise. From Chapter I we have the symbolic inversion formula

$$\varphi(t) = \cos(\pi D) f(t).$$

Replacing s by D in equation (5) we find that

$$\varphi(t) = \cos(\pi D) f(t) = \frac{1}{2\pi i} \int_C \frac{z}{z^2 + \pi^2} e^{Dz} \, dz \, f(t),$$

$$\varphi(t) = \frac{1}{2\pi i} \int_C \frac{z}{z^2 + \pi^2} f(t + z) \, dz. \tag{6}$$

We have made use here of the familiar operational formula $e^{zD} f(t) = f(t + z)$. The formula (6) is only formally true since $f(t + z)$ is not in general defined for z on C. A simple way of avoiding this difficulty is to replace the formula (6) by the formula

$$\varphi(t) = \lim_{\rho \to 1-} \frac{1}{2\pi i} \int_{C_\rho} \frac{z}{z^2 + \pi^2} f(t + \rho z) \, dz, \tag{7}$$

where C_ρ $(0 < \rho < 1)$ is a rectifiable closed curve containing $-i\pi$ and $i\pi$ in its interior, lying in the strip $|\operatorname{Im} z| < \pi/\rho$, and proceeding counterclockwise. Employing the calculus of residues in (7) we obtain

$$\varphi(t) = \lim_{\rho \to 1-} \tfrac{1}{2}[f(t + i\pi\rho) + f(t - i\pi\rho)]. \tag{8}$$

After a logarithmic change of variables this becomes

$$\Phi(t) = \lim_{\rho \to 1-} \frac{1}{2\pi} [F(te^{i\pi\rho}) e^{i\pi\rho/2} + F(te^{-i\pi\rho}) e^{-i\pi\rho/2}] \tag{8'}$$

which is substantially the same as

$$\Phi(t) = \lim_{\epsilon \to 0+} \frac{1}{2\pi i} [F(-t - i\epsilon) - F(-t + i\epsilon)],$$

the classical complex inversion formula for (3) due to Stieltjes; see D. V. Widder [1946; 338].

This example suggests precedure for the general case. We shall show that if G is defined by equations (1) and (2) then

$$f(z) = \int_{-\infty}^{\infty} G(z - t) \varphi(t) \, dt \tag{9}$$

is analytic for $|\operatorname{Im} z| < \pi\Omega$. Let

$$K(z) = \int_0^\infty E(s) e^{-sz} \, ds.$$

We shall prove that $K(z)$ is analytic and single valued in the z-plane slit along the imaginary axis from $-i\Omega\pi$ to $i\Omega\pi$, and further that

$$E(s) = \frac{1}{2\pi i} \int_C K(z) e^{sz}\, dz$$

where C is a closed rectifiable curve going counterclockwise around the segment $[-i\Omega\pi, i\Omega\pi]$. If $f(z)$ is defined by (9) then we have the symbolic inversion formula

$$\varphi(t) = E(D) f(t).$$

Just as before we obtain

$$\varphi(t) = \frac{1}{2\pi i} \int_C K(z) e^{zD} dz f(t),$$

$$\varphi(t) = \frac{1}{2\pi i} \int_C K(z) f(t + z)\, dz.$$

Again this is not in general meaningful and it must be replaced by

$$\varphi(t) = \lim_{\rho \to 1-} \frac{1}{2\pi i} \int_{C_\rho} K(z) f(t + \rho z)\, dz \tag{10}$$

where C_ρ is a closed rectifiable curve going counterclockwise around the segment $[-i\Omega\pi, i\Omega\pi]$ and lying in the strip $|\text{Im } z| < \pi\Omega/\rho$. The present chapter is devoted to establishing the validity of this inversion formula.

2. TRANSFORMS IN THE COMPLEX DOMAIN

2.1. The present section is concerned with the kernels $G(t)$ and the corresponding convolution transforms in the complex plane. We suppose that

$$E(s) = \prod_1^\infty \left(1 - \frac{s^2}{a_k^2}\right) \tag{1}$$

where

$$0 < a_1 \leqq a_2 \leqq a_3 \leqq \cdots, \tag{2}$$

$$\lim_{n \to \infty} n/a_n = \Omega.$$

We define

$$G(z) = \frac{1}{2\pi i} \int_{-i\infty}^{i\infty} [E(s)]^{-1} e^{sz}\, ds, \qquad z = x + iy. \tag{3}$$

In order to study $G(z)$ we must first investigate $E(s)$. The following result shows that $E(s)$ behaves very much like $\cos \pi\Omega s$.

LEMMA 2.1. *If $E(s)$ is defined by equations* (1) *and* (2) *then*

$$\lim_{r\to\infty} r^{-1} \log |E(re^{i\theta})| = \pi\Omega |\sin \theta|$$

uniformly for θ in any closed interval not containing an integral multiple of π.

Let $N(a)$ be the function which counts the a_k's. By equation (2) $N(a) = \Omega a + \epsilon(a) \cdot a$ where $\epsilon(a) = o(1)$ as $a \to \infty$. We have

$$\log |E(s)| = \mathrm{Rl} \int_0^\infty \log (1 - s^2a^{-2})\, dN(a).$$

Integrating by parts

$$\begin{aligned}\log |E(s)| &= \mathrm{Rl} \int_0^\infty \frac{N(a)}{a} \frac{-2s^2}{a^2 - s^2}\, da,\\ &= \mathrm{Rl} \int_0^\infty \Omega \frac{-2s^2}{a^2 - s^2}\, da + Rl \int_0^\infty \epsilon(a) \frac{-2s^2}{a^2 - s^2}\, da,\\ &= I_1(s) + I_2(s).\end{aligned}$$

If $s = re^{i\theta}$ then

$$I_1(s) = \pi\Omega r |\sin \theta| \qquad \theta \neq n\pi.$$

If θ lies in an interval $\theta_0 \leqq \theta \leqq \theta_1$ which does not include an integral multiple of π, then there exists a constant $A > 0$ independent of θ such that $|a \pm s| \geqq A(a + r)$. Thus

$$\begin{aligned}|I_2(s)| &\leqq 2A^{-2} \int_0^\infty \frac{|\epsilon(a)|\, r^2}{(a + r)^2}\, da,\\ &\leqq 2rA^{-2} \int_0^\infty \frac{|\epsilon(rb)|}{(1 + b)^2}\, db.\end{aligned}$$

Since

$$\lim_{r\to\infty} \int_0^\infty \frac{|\epsilon(rb)|}{(1 + b)^2}\, db = 0$$

our lemma follows.

Lemma 2.1 has long been known, see V. Bernstein [1933; 271].

2.2. The following theorem represents an extension to the complex plane of the results of Chapter V.

THEOREM 2.2a. *If*

1. *$E(s)$ is defined by equations* 2.1 (1) *and* 2.1 (2),
2. *μ = [multiplicity of $s - a_1$ as a zero of $E(s)$],*
3. *$G(z)$ is defined by equation* 2.1 (3),

then:

A. *$G(z)$ is an analytic function in the strip $|y| < \pi\Omega$, $z = x + iy$;*

B. $G(z) = p(z)e^{-a_1 z} + R_+(z)$,

$G(z) = p(-z)e^{a_1 z} + R_-(z)$,

where $p(z)$ is a polynomial of degree $\mu - 1$ and where

$$R_+(z), \frac{d}{dz} R_+(z) = O(e^{-(a_1+\epsilon)x}) \qquad x \to +\infty,$$

$$R_-(z), \frac{d}{dz} R_-(z) = O(e^{(a_1+\epsilon)x}) \qquad x \to -\infty,$$

for some $\epsilon > 0$, uniformly in every proper substrip $|y| < \pi(\Omega - \eta)$ of the strip $|y| < \pi\Omega$.

By Lemma 2.2a if $\eta > 0$ then

$$(1) \qquad |1/E(s)| = 0(e^{-\pi(\Omega-\eta)|\tau|}) \qquad \tau \to \pm\infty$$

uniformly for σ in any finite interval. It follows that the integral 2.1 (3) defining $G(z)$ converges uniformly in the strip $|y| \leqq \pi(\Omega - 2\eta)$ and defines an analytic function there. Since η is arbitrary conclusion A follows. To establish conclusion B let us choose $\epsilon > 0$ so small that no zeros of $E(s)$ lie in the interval $-a_1 - \epsilon \leqq \sigma < -a_1$. Integrating about the rectangular contour with vertices at $\pm iT$, $-a_1 - \epsilon \pm iT$ and letting T increase without limit we obtain, as in § 2 of Chapter V,

$$G(z) = p(z)e^{-a_1 z} + R_+(z),$$

$$R_+(z) = \frac{1}{2\pi i}\int_{-a\ -\epsilon-i\infty}^{-a_1-\epsilon+i\infty} [E(s)]^{-1} e^{sz}\, ds.$$

Using equation (1) it is easily seen that

$$(2) \qquad \left(\frac{d}{dz}\right)^n R_+(z) = 0(e^{-(a_1+\epsilon)x}) \qquad x \to +\infty$$

uniformly for $|y| \leqq \pi(\Omega - 2\eta)$. The second part of conclusion B is established similarly.

THEOREM 2.2b. *If*

1. *$E(s)$ is defined by equations* 2.1 (1) *and* 2.1 (2),

2. *$G(z)$ is defined by equation* 2.1 (3),

3. $f(z) = \int_{-\infty}^{\infty} G(z - t)d\alpha(t)$,

and if the transform 3 converges for any value of z in the strip $(-\infty < x < \infty;\ |y| < \pi\Omega)$, *it converges for all such z, uniformly in any compact set, so that* $f(z)$ *is analytic for* $(-\infty < x < \infty;\ |y| < \pi\Omega)$.

Suppose that the transform 3 converges for $z = z_0$, $|\operatorname{Im} z_0| < \pi\Omega$. Let R be any compact subset of the strip $|\operatorname{Im} z| < \pi\Omega$; we must show that

$$\lim_{A,B\to+\infty} \int_A^B G(z-t)\, d\alpha(t) = 0, \tag{3}$$

$$\lim_{A,B\to-\infty} \int_A^B G(z-t)\, d\alpha(t) = 0, \tag{3'}$$

uniformly for z in R. By Theorem 2.2a

$$G(z-t)/G(z_0-t) = O(1) \qquad t \to +\infty,$$

$$\frac{d}{dt}[G(z-t)/G(z_0-t)] = O\left(\frac{1}{t^2}\right) \qquad t \to +\infty, \tag{4}$$

uniformly for z in R. If $L(t) = \int_t^\infty G(z_0 - t)\, d\alpha(t)$ then

$$L(t) = o(1) \qquad t \to +\infty. \tag{5}$$

We have

$$\int_A^B G(z-t)\, d\alpha(t) = \int_A^B \frac{G(z-t)}{G(z_0-t)} G(z_0-t)\, d\alpha(t),$$

$$= \left[-\frac{G(z-t)}{G(z_0-t)} L(t)\right]_A^B + \int_A^B \left(\frac{d}{dt}\frac{G(z-t)}{G(z_0-t)}\right) L(t)\, dt.$$

Using equations (4) and (5) we see that equation (3) holds uniformly for z in R. We may similarly establish (3′).

2.3. Let us consider some examples. The nth iterate of the Stieltjes kernel $S_n(t)$ is defined by the formula

$$S_n(t) = \frac{1}{2\pi i}\int_{-i\infty}^{i\infty} \frac{e^{st}}{(\cos \pi s)^n}\, ds. \tag{1}$$

We shall compute $S_n(t)$ explicitly. We have

$$S_n(t) = 2^{n-1}\pi^{-1}\int_{-\infty}^{\infty} e^{it\tau}[e^{\pi\tau} + e^{-\pi\tau}]^{-n}\, d\tau.$$

Making the change of variable $e^{2\pi\tau} = y$ we obtain

$$S_n(t) = 2^{n-2}\pi^{-2}\int_0^\infty (1+y)^{-n} y^{\frac{n}{2}+\frac{it}{2\pi}-1}\, dy.$$

Since

(2) $$\int_0^\infty y^{s-1}(1+y)^{-\nu}\,dy = \Gamma(s)\Gamma(\nu-s)/\Gamma(\nu) \qquad 0 < \mathrm{Rl}\, s < \nu,$$

it follows that

$$S_n(t) = 2^{n-2}\pi^{-2}\Gamma\left(\frac{n}{2}+\frac{it}{2\pi}\right)\Gamma\left(\frac{n}{2}-\frac{it}{2\pi}\right)\Big/(n-1)!.$$

In particular

$$S_1 = \frac{1}{2\pi}\operatorname{sech}\frac{t}{2},$$

$$S_2 = \frac{t}{2\pi^2}\operatorname{cosech}\frac{t}{2}.$$

Making use of the relation $\Gamma(1+s) = s\Gamma(s)$ we obtain the recurrence formula

$$S_n(t) = [(n-1)\,(n-2)]^{-1}[(n-2)^2 + (t/\pi)^2]S_{n-2}(t)$$

from which it follows that

(3)
$$S_{2n+1}(t) = \frac{1}{2\pi}\operatorname{sech}\frac{t}{2}\frac{1}{(2n)!}\prod_{r=1}^{n}[(t/\pi)^2 + (2r-1)^2],$$

$$S_{2n}(t) = \frac{t}{2\pi^2}\operatorname{cosech}\frac{t}{2}\frac{1}{(2n-1)!}\prod_{r=1}^{n-1}[(t/\pi)^2 + (2r)^2].$$

These formulas are due to Barrucand [1950]; see also D. V. Widder [1946; 259–265]. $S_1(t)$ and $S_2(t)$ have been previously evaluated in § 9 of III.

The function sech $z/2$ has simple poles at $z = \pm(2r-1)i\pi, r = 1, 2, \cdots$ while the product

$$\prod_{r=1}^{n}[(z/\pi)^2 + (2r-1)^2]$$

vanishes for $z = \pm(2r-1)i\pi$, $r = 1, 2, \cdots, n$. It follows that $S_{2n+1}(z)$ is analytic in the strip $|\mathrm{Im}\, z| < (2n+1)\pi$. Similarly cosech $z/2$ has the simple poles at $z = \pm 2ri\pi$, $r = 0, 1, \cdots$. The product

$$z\prod_{r=1}^{n-1}[(z/\pi)^2 + (2r)^2]$$

vanishes for $z = \pm 2ri\pi$, $r = 0, 1, \cdots n-1$ from which it follows that $S_{2n}(z)$ is analytic in the strip $|\mathrm{Im}\, z| < 2n\pi$. We have thus verified directly conclusion A of Theorem 2.2a for these cases.

A second example is given by the kernels

$$T_\nu(t) = \frac{1}{2\pi i}\int_{-i\infty}^{i\infty} \frac{\Gamma\left(\frac{\nu}{2} - s\right)\Gamma\left(\frac{\nu}{2} + s\right)}{\left[\Gamma\left(\frac{\nu}{2}\right)\right]^2} e^{st}\, ds \qquad \nu > 0.$$

We have previously shown, see § 9.5 of III, that

$$T_\nu(t) = \frac{\Gamma(\nu)2^{-\nu}}{\left[\Gamma\left(\frac{\nu}{2}\right)\right]^2}\left(\operatorname{sech}\frac{t}{2}\right)^\nu.$$

3. BEHAVIOUR AT INFINITY

3.1. The results of the present section are an extension to the complex domain of the results of § 2 of Chapter VII.

LEMMA 3.1. If

1. $0 < a_1 < a$,
2. $A(t) = o(e^{a_1|t|})$, $\qquad t \to \pm\infty$,

then

$$\int_{-\infty}^{\infty} e^{-a|x-t|} A(t)\, dt = o(e^{a_1|x|}), \qquad x \to \pm\infty.$$

The proof is left to the reader.

THEOREM 3.1. *If*

1. *$E(s)$ is defined by equations* 2.1 (1) and 2.2 (2),
2. *$G(z)$ is defined by equation* 2.1 (3),
3. $f(z) = \int_{-\infty}^{\infty} G(z-t)\, d\alpha(t) \qquad z = x + iy,$

then $f(z) = o(e^{a_1|x|})$ as $x \to \pm\infty$, uniformly in every substrip $y \leq \pi(\Omega - \eta)$.

Let

$$F_{2\mu}(s) = \prod_1^\mu \left(1 - \frac{s^2}{a_k^2}\right), \quad E_{2\mu}(s) = \prod_{\mu+1}^\infty \left(1 - \frac{s^2}{a_k^2}\right),$$

and let $H_{2\mu}(t)$ and $G_{2\mu}(t)$ be the corresponding kernels. It follows from Theorem 8.1a of VI that if

$$A(t) = \int_{-\infty}^{\infty} H_{2\mu}(t-u)\, d\alpha(u)$$

then

(1) $$f(x) = \int_{-\infty}^{\infty} G_{2\mu}(x - t)A(t)\, dt \qquad -\infty < x < \infty.$$

Since $f(z)$ and $\int_{-\infty}^{\infty} G_{2\mu}(z - t)A(t)\, dt$ are analytic in the strip $|\text{Im } z| < \pi\Omega$ it follows that

(2) $$f(z) = \int_{-\infty}^{\infty} G_{2\mu}(z - t)A(t)\, dt \qquad |\text{Im } z| < \pi\Omega.$$

By Theorem 2.1 of Chapter VII

(3) $$A(t) = o(e^{a_1|t|}) \qquad t \to \pm\infty.$$

If $a_1 < a < a_{\mu+1}$ then Theorem 2.2a implies that

(4) $$G_{2\mu}(x + iy) = O(e^{-a|x|}) \qquad x \to \pm\infty$$

uniformly for $|y| \leqq \pi(\Omega - \eta)$, $\eta > 0$. Applying Lemma 3.1 we obtain our desired result.

As an example of this result we find, after a logarithmic change of variables, that if

$$F(z) = \int_{0+}^{\infty} \frac{dA(t)}{z + t} \qquad z = re^{i\theta},$$

then

$$F(z) = o(r) \qquad z \to \infty$$

$$= o(1/r) \qquad z \to 0,$$

uniformly in every sector $|\arg z| \leqq \pi - \epsilon$.

4. AUXILIARY KERNELS

4.1. Let $0 < \rho < 1$ and let $a > 0$. We define

(1) $$h(\rho, a, t) = (1 - \rho^2)\frac{1}{2}a \int_{-\infty}^{t} e^{-a|u|}\, du + \rho^2 j(t)$$

where $j(t) = 0$ for $t < 0$, $= \frac{1}{2}$ for $t = 0$, and $= 1$ for $t > 0$. It is easily verified that $h(\rho, a, t)$ is a distribution function with mean 0 and variance $2(1 - \rho^2)a^{-2}$, and that

(2) $$\int_{-\infty}^{\infty} e^{-st}\, dh(\rho, a, t) = \frac{1 - \dfrac{\rho^2 s^2}{a^2}}{1 - \dfrac{s^2}{a^2}},$$

the bilateral Laplace transform converging absolutely for $-a < \text{Rl } s < a$.

THEOREM 4.1. *If*

1. *$E(s)$ is defined equations* 2.1 (1) *and* 2.1 (2),

2. $0 < \rho < 1$,

3. $G(\rho, t) = \dfrac{1}{2\pi i} \displaystyle\int_{-i\infty}^{i\infty} \frac{E(\rho s)}{E(s)} e^{st}\, ds \qquad (-\infty < t < \infty),$

then:

A. *$G(\rho, t)$ is a frequency function with mean* 0 *and variance*

$$2(1 - \rho^2) \sum_1^\infty a_k^{-2};$$

B. $\displaystyle\int_{-\infty}^{\infty} G(\rho, t)e^{-st}\, dt = E(\rho s)/E(s)$ *the bilateral Laplace transform converging absolutely in the strip* $-a_1 < \text{Rl}\, s < a_1$;

C. *$G(\rho, z)$ is an analytic function in the strip* $|y| < \pi(1 - \rho)\Omega$, $z = x + iy$;

D. $G(\rho, z) = p(\rho, z)e^{-a_1 z} + R_+(z)$
$G(\rho, z) = p(\rho, -z)e^{a_1 z} + R_-(z)$

where $p(\rho, z)$ is a polynomial of degree $\mu - 1$ and where

$$\left(\frac{d}{dz}\right)^n R_+(z) = O(e^{-(a_1+\epsilon)x}) \qquad x \to +\infty \quad (n = 0, 1, \cdots),$$

$$\left(\frac{d}{dz}\right)^n R_-(z) = O(e^{+(a_1+\epsilon)x}) \qquad x \to -\infty \quad (n = 0, 1, \cdots),$$

for some $\epsilon > 0$, uniformly in every proper substrip $|y| < \pi(1 - \rho)(\Omega - \eta)$ of the strip $|y| < \pi(1 - \rho)\Omega$.

By the convolution theorem if

$$H_n(\rho, t) = h(a_1, \rho, t) \# h(a_2, \rho, t) \# \cdots \# h(a_n, \rho, t)$$

then $H_n(\rho, t)$ is a distribution function with characteristic function

$$\int_{-\infty}^{\infty} e^{-st}\, dH_n(\rho, t) = \prod_{k=1}^{n} \frac{1 - \dfrac{\rho^2 s^2}{a_k^2}}{1 - \dfrac{s^2}{a_k^2}}.$$

We have

$$\lim_{n\to\infty} \prod_{k=1}^{n} \frac{1 - \dfrac{\rho^2 s^2}{a_k^2}}{1 - \dfrac{s^2}{a_k^2}} = \frac{E(\rho s)}{E(s)},$$

uniformly for s in any compact set of the s-plane punctured at $\pm a_1$, $\pm a_2, \cdots$. By Corollary 2.3 of III $E(\rho s)/E(s)$ is the characteristic function of a distribution function $H(\rho, t)$,

$$\int_{-\infty}^{\infty} e^{-i\tau t}\, dH(\rho, t) = E(\rho i\tau)/E(i\tau).$$

Moreover, by Theorem 5.2 of III,

$$H(\rho, t_1) - H(\rho, t_2) = \frac{1}{2\pi i} \int_{-i\infty}^{i\infty} \frac{E(\rho s)}{E(s)} \frac{e^{st_1} - e^{st_2}}{s}\, ds.$$

Since by Lemma 2.1

$$\text{(3)} \qquad \log E(\rho i\tau)/E(i\tau) \sim -\pi(1-\rho)\Omega\, |\tau| \qquad \tau \to \pm\infty,$$

it follows that $H(\rho, t)$ is infinitely differentiable. If $G(\rho, t) = \dfrac{d}{dt} H(\rho, t)$ then $G(\rho, t)$ is a frequency function, and

$$\text{(4)} \qquad G(\rho, t) = \frac{1}{2\pi i} \int_{-i\infty}^{i\infty} \frac{E(\rho s)}{E(s)} e^{st}\, ds.$$

Conclusion C follows from (4). To demonstrate conclusion D the line of integration in (4) must be deformed to $\text{Rl}\, s = \pm(a_1 + \epsilon)$ as in § 2. From D we see that

$$\int_{-\infty}^{\infty} e^{-st} G(\rho, t)\, dt$$

converges absolutely for $|\text{Rl}\, s| < a_1$ and defines in this strip an analytic function. Since

$$\int_{-\infty}^{\infty} e^{-i\tau t} G(\rho, t)\, dt = E(\rho i\tau)/E(i\tau)$$

we have demonstrated conclusion B; that is, for $|\text{Rl}\, s| < a_1$

$$\text{(5)} \qquad \int_{-\infty}^{\infty} e^{-st} G(\rho, t)\, dt = E(\rho s)/E(s).$$

Differentiating equation (5) with respect to s, and setting $s = 0$ we obtain

$$\int_{-\infty}^{\infty} -tG(\rho, t)\, dt = \frac{d}{ds} \frac{E(\rho s)}{E(s)}\bigg|_{s=0} = 0,$$

$$\int_{-\infty}^{\infty} t^2 G(\rho, t)\, dt = \left(\frac{d}{ds}\right)^2 \frac{E(\rho s)}{E(s)}\bigg|_{s=0} = 2(1-\rho^2) \sum_1^{\infty} a_k^{-2}.$$

Thus conclusion A has been established.

4.2. Since the functions $G(\rho, t)$ are not variation diminishing the changes of sign of their derivatives cannot be studied by the methods of Chapter IV. However some important information may be obtained by an elementary argument. See A. Wintner [1938].

THEOREM 4.2. *If $G(\rho, t)$ is defined as in Theorem 4.1 then*

$$\operatorname{sgn} dG(\rho, t)/dt = -\operatorname{sgn} t.$$

A distribution function $h(t)$ is said to be convex if

$$h(t) + h(-t) = 1 \tag{1}$$

and if for $r > 0$

$$h(t \mid r) - h(t - r) \in \begin{cases} \uparrow & t \leqq 0 \\ \downarrow & t \geqq 0. \end{cases} \tag{2}$$

We will show that if $h_1(t)$ and $h_2(t)$ are convex distribution functions then $h = h_1 \# h_2$ is again a convex distribution function. There is no difficulty in showing that $h(t)$ satisfies equation (1). To establish equation (2) we note that

$$\begin{aligned} h(t + r) - h(t - r) &= \int_{-\infty}^{\infty} [h_1(t + r - u) - h_1(t - r - u)]\, dh_2(u), \\ &= \int_{-\infty}^{\infty} h_1(t - u)\, d[h_2(u + r) - h_2(u - r)]. \end{aligned}$$

Making use of the relation $h_2(u) + h_2(-u) = 1$ we see that

$$\begin{aligned} &\int_{-\infty}^{0} h_1(t - u)\, d[h_2(u + r) - h_2(u - r)] \\ &\qquad = \int_{-\infty}^{0} h_1(t - u)\, d[-h_2(-u - r) + h_2(-u + r)], \\ &\qquad = \int_{0}^{\infty} h_1(t + u)\, d[+h_2(u - r) - h_2(u + r)], \end{aligned}$$

and thus

$$h(t + r) - h(t - r) = \int_{0}^{\infty} [h_1(t + u) - h_1(t - u)]\, d[h_2(u - r) - h_2(u + r)].$$

The function $h_2(u - r) - h_2(u + r)$ is non-decreasing for $0 \leqq u < \infty$. For $t \geqq 0$ and $0 \leqq u < \infty$ the function $h_1(t + u) - h_1(t - u)$ is non-negative and non-increasing as t increases. It follows that for $t \geqq 0$ $h(t + r) - h(t - r)$ is non-increasing. The case $t \leqq 0$ may be dealt with similarly.

Let $k(t)$, $\{k_n(t)\}_{n=1}^{\infty}$ be normalized distribution functions and let $\lim_{n\to\infty} k_n(t) = k(t)$ at all points of continuity of $k(t)$. Then if the functions $k_n(t)$ are convex so is $k(t)$.

The functions $h(a_k, \rho, t)$ are convex by inspection. It follows that for each n $H_n(\rho, t)$ is convex and finally that $H(\rho, t)$ is convex. Our theorem is an immediate consequence of this fact.

COROLLARY 4.2. If $G(\rho, t)$ is defined as in Theorem 5.1 then $\lim_{\rho\to 1-} G(\rho, t) = 0$ for $t \neq 0$.

We may suppose that $t > 0$. We have

$$(\tfrac{1}{2}t)^3 G(\rho, t) \leqq \int_{t/2}^{t} u^2 G(\rho, u)\, du \leqq \int_{-\infty}^{\infty} u^2 G(\rho, u)\, du \leqq 2(1-\rho^2) \sum_{1}^{\infty} a_k^{-2},$$

from which our assertion follows.

The functions $G(\rho, t)$ have in the complex inversion theory a role analogous to the functions $G_n(t)$ in the real inversion theory. The continuous parameter ρ corresponds to the integral parameter n, and $\rho \to 1-$ corresponds to $n \to \infty$. As $n \to \infty$ the variance of $G_n(t)$ decreases to 0; similarly as $\rho \to 1-$ the variance of $G(\rho, t)$ decreases to 0, etc.

As an example let us compute explicitly the one parameter family of kernels associated with the Stieltjes transform. If

$$S_1(\rho, t) = \frac{1}{2\pi i} \int_{-i\infty}^{i\infty} \frac{\cos \pi\rho s}{\cos \pi s} e^{st}\, ds \qquad 0 < \rho < 1,$$

we have

$$S_1(\rho, t) = \frac{1}{2\pi} \int_{-\infty}^{\infty} \frac{e^{\pi\rho\tau} + e^{-\pi\rho\tau}}{e^{\pi\tau} + e^{-\pi\tau}} e^{it\tau}\, d\tau.$$

Making the change of variable $e^{2\pi\tau} = y$

$$S_1(\rho, t) = \frac{1}{(2\pi)^2} \int_0^{\infty} \left(y^{\frac{1+\rho}{2} + \frac{it}{2\pi} - 1} + y^{\frac{1-\rho}{2} + \frac{it}{2\pi} - 1} \right) (1+y)^{-1}\, dy.$$

Since

$$\int_0^{\infty} y^{s-1}(1+y)^{-1}\, dy = \frac{\pi}{\sin \pi s} \qquad 0 < \mathrm{Rl}\, s < 1,$$

we obtain

$$S_1(\rho, t) = \left[4\pi \sin\left(\frac{\pi}{2}(1+\rho) + \frac{it}{2} \right) \right]^{-1} + \left[4\pi \sin\left(\frac{\pi}{2}(1-\rho) + \frac{it}{2} \right) \right]^{-1}.$$

After a few simplifications this becomes

$$S_1(\rho, t) = \frac{1}{\pi} \frac{\cos \frac{\pi\rho}{2} \cosh \frac{t}{2}}{\cos \pi\rho + \cosh t}.$$

By inspection $S_1(\rho, z)$ is analytic in the strip $|y| < (1-\rho)\pi$ thus verifying conclusion C of Theorem 4.1.

5. THE INVERSION FUNCTION

5.1. In this section we shall determine the properties of $K(z)$.

THEOREM 5.1. *If $E(s)$ is defined by equations* 2.1 (1) and 2.1 (2) *and if*

$$K(z) = \sum_{k=0}^{\infty} E^{(k)}(0) z^{-k-1} \tag{1}$$

then $K(z)$ is analytic and single valued in the z plane except on the segment $[-i\pi\Omega, i\pi\Omega]$. Moreover

$$E(s) = \frac{1}{2\pi i} \int_C K(z) e^{sz}\, dz \tag{2}$$

where C is a closed rectifiable curve going around $[-i\pi\Omega, i\pi\Omega]$ in the positive direction.

The inequality $|E(s)| \leqq E(i|s|)$ together with Lemma 2.1 shows that $E(s)$ is of order 1 type $\pi\Omega$. Hence $\overline{\lim}_{k\to\infty} |E^{(k)}(0)|^{1/k} \leqq \pi\Omega$. Indeed if $\epsilon > 0$ then there exists $A(\epsilon)$ such that $|E(s)| \leqq A(\epsilon) \exp[\pi(\Omega + \epsilon)|s|]$ for all s. We have

$$E^{(k)}(0) = \frac{k!}{2\pi i} \int_{|s|=r} E(s) s^{-k-1}\, ds .$$

Choosing $r = k[\pi(\Omega + \epsilon)]^{-1}$ we find that

$$|E^{(k)}(0)| \leqq \frac{A(\epsilon) k! e^k}{k^k} [\pi(\Omega + \epsilon)]^k.$$

Thus

$$\overline{\lim_{k\to\infty}} |E^{(k)}(0)|^{1/k} \leqq \pi(\Omega + \epsilon),$$

or since ϵ is arbitrary

$$\overline{\lim_{k\to\infty}} |E^{(k)}(0)|^{1/k} \leqq \pi\Omega.$$

It follows that the series (1) converges for $|z| > \pi\Omega$.

Using term by term integration of the series $E(s) = \sum_{0}^{\infty} E^{(k)}(0) s^k/k!$ one may verify that for $\arg s = \theta$

$$\int_0^{\infty} e^{-sz} E(s)\, ds = K(z),$$

the integral converging because of Lemma 2.1 in the half plane $\mathrm{Rl}\, e^{i\theta} z > \pi\Omega|\sin\theta|$. That $K(z)$ is analytic except possibly on the segment $[-i\pi\Omega, i\pi\Omega]$ is now evident. Equation (2) may be demonstrated by first taking C as a circle $|z| = \pi\Omega(1 + \epsilon)$ and applying the calculus of residues to the integral (2). Cauchy's theorem then shows that C may be deformed to the more general curve described above.

Theorem 5.1 is well known, see V. Bernstein [1933; 294–311].

EXAMPLE 1. $E_n(s) = (\cos \pi s)^n$. We have

$$K_n(z) = \int_0^\infty (\cos \pi s)^n e^{-sz}\, ds$$

$$= 2^{-n} \int_0^\infty \sum_{k=0}^{n} \binom{n}{k} e^{(n-2k)i\pi s} e^{-sz}\, ds,$$

$$K_n(z) = 2^{-n} \sum_0^n \binom{n}{k} [z - (n-2k)i\pi]^{-1}. \tag{3}$$

EXAMPLE 2. $E_\nu(s) = \left[\Gamma\left(\frac{\nu}{2}\right)\right]^2 \Big/ \Gamma\left(\frac{\nu}{2} - s\right)\Gamma\left(\frac{\nu}{2} + s\right) \qquad (0 < \nu).$

Consider the function $\varphi_\rho(x)$ which is $e^{-x}x^{\rho-1}$ for $x \geqq 0$ and 0 for $x < 0$ and let

$$\Phi_\rho(t) = \frac{1}{\sqrt{2\pi}} \int_{-\infty}^{\infty} \varphi_\rho(x) e^{itx}\, dx = \frac{1}{\sqrt{2\pi}} \frac{\Gamma(\rho)}{(1 - it)^\rho}.$$

(This formula is valid for $\text{Rl}\, \rho > 0$.) If $\text{Rl}\, \rho > \frac{1}{2}$ then

$$\varphi_\rho(x) \in L^2(-\infty, \infty);$$

thus, by Parseval's equality, if $\text{Rl}\, \rho > \frac{1}{2}$, $\text{Rl}\, \sigma > \frac{1}{2}$, we have

$$\int_{-\infty}^{\infty} \Phi_\rho(t)\overline{\Phi_\sigma(t)}\, dt = \int_{-\infty}^{\infty} \varphi_\rho(x)\overline{\varphi_\sigma(x)}\, dx,$$

$$\frac{\Gamma(\rho)\Gamma(\sigma)}{2\pi} \int_{-\infty}^{\infty} \frac{dt}{(1 - it)^\rho (1 + it)^\sigma} = \int_0^\infty e^{-2x} x^{\rho+\sigma-2}\, dx = \frac{\Gamma(\rho + \sigma - 1)}{2^{\rho+\sigma-1}}.$$

Appealing to the principle of analytic continuation we see that this formula holds for $\text{Rl}\, \rho + \text{Rl}\, \sigma > 1$. Let $t = \tan \frac{x}{2}$, $\rho + \sigma = \nu$, $(\rho - \sigma)/2 = s$; we obtain

$$\frac{\left[\Gamma\left(\frac{\nu}{2}\right)\right]^2 2^{\nu-3}}{\pi\Gamma(\nu - 1)} \int_{-\pi}^{\pi} \left[\cos\left(\frac{x}{2}\right)\right]^{\nu-2} e^{ixs} dx$$

$$= \left[\Gamma\left(\frac{\nu}{2}\right)\right]^2 \Big/ \Gamma\left(\frac{\nu}{2} + s\right)\Gamma\left(\frac{\nu}{2} - s\right),$$

this result being valid for $\text{Rl}\, \nu > 1$ and all (complex) s. Since

$$\int_0^\infty e^{-sz}\, ds \int_{-\pi}^{\pi} \left[\cos\left(\frac{t}{2}\right)\right]^{\nu-2} e^{its}\, dt = \int_{-\pi}^{\pi} \left[\cos\left(\frac{t}{2}\right)\right]^{\nu-2} \frac{dt}{z - it}$$

we have

$$(4)\qquad K_\nu(z) = \frac{\left[\Gamma\left(\frac{\nu}{2}\right)\right]^2 2^{\nu-3}}{\pi\Gamma(\nu-1)} \int_{-\pi}^{\pi} \frac{\left[\cos\left(\frac{t}{2}\right)\right]^{\nu-2}}{z-it}\, dt \qquad (\nu > 1).$$

The case $\nu = 1$ is included under Example 1. We have

$$(4')\qquad K_1(z) = \frac{z}{\pi^2 + z^2}\,.$$

To obtain a formula valid for $0 < \nu < 1$, we note that

$$1\Big/\left[\Gamma\left(\frac{\nu}{2} - s\right)\Gamma\left(\frac{\nu}{2} + s\right)\right] = \left(\frac{\nu^2}{4} - s^2\right)\Big/\left[\Gamma\left(\frac{\nu}{2} + 1 - s\right)\Gamma\left(\frac{\nu}{2} + 1 + s\right)\right].$$

It follows that

$$\int_0^\infty \left[\Gamma\left(\frac{\nu}{2} - s\right)\Gamma\left(\frac{\nu}{2} + s\right)\right]^{-1} e^{-sz}\, ds = \left[\frac{\nu^2}{4} - \left(\frac{d}{dz}\right)^2\right] \int_0^\infty \left[\Gamma\left(\frac{\nu}{2} + 1 - s\right)\Gamma\left(\frac{\nu}{2} + 1 + s\right)\right]^{-1} e^{-sz}\, ds,$$

$$K_\nu(z) = \left[1 - \left(\frac{2}{\nu}\frac{d}{dz}\right)^2\right] K_{\nu+2}(z).$$

Thus

$$K_\nu(z) = \frac{\Gamma\left(\frac{\nu}{2} + 1\right)^2 2^{\nu-1}}{\pi\Gamma(\nu+1)} \int_{-\pi}^{\pi} \left[\cos\left(\frac{t}{z}\right)\right]^\nu \left[\frac{1}{z-it} - \frac{(8/\nu^2)}{(z-it)^3}\right] dt.$$

Integration by parts gives

$$\int_{-\pi}^{\pi} \left[\cos\left(\frac{t}{2}\right)\right]^\nu \frac{dt}{(z-it)^3} = +\frac{\nu}{4i}\int_{-\pi}^{\pi} \left[\cos\left(\frac{t}{2}\right)\right]^{\nu-1} \sin\left(\frac{t}{2}\right) \frac{dt}{(z-it)^2}\,.$$

Thus finally

$$(4'')\qquad K_\nu(z) = \frac{\Gamma\left(\frac{\nu}{2} + 1\right)^2 2^{\nu-1}}{\pi\Gamma(\nu+1)} \cdot \int_{-\pi}^{\pi} \left[\cos\left(\frac{t}{2}\right)\right]^{\nu-1} \left[\frac{\cos\left(\frac{t}{2}\right)}{z-it} + \frac{\frac{2i}{\nu}\sin\left(\frac{t}{2}\right)}{(z-it)^2}\right] dt \qquad (0 < \nu < 1).$$

This formula is of course valid for $\nu \geqq 1$; however if $\nu \geqq 1$ it may be reduced to the simpler formulas given above. This example is due to D. B. Sumner [1949].

6. APPLICATION OF THE INVERSION OPERATOR

6.1. We consider here the application of the complex inversion operator.

THEOREM 6.1a. *Let*

1. *$G(z)$ be defined by equation* 2.1 (3),

2. $f(z) = \int_{-\infty}^{\infty} G(z - t)e^{ct}\, d\alpha(t) \qquad (c \text{ real}),$

3. *$K(z)$ be defined as in Theorem* 5.1,

4. *$G(\rho, t)$ be defined as in Theorem* 4.1,

5. *C_ρ be a closed rectifiable curve going around the segment $[-i\pi\Omega, i\pi\Omega]$ in the positive direction and lying in the strip $|\operatorname{Im} z| < \pi\Omega/\rho$, then for $-\infty < u < \infty$*

$$(1/2\pi i) \int_{C_\rho} f(u + \rho z)K(z)\, dz = \int_{-\infty}^{\infty} G(\rho, u - t)e^{ct}\, d\alpha(t).$$

We have

$$\frac{1}{2\pi i} \int_{C_\rho} f(u + \rho z)K(z)\, dz = \frac{1}{2\pi i} \int_{C_\rho} K(z)\, dz \int_{-\infty}^{\infty} G(u + \rho z - t)e^{ct}\, d\alpha(t).$$

The inner integral converges uniformly for z on C_ρ by Theorem 2.2b, and we may therefore interchange the order of the integrations to obtain

$$\frac{1}{2\pi i} \int_{C_\rho} f(u + \rho z)K(z)\, dz = \int_{-\infty}^{\infty} e^{ct}\, d\alpha(t) \frac{1}{2\pi i} \int_{C_\rho} G(u + \rho z - t)K(z)\, dz.$$

We have seen in Theorem 2.2a that the integral

$$(1/2\pi i) \int_{-i\infty}^{i\infty} [E(s)]^{-1} e^{s(u + \rho z - t)}\, ds = G(u + \rho z - t)$$

converges uniformly for z on C_ρ. Inserting this integral and again inverting the order of integration we find that

$$\frac{1}{2\pi i} \int_{C_\rho} f(u + \rho z)K(z)\, dz$$

$$= \int_{-\infty}^{\infty} e^{ct}\, d\alpha(t) \frac{1}{2\pi i} \int_{-i\infty}^{i\infty} \frac{e^{s(u-t)}}{E(s)}\, ds \frac{1}{2\pi i} \int_{C_\rho} e^{\rho s z} K(z)\, dz,$$

$$= \int_{-\infty}^{\infty} e^{ct}\, d\alpha(t) \frac{1}{2\pi i} \int_{-i\infty}^{i\infty} \frac{E(\rho s)}{E(s)} e^{s(u-t)}\, ds = \int_{-\infty}^{\infty} G(\rho, u - t)e^{ct}\, d\alpha(t),$$

as desired.

COROLLARY 6.1. Under the assumptions of Theorem 6.1a the integral

$$\int_{-\infty}^{\infty} G(\rho, u - t)e^{ct}\, d\alpha(t)$$

converges uniformly for u in any finite interval.

If A and B are sets in the z plane and a and b are (complex) numbers then by $aA + bB$ we mean the set of all z of the form $z = az_1 + bz_2$ with $z_1 \in A$, $z_2 \in B$.

Let I be any finite interval $-\infty < u_1 \leqq x \leqq u_2 \leqq \infty$. We set $J = I + \rho C_\rho$. It is easily seen that J is a compact set in the z plane lying in the strip $|\text{Im } y| < \pi\Omega$. Given $\epsilon > 0$ we can find T so large that $|t_1|, |t_2| \geqq T$, sgn $t_1 =$ sgn t_2, implies that

$$\left| \int_{t_1}^{t_2} G(w - t)e^{ct}\, d\alpha(t) \right| \leqq \epsilon \qquad w \in J.$$

Let $M =$ l.u.b. $|K(z)|$ for z on C_ρ and let L be the length of C_ρ. For t_1 and t_2 as above we have

$$\int_{t_1}^{t_2} G(\rho, u - t)e^{ct}\, d\alpha(t) = \frac{1}{2\pi i} \int_{C_\rho} K(z)\, dz \int_{t_1}^{t_2} G(u + \rho z - t)e^{ct}\, d\alpha(t),$$

$$\left| \int_{t_1}^{t_2} G(\rho, u - t)e^{ct}\, d\alpha(t) \right| \leqq ML\epsilon(2\pi)^{-1}$$

This establishes our assertion.

THEOREM 6.1b. *Under the assumptions of Theorem* 6.1a.:

A. $-a_1 < c < a_1$ *implies that*

$$\int_{x_1}^{x_2} e^{-cu}\, du \frac{1}{2\pi i} \int_{C_\rho} f(u + \rho z)K(z)\, dz$$
$$= \int_{-\infty}^{\infty} G(\rho, t)e^{-ct}\alpha(x_2 - t)\, dt - \int_{-\infty}^{\infty} G(\rho, t)e^{-ct}\alpha(x_1 - t)\, dt;$$

B. $c = a_1, -a_1$, *implies that*

$$\int_{x_1}^{x_2} e^{-cu}\, du \frac{1}{2\pi i} \int_{C_\rho} f(u + \rho z)K(z)\, dz$$
$$= \int_{-\infty}^{\infty} G(\rho, t)e^{-ct}[\alpha(x_2 - t) - \alpha(x_1 - t)]\, dt;$$

C. $c > a_1$ *implies that* $\alpha(+\infty)$ *exists and*

$$\int_{x_1}^{\infty} e^{-cu}\, du \frac{1}{2\pi i} \int_{C_\rho} f(u + \rho z)K(z)\, dz$$
$$= \int_{-\infty}^{\infty} G(\rho, t)e^{-ct}[\alpha(+\infty) - \alpha(x_1 - t)]\, dt;$$

D. $c < -a_1$ *implies that* $\alpha(-\infty)$ *exists and*

$$\int_{-\infty}^{x_2} e^{-cu}\, du \frac{1}{2\pi i} \int_{C_\rho} f(u + \rho z)K(z)\, dz$$

$$= \int_{-\infty}^{\infty} G(\rho, t)e^{-ct}[\alpha(x_2 - t) - \alpha(-\infty)]\, dt.$$

Let us first consider the case $-a_1 < c < a_1$. Lemma 2.1c of Chapter VI implies that $\alpha(t)G(\rho, -t)e^{ct} = o(1)$ as $t \to \pm\infty$, from which it follows that

$$\lim_{t\to\pm\infty} G(\rho, u - t)e^{ct}\alpha(t) = 0 \tag{1}$$

uniformly for u in any finite interval. We have

$$\int_{-\infty}^{\infty} \alpha(t) \frac{d}{dt}[G(\rho, u - t)e^{ct}]\, dt = [\alpha(t)G(\rho, u - t)e^{ct}]_{-\infty}^{\infty} \tag{2}$$

$$- \int_{-\infty}^{\infty} G(\rho, u - t)e^{ct}\, d\alpha(t).$$

Equations (1) and (2) and Corollary 6.1 show that

$$e^{-cu} \frac{1}{2\pi i} \int_{C_\rho} f(u + \rho z)K(z)\, dz = - \int_{-\infty}^{\infty} \alpha(t) \frac{d}{dt}[G(\rho, u - t)\, e^{-c(u-t)}]\, dt,$$

the integral on the left converging uniformly for u in any finite interval. This may be rewritten as

$$e^{-cu} \frac{1}{2\pi i} \int_{C_\rho} f(u + \rho z)K(z)\, dz = + \int_{-\infty}^{\infty} \alpha(t) \frac{d}{du}[G(\rho, u - t)e^{-c(u-t)}]\, dt.$$

Because of the uniform convergence we may integrate under the integral sign to obtain

$$\int_{x_1}^{x_2} e^{-cu}\, du \frac{1}{2\pi i} \int_{C_\rho} f(u + \rho z)K(z)\, dz \tag{3}$$

$$= \int_{-\infty}^{\infty} [G(\rho, x_2 - t)e^{-c(x_2-t)} - G(\rho, x_1 - t)e^{-c(x_1-t)}]\alpha(t)\, dt.$$

We have

$$\int_{-T}^{T} G(\rho, x - t)e^{-c(x-t)}\alpha(t)\, dt - \int_{-T}^{T} G(\rho, t)e^{-ct}\alpha(x - t)\, dt$$

$$= - \int_{T}^{x+T} G(\rho, x - t)e^{-c(x-t)}\alpha(t)\, dt + \int_{-T}^{x-T} G(\rho, x - t)e^{-c(x-t)}\, \alpha(t)\, dt.$$

Making use of equation (1) we see that

$$\int_{-T}^{T} G(\rho, x - t)e^{-c(x-t)}\alpha(t)\, dt - \int_{-T}^{T} G(\rho, t)e^{-ct}\alpha(x - t)\, dt = o(1) \quad (T \to \infty).$$

Combining this result with equation (3) we obtain

$$\int_{x_1}^{x_2} e^{-cu}\,du \frac{1}{2\pi i}\int_{C_\rho} f(u+\rho z)K(z)\,dz = \int_{-\infty}^{\infty} G(\rho,t)e^{-ct}[\alpha(x_2-t)-\alpha(x_1-t)]\,dt. \tag{4}$$

Using Theorem 4.1 we may show that if c is not equal to $\pm a_1$ then

$$[G(\rho,t)e^{-ct}]/[G(\rho,t)e^{-ct}]' = O(1) \qquad t\to\pm\infty$$

$$\frac{d}{dt}\{[G(\rho,t)e^{-ct}]/[G(\rho,t)e^{-ct}]'\} = O\left(\frac{1}{t^2}\right) \qquad t\to\pm\infty. \tag{5}$$

We have shown in the course of our proof that the integral

$$\int_{-\infty}^{\infty} \alpha(x-t)\,[G(\rho,t)e^{-ct}]'\,dt$$

is convergent. Employing the relations (5) as in the demonstration of Theorem 2.2b we find that the integral

$$\int_{-\infty}^{\infty} \alpha(x-t)\,[G(\rho,t)e^{-ct}]\,dt$$

is also convergent. Thus

$$\int_{x_1}^{x_2} e^{-cu}\,du \frac{1}{2\pi i}\int_{C_\rho} f(u+\rho z)K(z)\,dz = \int_{-\infty}^{\infty} G(\rho,t)e^{-ct}\alpha(x_2-t)\,dt - \int_{-\infty}^{\infty} G(\rho,t)e^{-ct}\alpha(x_1-t)\,dt.$$

We have established conclusion A. If $c > a_1$ almost exactly the same argument shows that

$$\begin{aligned}\int_{x_1}^{x_2} e^{-cu}\,du \frac{1}{2\pi i}\int_{C_\rho} f(u+\rho z)K(z)\,dz &= \int_{-\infty}^{\infty} G(\rho,t)e^{-ct}[\alpha(x_2-t)-\alpha(+\infty)]\,dt \\ &\quad - \int_{-\infty}^{\infty} G(\rho,t)e^{-ct}[\alpha(x_1-t)-\alpha(+\infty)]\,dt.\end{aligned} \tag{6}$$

We have

$$\begin{aligned}\int_{-\infty}^{\infty} G(\rho,t)e^{-ct}[\alpha(x_2-t)-\alpha(+\infty)]\,dt &= \int_{-\infty}^{\infty} G(\rho,x_2-t)e^{-c(x_2-t)}[\alpha(t)-\alpha(+\infty)]\,dt, \\ &= e^{-cx_2}\frac{1}{2\pi i}\int_{C_\rho} K(z)\,dz \int_{-\infty}^{\infty} G(x_2+\rho z-t)e^{ct}[\alpha(t)-\alpha(+\infty)]\,dt.\end{aligned}$$

Using Theorem 3.1 we find that

$$\int_{-\infty}^{\infty} G(\rho, t)e^{-ct}[\alpha(x_2 - t) - \alpha(+\infty)]\, dt = o(e^{-cx_2+a_1x_2}),$$

$$= o(1) \qquad (x_2 \to +\infty).$$

Letting x_2 increase without limit in equation (6) and using the above result we obtain

$$\int_{x_1}^{\infty} e^{-cu}\, du \frac{1}{2\pi i} \int_{C_\rho} f(u + \rho z)K(z)\, dz$$

$$= \int_{-\infty}^{\infty} G(\rho, t)e^{-ct}[\alpha(+\infty) - \alpha(x_1 - t)]\, dt.$$

We have established conclusion C. The same arguments suffice to establish conclusions B and D.

7. THE INVERSION THEOREMS

7.1. We are now in position to prove our principal theorems. It is convenient to establish them both at the same time.

THEOREM 7.1a. *If*

1. *$G(z)$ is defined by equation* 2.1 (3),

2. $f(z) = \int_{-\infty}^{\infty} G(z - t)e^{ct}\, d\alpha(t)$, *where $\alpha(t)$ is normalized,*

3. *$K(z)$ is defined as in Theorem* 5.1,

4. *C_ρ is defined as in Theorem* 6.1a,

then:

A. *$-a_1 \leqq c \leqq a_1$ implies that*

$$\lim_{\rho\to 1-} \int_{x_1}^{x_2} e^{-cu}\, du(1/2\pi i) \int_{C_\rho} f(u + \rho z)K(z)\, dz = \alpha(x_2) - \alpha(x_1);$$

B. *$c > a_1$ implies that $\alpha(+\infty)$ exists and*

$$\lim_{\rho\to 1-} \int_{x_1}^{\infty} e^{-cu}\, du(1/2\pi i) \int_{C_\rho} f(u + \rho z)K(z)\, dz = a(+\infty) - \alpha(x_1);$$

C. *$c < -a_1$ implies that $\alpha(-\infty)$ exists and*

$$\lim_{\rho\to 1-} \int_{-\infty}^{x_2} e^{-cu}\, du(1/2\pi i) \int_{C_\rho} f(u + \rho z)K(z)\, dz = \alpha(x_2) - \alpha(-\infty).$$

THEOREM 7.1b. *If*

1. $G(z)$ *is defined by equation* 2.1 (3),

2. $$f(z) = \int_{-\infty}^{\infty} G(z - t)\varphi(t)\,dt,$$

3. $K(z)$ *is defined as in Theorem* 5.1,

4. C_ρ *is defined as in Theorem* 6.1a,

5. $$\int_0^h [\phi(x + t) - \phi(x)]\,dt = o(h) \qquad g \to 0,$$

then

$$\lim_{\rho \to 1-} \int_{C_\rho} f(x + \rho z)K(z)\,dz = \varphi(x).$$

In view of Theorems 6.1a and 6.1b it is sufficient to show that if the integral

$$(1) \qquad \int_{-\infty}^{\infty} G(t)\psi(t)\,dt$$

is convergent, and if either

$$(2) \qquad [\psi(0+) + \psi(0-] = 0,$$

or

$$(2') \qquad \int_0^h \psi(t)\,dt = o(h) \qquad h \to 0,$$

holds, then

$$\lim_{\rho \to 1-} \int_{-\infty}^{\infty} G(\rho, t)\psi(t)\,dt = 0.$$

Let

$$E_0(\rho, s) = \prod_{k > \mu} \left\{ \frac{1 - \dfrac{\rho^2 s^2}{a_k^2}}{1 - \dfrac{s^2}{a_k^2}} \right\}$$

and let

$$G_k(\rho, t) = \frac{1}{2\pi i} \int_{-i\infty}^{i\infty} \left(1 - \frac{s^2}{a_1^2}\right)^{-k} E_0(\rho, s) e^{st}\,ds$$

for $k = 0, 1, \cdots, \mu$. If we define

$$H_k(t) = \frac{1}{2\pi i} \int_{-i\infty}^{i\infty} \left(1 - \frac{s^2}{a_1^2}\right)^{-k} e^{st}\,ds$$

then evidently

$$G_k(\rho, t) = H_k(t) * G_0(\rho, t).$$

It follows from this that $G_k(\rho, t)$ is a frequency function with mean 0 and with variance $2ka_1^{-2} + 2(1 - \rho^2) \sum_{k>\mu} a_k^{-2}$. We have

$$\int_{-\infty}^{\infty} G_k(\rho, t)e^{-st}\, dt = \left(1 - \frac{s^2}{a_1^2}\right)^{-k} E_0(\rho, s), \tag{3}$$

the bilateral Laplace transform converging absolutely in the strip

$$-a_1 < \text{Rl}\, s < a_1 \qquad k = 1, \cdots, \mu,$$

$$-a_{\mu+1} < \text{Rl}\, s < a_{\mu+1} \qquad k = 0.$$

Further if $k = 1, \cdots, \mu$, then

$$\left(\frac{d}{dt}\right)^n G_k(\rho, t) = \left(\frac{d}{dt}\right)^n p_k(\rho, t)e^{-a_1 t} + O(e^{-(a_1+\epsilon)t}) \qquad t \to +\infty,$$

$$= \left(\frac{d}{dt}\right)^n p_k(\rho, -t)e^{a_1 t} + O(e^{(a_1+\epsilon)t}) \qquad t \to -\infty,$$

where $p_k(\rho,t)$ is a real polynomial of degree $k - 1$ and where $\epsilon > 0$; and

$$\left(\frac{d}{dt}\right)^n G_0(\rho, t) = \left(\frac{d}{dt}\right)^n p(\rho, t)e^{-a_{\mu+1}t} + O(e^{-(a_{\mu+1}+\epsilon)t}) \qquad t \to +\infty,$$

$$= \left(\frac{d}{dt}\right)^n p(\rho, -t)e^{a_{\mu+1}t} + O(e^{(a_{\mu+1}+\epsilon)t}) \qquad t \to -\infty,$$

where $p(\rho, t)$ is a real polynomial and where $\epsilon > 0$. From these asymptotic expansions we may first verify that

$$G_k(\rho, t)/G(t) = O(1) \qquad t \to \pm\infty,$$

$$\frac{d}{dt}[G_k(\rho, t)/G(t)] = O\left(\frac{1}{t^2}\right) \qquad t \to \pm\infty,$$

and from this deduce that the integrals

$$\int_{-\infty}^{\infty} G_k(\rho, t)\psi(t)dt \qquad k = 0, \cdots, \mu \tag{4}$$

are convergent. If the constants $u_k(\rho, t)$ are defined by the equation

$$\left[\frac{1 - \dfrac{\rho^2 s^2}{a_1^2}}{1 - \dfrac{s^2}{a_1^2}}\right]^{\mu} = \sum_{k=0}^{\mu} \frac{u_k(\rho)}{\left(1 - \dfrac{s^2}{a_1^2}\right)^k}$$

then we have

$$G(\rho, t) = \sum_{k=0}^{\mu} u_k(\rho) G_k(\rho, t).$$

Note that

(5) $$\lim_{\rho \to 1-} u_k(\rho) = 0 \qquad (k = 1, \cdots, \mu); \qquad \lim_{\rho \to 1-} u_0(\rho) = 1.$$

Because the integrals (4) are convergent we obtain

(6) $$\int_{-\infty}^{\infty} G(\rho, t)\psi(t)\, dt = \sum_{k=0}^{\mu} u_k(\rho) \int_{-\infty}^{\infty} G_k(\rho, t)\psi(t)\, dt.$$

By arguments like those given in § 8 of Chapter VI one may easily show that if

$$B_k(t) = \int_{-\infty}^{\infty} H_k(u - t)\psi(u)\, du$$

then

$$\int_{-\infty}^{\infty} G_k(\rho, t)\psi(t)\, dt = \int_{-\infty}^{\infty} G_0(\rho, t) B_k(t)\, dt \qquad (k = 1, \cdots, \mu).$$

By Theorem 2.1 of Chapter VII

$$B_k(t) = o(e^{a_1|t|}) \qquad t \to \pm\infty$$

and thus if $1 \leqq k \leqq \mu$ then here exists a constant $O(1)$ such that $|B_k(t)| \leqq O(1) \cosh a_1 t$ $(-\infty < t < \infty)$. Now

$$\int_{-\infty}^{\infty} G_0(\rho, t) \cosh a_1 t\, dt = \tfrac{1}{2}[E_0(\rho, -a_1) + E_0(\rho, a_1)] = O(1) \qquad \rho \to 1-.$$

Making use of (5) we see that

$$\lim_{\rho \to 1-} u_k(\rho) \int_{-\infty}^{\infty} G_k(\rho, t)\psi(t)\, dt = 0 \quad (k = 1, \cdots, \mu).$$

It remains to show that if condition (2) or condition (2′) is satisfied then

(7) $$\lim_{\rho \to 1-} \int_{-\infty}^{\infty} G_0(\rho, t)\psi(t)\, dt = 0.$$

We first assert if $\delta > 0$ then

(8) $$\lim_{\rho \to 1-} \int_{|t| \geqq \delta} G_0(\rho, t)\psi(t)\, dt = 0.$$

Let $\Psi(t) = \int_0^t \psi(u)\, du$. Lemma 2.1c of Chapter VI and the convergence of the integral (1) imply that $\Psi(t) = o(e^{a_1|t|})$ $t \to \pm\infty$. We have

$$\int_{|t| \geqq \delta} G_0(\rho, t)\psi(t)\, dt = -\Big[G_0(\rho, t)\Psi(t)\Big]_{-\delta}^{\delta} - \int_{|t| \geqq \delta} G_0'(\rho, t)\Psi(t)\, dt.$$

By Corollary 4.2

$$\lim_{\rho\to 1-} [G_0(\rho, t)\Psi(t)]_{-\delta}^{\delta} = 0.$$

For $|t| \geqq \delta$ we have $|\Psi(t)| \leqq O(1) \left|\sinh^3 \frac{a_1 t}{3}\right|$. Since by Theorem 4.2 $\operatorname{sgn} G_0'(\rho, t) = -\operatorname{sgn} t$ we have

$$\int_{|t|\geqq\delta} G_0'(\rho, t) \left|\Psi(t)\, dt\right| \leqq -O(1) \int_{-\infty}^{\infty} \left[\sinh^3 \frac{a_1 t}{3}\right] G_0'(\rho, t)\, dt.$$

Integrating by parts we find that

$$-\int_{-\infty}^{\infty} \sinh^3 \left(\frac{a_1 t}{3}\right) G_0'(\rho, t)\, dt = a_1 \int_{-\infty}^{\infty} \sinh^2 \frac{a_1 t}{3} \cosh \frac{a_1 t}{3} G_0(\rho, t)\, dt,$$

$$= \frac{a_1}{8}\left[E_0(\rho, a_1) - E_0\left(\rho, \frac{a_1}{3}\right) - E_0\left(\rho, \frac{-a_1}{3}\right) + E_0(\rho, -a_1)\right],$$

$$= o(1) \qquad \rho \to 1-,$$

since $\lim_{\rho\to 1-} E_0(\rho, \sigma) = 1$ for $-a_{\mu+1} < \sigma < a_{\mu+1}$. Combining our results, we have established equation (8).

Let condition (2) hold. Given $\epsilon > 0$ we choose $\delta > 0$ so small that $|\psi(-t) + \psi(t)| \leqq \epsilon$ for $0 < t \leqq \delta$; then

$$\left|\int_{-\delta}^{\delta} G_0(\rho, t)\psi(t)\, dt\right| = \left|\int_0^{\delta} G(\rho, t)\, [\psi(t) + \psi(-t)]\, dt\right| \leqq \epsilon.$$

Using equation (8) we obtain

$$\varlimsup_{\rho\to 1-} \left|\int_{-\infty}^{\infty} G_0(\rho, t)\psi(t)\, dt\right| \leqq \epsilon.$$

Since ϵ is arbitrary, equation (7) follows.

Let condition (2′) hold. Given $\epsilon > 0$ we choose $\delta > 0$ so small that if $\Psi(t) = \int_0^t \psi(u)\, du$ then $|\Psi(t)| \leqq \epsilon |t|$ for $|t| \leqq \delta$. Employing Theorem 4.2 we find that

$$\int_{-\delta}^{\delta} G_0(\rho, t)\psi(t)\, dt = [G_0(\rho, t)\Psi(t)]_{-\delta}^{\delta} - \int_{-\delta}^{\delta} \Psi(t) G_0'(\rho, t)\, dt,$$

$$\left|\int_{-\delta}^{\delta} \Psi(t) G_0'(\rho, t)\, dt\right| \leqq -\epsilon \int_{-\delta}^{\delta} t G_0'(\rho, t)\, dt,$$

$$\leqq -\epsilon [t G_0(\rho, t)]_{-\delta}^{\delta} + \epsilon \int_{-\delta}^{\delta} G_0(\rho, t)\, dt.$$

Recalling Corollary 4.2 we see that

$$\overline{\lim_{\rho\to 1-}} \left| \int_{-\delta}^{\delta} G_0(\rho, t)\psi(t)\, dt \right| \leqq \epsilon.$$

Using equation (8) we obtain

$$\overline{\lim_{\rho\to 1-}} \left| \int_{-\infty}^{\infty} G_0(\rho, t)\psi(t)\, dt \right| \leqq \epsilon.$$

Since ϵ is arbitrary equation (7) follows.

As an example we see that if

$$f(z) = \int_{-\infty}^{\infty} \left[\operatorname{sech}\left(\frac{z-t}{2}\right)\right]^{\nu} \varphi(t)\, dt$$

then

$$\varphi(t) = \lim_{\rho\to 1-} \frac{\nu - 1}{8\pi} \int_{-\pi}^{\pi} f(t + i\rho y)\, (\cos \tfrac{1}{2}y)^{\nu-2}\, dy \qquad (\nu > 1),$$

$$\varphi(t) = \lim_{\rho\to 1-} \frac{1}{4\pi} [f(t + i\rho\pi) + f(t - i\rho\pi)] \qquad (\nu = 1),$$

$$\varphi(t) = \lim_{\rho\to 1-} \frac{\nu}{8\pi} \left[\int_{-\pi}^{\pi} f(t + i\rho y) \left(\cos\frac{y}{2}\right)^{\nu} dy \right.$$

$$\left. + \frac{2i}{\nu} \int_{-\pi}^{\pi} f'(t + i\rho y) \left(\cos\frac{y}{2}\right)^{\nu-1} \sin\frac{y}{2}\, dy\right] \qquad (0 < \nu < 1).$$

8. A GENERAL REPRESENTATION THEOREM

8.1. In the remaining sections of this chapter we shall construct a representation theory corresponding to our complex inversion operator. The following result is analogous to Theorem 3.1 of VII.

THEOREM 8.1. *If*

1. *$G(t)$ is defined by equation 2.1 (3),*
2. *$K(z)$ is defined as in Theorem 5.1,*
3. *$f(z)$ is analytic for $|\operatorname{Im} z| < \pi\Omega/\theta$* $\qquad 0 < \theta < 1,$
4. *$f(z) = o(e^{a_1\theta|x|})$ $x \to \pm\infty$ uniformly in every substrip*
   $$|\operatorname{Im} z| < \pi\Omega/\theta', \qquad \theta' > \theta,$$

then

$$\int_{-\infty}^{\infty} G(x - t)\, dt (1/2\pi i) \int_{C_\theta} f(t + z) K(z)\, dz = f(x)$$

where C_θ is a rectifiable closed curve going around $[-i\pi\Omega, i\pi\Omega]$ in the positive direction and lying in the strip $|\operatorname{Im} z| < \pi\Omega/\theta$.

By assumption 4 and Theorem 2.2a we see that

$$\int_{-\infty}^{\infty} G(x-t)\,dt(1/2\pi i)\int_{C_\theta} f(t+z)K(z)\,dz$$
$$= \lim_{\rho\to 1-}\int_{-\infty}^{\infty} G(x-t)\,dt\int_{C_\theta} f(t+\rho z)K(z)\,dz.$$

For $\theta<\rho<1$ let C_ρ be a closed rectifiable curve going around $[-i\pi\Omega, i\pi\Omega]$ in the positive direction and lying in the strip $|\operatorname{Im} z|<\pi\Omega/\rho$. By Cauchy's theorem

$$(1/2\pi i)\int_{C_\theta} f(t+\rho z)K(z)\,dz = (1/2\pi i)\int_{C_\rho} f(t+\rho z)K(z)\,dz.$$

Assumption 4 and Theorem 2.2a show that the iterated integral

$$\int_{-\infty}^{\infty} G(x-t)\,dt(1/2\pi i)\int_{C_\rho} f(t+\rho z)K(z)\,dz$$

is absolutely convergent and so may be inverted to give

$$(1/2\pi i)\int_{C_\rho} K(z)\,dz\int_{-\infty}^{\infty} f(t+\rho z)G(x-t)\,dt.$$

Using Theorem 2.2a and assumptions 3 and 4 one may, for z on C_ρ, verify that

$$\int_{C_\rho} f(t+\rho z)G(x-t)\,dt = \int_{-\infty}^{\infty} G(x+\rho z-t)f(t)\,dt$$

by integrating around the rectangular contour $\operatorname{Im} t=0$, $\operatorname{Im} t=-\operatorname{Im}\rho z$, $\operatorname{Rl} t=\pm T$, and allowing T to increase without limit. Inserting this we obtain

$$(1/2\pi i)\int_{C_\rho} K(z)\,dz\int_{-\infty}^{\infty} G(x+\rho z-t)f(t)\,dt.$$

By Theorem 7.1b

$$\lim_{\rho\to 1-}(1/2\pi i)\int_{C_\rho} K(z)\,dz\int_{-\infty}^{\infty} G(x+\rho z-t)f(t)\,dt = f(x).$$

Our theorem follows.

9. DETERMINING FUNCTION NON-DECREASING

9.1.

THEOREM 9.1. *Let $G(t)$ be defined as in equation 2.1 (3) and $K(z)$ as in Theorem 5.1. Necessary and sufficient conditions that* $f(z)=\int_{-\infty}^{\infty} G(z-t)\,d\alpha(t)$ *with $\alpha(t)\in\uparrow$ are:*

1. *$f(z)$ is analytic in the strip $|\operatorname{Im} z|<\pi\Omega$,*
2. *$f(z)=o(e^{a_1|x|})$ for $x\to\pm\infty$, uniformly in every proper substrip $|y|\leq\pi(\Omega-\eta)$,*

3. $(1/2\pi i)\int_{C_\rho} f(x+\rho z)K(z)\,dz \geqq 0$ *for* $0 \leqq \rho < 1,\ -\infty < x < \infty$,

where C_ρ *is defined as in Theorem* 6.1a.

The necessity of conditions 1 and 2 follows from Theorem 2.2b and Theorem 3.1. The necessity of condition 3 is a consequence of Theorem 6.1a which implies that for $0 \leqq \rho < 1$, $-\infty < x < \infty$

$$(1/2\pi i)\int_{C_\rho} f(x+\rho z)K(z)\,dz = \int_{-\infty}^{\infty} G(\rho, x-t)\,d\alpha(t) > 0.$$

Let $0 < \theta < 1$. It is evident from conditions 1 and 2 that $f(\theta z)$ satisfies the assumptions of Theorem 8.1. Thus $f(\theta x) = \int_{-\infty}^{\infty} G(x-t)\varphi_\theta(t)\,dt$ where $\varphi_\theta(t) = (1/2\pi i)\int_{C_\theta} f(\theta t + \theta z)K(z)\,dz$. By condition 3, $\varphi_\theta(t) \geqq 0$.

By Theorem 5.1a of Chapter VI we have

$$\prod_1^n \left(1 - \frac{D^2}{a_k^2}\right)\int_{-\infty}^{\infty} G(x-t)\varphi_\theta(t)\,dt = \int_{-\infty}^{\infty} G_n(x-t)\varphi_\theta(t)\,dt,$$

where $G_n(t) = \prod_1^n (1 - D^2 a_k^{-2})G(t) \geqq 0$ for $-\infty < t < \infty$. It follows that

$$\prod_1^n \left(1 - \frac{D^2}{a_k^2}\right) f(\theta t) \geqq 0 \qquad (-\infty < t < \infty).$$

Allowing θ to approach 1 we obtain

$$\prod_1^n \left(1 - \frac{D^2}{a_k^2}\right) f(t) \geqq 0 \qquad (-\infty < t < \infty;\quad n = 0, 1, \cdots). \tag{1}$$

Equation (1) and condition 2 for z real together with Theorem 6.1 of Chapter VII now imply our desired result.

As an example we see that necessary and sufficient conditions that

$$f(z) = \frac{1}{2\pi}\int_{-\infty}^{\infty} \operatorname{sech}\frac{1}{2}(z-t)\,d\alpha(t)$$

with $\alpha(t) \in \uparrow$ are:

1. $f(z)$ analytic in the strip $|\operatorname{Im} z| < \pi$,
2. $f(z) = o(e^{|x|/2})$ $x \to \pm\infty$ uniformly in every substrip $|\operatorname{Im} z| < \pi - \eta$,
3. $f(x + i\rho\pi) + f(x - i\rho\pi) \geqq 0 \quad 0 \leqq \rho < 1, \quad -\infty < x < \infty$.

After a logarithmic change of variable we find that necessary and sufficient conditions that

$$F(z) = \int_{0+}^{\infty} \frac{dA(t)}{z+t}$$

with $A(t) \in \uparrow$ are:

1. $F(z)$ analytic in the sector $|\arg z| < \pi$,

2. $F(z) = o(1)$ for $|z| \to \infty$, $F(z) = o(|z|^{-1})$ for $|z| \to 0$, uniformly in every sector $|\arg z| \leqq \pi - \eta$, $\eta > 0$,

3. $e^{i\theta/2}F(re^{i\theta}) + e^{-i\theta/2}F(re^{-i\theta}) \geqq 0 \qquad 0 < r < \infty, \quad |\theta| < \pi.$

This result is evidently related to the theorem of Herglotz, concerning functions with positive imaginary part. See J. A. Shohat and J. D. Tamarkin [1943].

10. DETERMINING FUNCTION IN L^p

10.1. We conclude by establishing one other typical representation theorem. Further results may be obtained by essentially the same arguments.

THEOREM 10.1. *Let $G(t)$ be defined as in equation 2.1 (3) and $K(z)$ as in Theorem 5.1. Necessary and sufficient conditions that $f(z) = \int_{-\infty}^{\infty} G(z-t)\varphi(t)\,dt$ with $\varphi(t) \in L^p(-\infty, \infty)$, $1 < p < \infty$, are:*

1. *$f(z)$ is analytic in the strip $|\operatorname{Im} z| < \pi\Omega$;*

2. *$f(z) = o(e^{a_1|x|})$ for $x \to \pm\infty$, uniformly in every substrip $|\operatorname{Im} z| \leqq \pi(\Omega - \eta)$, $\eta > 0$;*

3. *$\left\| (1/2\pi i) \int_{C_\rho} f(x + \rho z)K(z)\,dz \right\|_p \leqq M$, $0 \leqq \rho < 1$ for some constant M independent of ρ. Here C_ρ is defined as in Theorem 6.1a.*

The necessity of conditions 1 and 2 follows from Theorem 2.2b and Theorem 3.1. Assuming that $f(z)$ has the desired representation we have by Theorem 6.1a

$$(1/2\pi i) \int_{C_\rho} f(x + \rho z)K(z)\,dz = \int_{-\infty}^{\infty} G(\rho, x - t)\varphi(t)\,dt.$$

By Hölder's inequality if $p^{-1} + q^{-1} = 1$, then

$$\left| \int_{-\infty}^{\infty} G(\rho, x-t)\varphi(t)\,dt \right|^p \leqq \int_{-\infty}^{\infty} G(\rho, x-t)\,|\varphi(t)|^p\,dt \left[\int_{-\infty}^{\infty} G(\rho, x-t)\,dt \right]^{p/q},$$

from which it follows that

$$\left\| (1/2\pi i) \int_{C_\rho}^{\infty} f(x + \rho z)K(z)\,dz \right\|_p \leqq \| \varphi(t) \|_p.$$

This establishes the necessity of condition 3.

To establish the sufficiency of our conditions let $0 < \theta < 1$. Using Theorem 8.1 we see that

$$f(\theta x) = \int_{-\infty}^{\infty} G(x - t)\varphi_\theta(t)\, dt$$

where $\varphi_\theta(t) = (1/2\pi i) \int_{C_\theta} f(\theta t + \theta z)K(z)\, dz$. Condition 3 implies that $\| \varphi_\theta(t) \|_p \leqq M\theta^{1/p}$. By Theorem 4.1 of VII there exists a sequence of values $\{\theta_n\}_{n=1}^{\infty}$, $\theta_n \to 1-$, and a function $\varphi(t)$, $\| \varphi(t) \|_p \leqq M$, such that if $\chi(t) \in L^q\,(-\infty, \infty)$ then

$$\lim_{n\to\infty} \int_{-\infty}^{\infty} \varphi_{\theta_n}(t)\chi(t)\, dt = \int_{-\infty}^{\infty} \varphi(t)\chi(t)\, dt.$$

Since for each x, $G(x - t) \in L^q\,(-\infty, \infty)$ we obtain in the limit

$$f(x) = \lim_{n\to\infty} f(\theta_n x) = \lim_{n\to\infty} \int_{-\infty}^{\infty} G(x - t)\varphi_{\theta_n}(t)\, dt,$$

$$f(x) = \int_{-\infty}^{\infty} G(x - t)\varphi(t)\, dt.$$

11. SUMMARY

11.1. In this chapter we have seen how to associate a complex inversion formula with each convolution transform whose kernel is of the form

$$G(t) = \frac{1}{2\pi i} \int_{-i\infty}^{i\infty} \left[\prod_1^\infty \left(1 - \frac{s^2}{a_k^2}\right)\right]^{-1} e^{st}\, ds,$$

where

$$0 < a_1 \leqq a_2 \leqq a_3 \leqq \cdots,$$

$$\lim_{k\to\infty} k/a_k = \Omega \qquad (0 < \Omega < \infty).$$

For $a_k = k - \frac{1}{2}$ this formula reduces to the familiar complex inversion of the Stieltjes transform.

CHAPTER X

Miscellaneous Topics

1. INTRODUCTION

1.1. The present chapter consists of four short topics which although connected with the remainder of the book are not related to each other. In section 2 we consider the theory of generalized Bernstein polynomials, in section 3 the behaviour of convolution transforms at infinity, in section 4 the analytic character of kernels of class II and class III, and in section 5 non-quasi-analytic classes of functions.

2. BERNSTEIN POLYNOMIALS

2.1. A classical theorem of Weierstrass asserts that a function $f(x)$ continuous for $0 \leqq x \leqq 1$ can be uniformly approximated by polynomials. S. Bernstein has given an explicit method of effecting this approximation in terms of the functions

$$\lambda_{n,m}(x) = \binom{n}{m} x^m(1-x)^{n-m} \qquad (m = 0, 1, \cdots, n;\ n = 0, 1, \cdots);$$

the Bernstein polynomial of order n corresponding to $f(x)$ being

$$B_n[f(x)] = \sum_{m=0}^{n} f(m/n)\lambda_{n,m}(x).$$

For $f(x)$ continuous on $0 \leqq x \leqq 1$ Bernstein has proved (see D. V. Widder [1946; 152]) that

$$\lim_{n\to\infty} B_n[f(x)] = f(x)$$

uniformly for $0 \leqq x \leqq 1$.

It is convenient to replace $x(0 \leqq x \leqq 1)$ by $e^t(-\infty \leqq t \leqq 0)$ in the formulas above. We find that if $F(t)$ is continuous for $-\infty \leqq t \leqq 0$ ($F(-\infty)$ exists) then

$$\lim_{n\to\infty} \sum_{m=0}^{n} F\left(\log \frac{m}{n}\right) \lambda_{n,m}(e^{-t}) = F(t)$$

uniformly for $-\infty \leqq t \leqq 0$. The formula defining Euler's constant γ,

$$\lim_{n\to\infty} \left[\sum_{j=1}^{n} \frac{1}{j} - \log n \right] = \gamma,$$

shows that the quantities

$$\log \frac{m}{n} \text{ and } -\sum_{j=m+1}^{n} \frac{1}{j}$$

are very nearly the same, at least when n and m are large. This implies that the following variant of Bernstein's result,

$$\lim_{n\to\infty} \sum_{m=0}^{n} F\left(-\sum_{j=m+1}^{n} \frac{1}{j}\right) \lambda_{n,m}(e^{-t}) = F(t) \tag{1}$$

uniformly for $-\infty \leqq t \leqq 0$, is true. Let us recall the following relation from § 10 of Chapter II.

$$\int_{-\infty}^{0} e^{-st} \lambda_{n,m}(e^{-t})\, dt = \left[(m - s) \prod_{j=m+1}^{n} \left(1 - \frac{s}{j}\right)\right]^{-1}. \tag{2}$$

Expressed in this form Bernstein's result suggests the following generalization: if

$$0 = a_0 < a_1 < a_2 \cdots,$$

where

$$\sum_{1}^{\infty} \frac{1}{a_k} = \infty, \lim_{k\to\infty} a_k = \infty, \tag{3}$$

and if

$$\int_{-\infty}^{0} e^{-st} \Lambda_{n,m}(t)\, dt = \left[(a_m - s) \prod_{m+1}^{n} \left(1 - \frac{s}{a_k}\right)\right]^{-1}, \tag{4}$$

then

$$\lim_{n\to\infty} \sum_{m=0}^{n} F\left(-\sum_{m+1}^{n} \frac{1}{a_k}\right) \Lambda_{n,m}(t) = F(t) \tag{5}$$

uniformly for $-\infty \leqq t \leqq 0$. Note that the conditions (3) are obviously necessary since otherwise no use would be made of value of $F(t)$ in certain intervals. In the following sections we shall establish this conjecture. See I. I. Hirschman and D. V. Widder [1949], and also A. O. Gelfond [1950].

2.2. We assume as given a sequence of real numbers

$$0 = a_0 < a_1 < a_2 < \cdots.$$

We set

$$D_{n,m}(s) = (s - a_m) \prod_{j=m+1}^{n} \left(1 - \frac{s}{a_j}\right), \tag{1}$$

and

$$\Lambda_{n,m}(t) = \sum_{k=m}^{n} e^{a_k t} / D'_{n,m}(a_k), \tag{2}$$

for $(m = 0, 1, \cdots, n;\ n = 0, 1, \cdots)$.

THEOREM 2.2a. *If* $D_{n,m}(s)$ *and* $\Lambda_{n,m}$ *are defined as above then:*

A. $\int_{-\infty}^{0} \Lambda_{n,m}(t)e^{-st}\,dt = -1/D_{n,m}(s) \qquad (-\infty < \mathrm{Rl}\, s < a_m);$

B. $\Lambda_{n,m}(t) = \frac{-1}{2\pi i}\int_{-1-i\infty}^{-1+i\infty} [D_{n,m}(s)]^{-1}e^{st}\,ds \qquad (-\infty < t < 0).$

Making use of the definition of $\Lambda_{n,m}(t)$ we find that

$$\int_{-\infty}^{0} \Lambda_{n,m}(t)e^{-st}\,dt = \sum_{k=m}^{n} [D'_{n,m}(a_k)\,(a_k - s)]^{-1} \qquad (-\infty < \mathrm{Rl}\, s < a_m).$$

Since

$$\sum_{k=m}^{n}[D'_{n,m}(a_k)\,(a_k - s)]^{-1} = -[D_{n,m}(s)]^{-1},$$

conclusion A is established. Conclusion B is a consequence of the complex inversion formula for the Laplace transform.

COROLLARY 2.2a. $\Lambda_{n,m}(t) \geqq 0 \; (-\infty \leqq t \leqq 0).$

This follows from Theorem 6.2 of Chapter II. Let

$$P_{n,m}(a) = \prod_{k=m+1}^{n}\left(1 - \frac{a}{a_k}\right).$$

THEOREM 2.2b. *If* $\Lambda_{n,k}$, $P_{n,m}$ *are defined as above then:*

A. $\sum_{k=0}^{n}\Lambda_{n,k}(t) = 1;$

B. $\sum_{k=m}^{n}P_{n,k}(a_m)\Lambda_{n,k}(t) = e^{a_m t}.$

Conclusion A is a special case of conclusion B which we proceed to prove. It is enough to show that

$$\int_{-\infty}^{0}\left[\sum_{k=m}^{n} P_{n,k}(a_m)\Lambda_{n,k}(t)\right] e^{-st}\,dt = \int_{-\infty}^{0} e^{a_m t}e^{-st}\,dt,$$

or equivalently that

$$\sum_{k=m}^{n} P_{n,k}(a_m)/D_{n,k}(s) = 1/(s - a_m). \tag{3}$$

We will establish this by induction on n. It is true if $n = m$. Suppose that it is true for a general $n > m$. We will prove that it is then true for $n + 1$. By our induction assumption equation (3) is valid. Multiplying both sides of (3) by $(1 - a_m a_{n+1}^{-1})/(1 - s a_{n+1}^{-1})$ we obtain

$$\sum_{k=m}^{n} P_{n+1,k}(a_m)/D_{n+1,k}(s) = (1 - a_m a_{n+1}^{-1})/[(1 - s a_{n+1}^{-1})(s - a_m)].$$

Adding $1/(s - a_{n+1})$ to both sides of this equation we have

$$\sum_{k=m}^{n+1} P_{n+1,k}(a_m)/D_{n+1,k}(s) = 1/(s - a_m)$$

as desired.

COROLLARY 2.2b. $\Lambda_{n,k}(t) \leqq 1 \qquad (-\infty \leqq t \leqq 0)$.

This is an immediate consequence of Corollary 2.2a and conclusion A of Theorem 2.2b.

If $0 \leqq a \leqq a_{m+1}$ then we set

$$D_{n,m}(a, s) = (s - a) \prod_{m+1}^{n} \left(1 - \frac{s}{a_j}\right),$$

$$h_{n,m}(a, t) = e^{at}/D'_{n,m}(a, a) + \sum_{m+1}^{n} e^{a_k t}/D'_{n,m}(a, a_k). \tag{4}$$

The function $h_{n,m}(a, t)$ is defined exactly as $\Lambda_{n,m}(t)$ except that a is used in place of a_m. As a consequence of this remark we have

THEOREM 2.2c. *If $h_{n,m}(a, t)$ is defined as above then:*

A. $0 \leqq h_{n,m}(a, t) \leqq 1 \qquad (-\infty \leqq t \leqq 0)$;

B. $e^{at} = P_{n,m}(a)h_{n,m}(a, t) + \sum_{k=m+1}^{n} P_{n,k}(a)\Lambda_{n,k}(t)$.

2.3. We shall need the following result

LEMMA 2.3. If $0 \leqq x_i \leqq 1 (i = 1, 2, \cdots, n)$ then

$$0 \leqq e^{-(x_1 + \cdots + x_n)} - (1 - x_1) \cdots (1 - x_n) \leqq \tfrac{1}{2}(x_1^2 + \cdots + x_n^2).$$

By Taylor's theorem we have

$$e^{-x} = 1 - x + \tfrac{1}{2}x^2 e^{\theta x} \qquad (0 < \theta < 1)$$

so that our lemma is true for $n = 1$. It is evident that

$$0 \leqq e^{-(x_1 + \cdots + x_n)} - (1 - x_1) \cdots (1 - x_n).$$

To show that

$$e^{-(x_1 + \cdots + x_n)} - (1 - x_1) \cdots (1 - x_n) \leqq \tfrac{1}{2}(x_1^2 + \cdots + x_n^2)$$

we proceed by induction. We know it is true for $n = 1$; assuming that it is true for n we will prove it for $n + 1$. If

$$I = e^{-(x_1 + \cdots + x_{n+1})} - (1 - x_1) \cdots (1 - x_{n+1})$$

then $I = I_1 + I_2$ where

$$I_1 = e^{-(x_1 + \cdots + x_{n+1})} - e^{-(x_1 + \cdots + x_n)}(1 - x_{n+1}),$$

$$I_2 = e^{-(x_1 + \cdots + x_n)}(1 - x_{n+1}) - (1 - x_1) \cdots (1 - x_{n+1}).$$

We have

$$I_1 = e^{-(x_1+\cdots+x_n)}[e^{-x_{n+1}} - (1 - x_{n+1})] \leqq \tfrac{1}{2}x_{n+1}^2,$$

$$I_2 = (1 - x_{n+1})[e^{-(x_1+\cdots+x_n)} - (1 - x_1)\cdots(1 - x_n)] \leqq \tfrac{1}{2}(x_1^2 + \cdots + x_n^2),$$

from which it follows that

$$I \leqq \tfrac{1}{2}(x_1^2 + \cdots + x_{n+1}^2)$$

as desired.

We set

$$\sigma_{n,m} = \sum_{k=m+1}^{n} a_k^{-1}. \tag{1}$$

The generalized Bernstein polynomial of order n corresponding to a function $F(t)$ $(-\infty \leqq t \leqq 0)$ is

$$B_n[F(t)] = \sum_{m=0}^{n} F(-\sigma_{n,m})\Lambda_{n,m}(t). \tag{2}$$

THEOREM 2.3. *If*

1. $\{a_k\}_0^\infty$ *satisfies the conditions* 2.1 (3),
2. $F(t)$ *is continuous for* $-\infty \leqq t \leqq 0$,

then

$$\lim_{n\to\infty} B_n[F(t)] = F(t)$$

uniformly for $-\infty \leqq t \leqq 0$.

We first note that if $|F(t)| \leqq M$ for $-\infty \leqq t \leqq 0$ then $|B_n[F(t)]| \leqq M$ in the same interval. This follows from Corollary 2.2a and Theorem 2.2b. We assert that it is sufficient to prove our theorem for the special functions $f(t) = e^{jt}$, $j = 0, 1, \cdots$. Indeed, let $F(t)$ be an arbitrary function continuous for $-\infty \leqq t \leqq 0$. By the Weierstrass approximation theorem given $\epsilon > 0$ there exists a polynomial $P(t) = \sum_{j=0}^{m} p_j e^{jt}$ such that

$$\|P(t) - F(t)\|_\infty \leqq \epsilon.$$

By the above remark

$$\|B_n[P(t)] - B_n[F(t)]\|_\infty \leqq \epsilon \qquad (n = 0, 1, \cdots).$$

Since we assumed our theorem valid for e^{jt} $j = 0, 1, \cdots$ $B_n[P(t)]$ tends uniformly to $P(t)$ as $n \to \infty$. That is, there exists an integer N such that

$$\|B_n[P(t) - P(t)\|_\infty \leqq \epsilon \qquad (n \geqq N).$$

Thus combining our results

$$\|B_n[F(t)] - F(t)\|_\infty \leqq 3\epsilon \qquad (n \geqq N).$$

Since ϵ is arbitrary our theorem follows.

We will now show that for $0 \leqq a < \infty$ $B_n[e^{at}]$ tends uniformly to e^{at} as $n \to \infty$, which will complete our proof. This is immediate for $a = 0$. Since $B_n[1] = 1$, $n = 0, 1, \cdots$. We suppose $a > 0$. By definition

$$B_n[e^{at}] = \sum_{k=0}^{n} e^{-a\sigma_{n,k}} \Lambda_{n,k}(t).$$

Choose m so that $a_{m+1} > a$; by conclusion B of Theorem 2.2c

$$e^{at} = P_{n,m}(a) h_{n,m}(a, t) + \sum_{k=m+1}^{n} P_{n,k}(a) \Lambda_{n,k}(t).$$

It follows that

$$B_n[e^{at}] - e^{at} = J_1(t) + J_2(t) + J_3(t)$$

where

$$J_1(t) = \sum_{k=0}^{m} \Lambda_{n,k}(t) e^{-a\sigma_{n,k}},$$

$$J_2(t) = -P_{n,m}(a) h_{n,m}(a, t),$$

$$J_3(t) = \sum_{k=m+1}^{n} \Lambda_{n,k}(t) [e^{-a\sigma_{n,k}} - P_{n,k}(a)].$$

Using Corollary 2.2a and Theorem 2.2b we see that

$$0 \leqq J_1(t) \leqq e^{-a\sigma_{n,m}}.$$

Using Theorem 2.2c and Lemma 2.3 we find that

$$0 \leqq -J_2(t) \leqq e^{-a\sigma_{n,m}}.$$

It follows that $\lim_{n\to\infty} J_1(t) = \lim_{n\to\infty} J_2(t) = 0$ uniformly for $(-\infty \leqq t \leqq 0)$. If ϵ and n are given let $r(n)$ be largest integer such that $e^{-a\sigma_{n,k}} \leqq \epsilon$ for $k \leqq r(n)$; $r(n)$ is defined for all sufficiently large n and $r(n) \to \infty$ as $n \to \infty$. We set

$$J_3(t) = J_4(t) + J_5(t)$$

where

$$J_4(t) = \sum_{k=m+1}^{r(n)} \Lambda_{n,k}(t) [e^{-a\sigma_{n,k}} - P_{n,k}(a)],$$

$$J_5(t) = \sum_{k=r(n)+1}^{n} \Lambda_{n,k}(t) [a^{-a\sigma_{n,k}} - P_{n,k}(a)].$$

Using Lemma 2.3 and conclusion A of Theorem 2.2b we see that

$$|J_4(t)| \leqq \sum_{k=m+1}^{r(n)} \Lambda_{n,k}(t) e^{-a\sigma_{n,k}} \leqq \epsilon.$$

For $r(n) + 1 \leqq k \leqq n$ we find, using Lemma 2.3, that

$$|e^{-a\sigma_{n,k}} - P_{n,k}(a)| \leqq \tfrac{1}{2} a^2 \sum_{j=k+1}^{n} a_j^{-2}.$$

If $R(n) = (a_{r(n)+1})^{-1}$ then evidently

$$\sum_{j=k+1}^{n} a_j^{-2} \leqq (a_{k+1})^{-1}\sigma_{n,k} \leqq R(n)\,\frac{1}{a}\log\left(\frac{1}{\epsilon}\right).$$

By conclusion A of Theorem 2.2b

$$|J_5(t)| \leqq \frac{1}{2} a \log\left(\frac{1}{\epsilon}\right) R(n)$$

from which it follows that $\lim_{n\to\infty} J_5(t) = 0$ uniformly for $-\infty \leqq t \leqq 0$. Combining our results we have shown that

$$\overline{\lim_{n\to\infty}}\; \| e^{at} - B_n[e^{at}] \|_\infty \leqq \epsilon.$$

Since ϵ is arbitrary our theorem is proved.

2.4. Instead of the points $\sigma_{n,m}$ we may use any sufficiently close set of points $\sigma_{n,m}^*$. We set

$$B_n^*[F(t)] = \sum_{m=0}^{n} F(-\sigma_{n,m}^*)\Lambda_{n,m}(t). \tag{3}$$

COROLLARY 2.4a. If

1. $\{a_k\}_0^\infty$ satisfies conditions 2.1 (3),
2. $F(t)$ is continuous $-\infty \leqq t \leqq 0$,
3. $e^{-\sigma_{n,m}} - e^{-\sigma_{n,m}^*} \to 0$ as $n \to \infty$ uniformly in m,

then

$$\lim_{n\to\infty} \| B_n^*[F(t)] - F(t) \|_\infty = 0.$$

We have

$$| B_n^*[F(t)] - B_n[F(t)] | \leqq \underset{0\leqq m\leqq n}{\text{l.u.b.}} | F(-\sigma_{n,m}) - F(-\sigma_{n,m}^*) |.$$

If $F(t)$ is continuous for $-\infty \leqq t \leqq 0$ and if condition 3 is satisfied then

$$\underset{0\leqq m\leqq n}{\text{l.u.b.}} | F(-\sigma_{n,m}) - F(-\sigma_{n,m}^*) | = o(1) \qquad n \to \infty.$$

3. BEHAVIOUR AT INFINITY

3.1. Let

$$E(s) = e^{bs} \prod_{k=1}^{\infty} \left(1 - \frac{s}{a_k}\right) e^{s/a_k} \tag{1}$$

where b, $\{a_k\}_1^\infty$ are real and

$$\sum_{k=1}^{\infty} a_k^{-2} < \infty,$$

and let

$$G(t) = (2\pi i)^{-1} \int_{-i\infty}^{i\infty} [E(s)]^{-1} e^{st}\, ds, \tag{2}$$

$$f(x) = \int_{-\infty}^{\infty} G(x-t)\, d\alpha(t). \tag{3}$$

We shall consider those kernels $G(t)$ for which

$$\sum_{1}^{\infty} |a_k|^{-1} = \infty. \tag{4}$$

In the following sections we shall show that if $f(x)$ is defined by (3) then $f(x)$ cannot approach zero too rapidly as x approaches infinity without being identically zero. Let us illustrate this by means of the Laplace and Stieltjes transforms. It will follow from our results that if

$$F(x) = \int_{0+}^{\infty} e^{-xt}\, dA(t) \qquad (x > \sigma_c)$$

and if

$$F(x) = O(e^{-rx}) \qquad (x \to +\infty, \text{ for every } r > 0)$$

then

$$F(x) \equiv 0.$$

Similarly if

$$F(x) = \int_{0+}^{\infty} (x+t)^{-1}\, dA(t)$$

and if

$$F(x) = O(e^{-rx^{1/2}}) \qquad (x \to +\infty, \text{ for every } r > 0)$$

or

$$F(x) = O(e^{-rx^{-1/2}}) \qquad (x \to 0+, \text{ for every } r > 0)$$

then

$$F(x) \equiv 0.$$

See I. I. Hirschman, Jr. [1951].

3.2. We first consider the case when $G(t) \in$ class II.

LEMMA 3.2. If

1. $G(t) \in$ class II,
2. $\varphi(t) \in L(\rho \leqq t < \infty)$,
3. $h(x) = \int_{\rho}^{\infty} G(x-t) e^{ct} \varphi(t)\, dt \qquad c < \alpha_2,$
4. $h(x) = O[G(x-\rho)] \qquad (x \to +\infty),$

then $\varphi(t) = 0$ almost everywhere $(\rho \leqq t < \infty)$.

We define

$$\varphi^*(s) = \int_{\rho}^{\infty} e^{ct} \varphi(t) e^{-st}\, dt. \tag{1}$$

This bilateral Laplace transform converges absolutely for $(c \leqq \sigma < \infty)$. We know that

$$(2) \qquad 1/E(s) = \int_{-\infty}^{\infty} G(t)e^{-st}\,dt,$$

the integral converging absolutely for $(-\infty < \sigma < \alpha_2)$. Since $G(t)e^{-ct}/E(c)$ has $E(s+c)/E(c)$ as its bilateral Laplace transform, and since $E(s+c)/E(c) \in E$, it follows from Theorem 5.1 of IV that $G(t)e^{-ct}$ is bell-shaped. Thus if x is sufficiently large and negative

$$\begin{aligned} |h(x)| &\leqq e^{cx}[\max_{\rho \leqq t \leqq \infty} G(x-t)e^{-cx+ct}] \int_{\rho}^{\infty} |\varphi(t)|\,dt \\ &= O[G(x-\rho)] \qquad (x \to -\infty). \end{aligned}$$

From this and from assumption 4 it follows that

$$(3) \qquad |h(x)| \leqq AG(x-\rho) \qquad (-\infty < x < \infty)$$

for some positive constant A. This implies that the bilateral Laplace transform

$$(4) \qquad h^*(s) = \int_{-\infty}^{\infty} h(x)e^{-sx}\,dx$$

converges absolutely for $-\infty < \sigma < \alpha_2$.

Since the integrals (1) and (2) have a common strip of absolute convergence $(c < \sigma < \alpha_2)$, the convolution theorem for the bilateral Laplace transform, D. V. Widder [1946; 257], tells us that in this strip

$$h^*(s) = \frac{1}{E(s)}\varphi^*(s),$$

or

$$(5) \qquad \varphi^*(s) = h^*(s)E(s).$$

Equation (5) provides a continuation of $\varphi^*(s)$ into the half-plane $\sigma \leqq c$ so that $\varphi^*(s)$ is an entire function.

By a change of variable in equation (1) we obtain

$$e^{\rho(s-c)}\varphi^*(s) = \int_0^{\infty} e^{ct}\varphi(t+\rho)e^{-st}\,dt.$$

It is clear that $e^{\rho s}\varphi^*(s)$ is bounded in the half plane $\sigma \geqq c$. In particular $e^{\rho s}\varphi^*(s)$ is bounded on the line $(\sigma = c,\ -\infty < \tau < \infty)$. Using inequality (3) we have

$$|h^*(s)| \leqq A[E(\sigma)e^{\rho\sigma}]^{-1} \qquad (-\infty < \sigma < \alpha_2),$$

from which we obtain

$$(6) \qquad |e^{\rho s}\varphi^*(s)| \leqq O(1)\ |E(s)/E(\sigma)| \qquad -\infty < \sigma < \alpha_2.$$

It follows that $e^{\rho s}\varphi^*(s)$ is bounded on the half line $(-\infty < \sigma \leqq c,\ \tau = 0)$.

We further assert that

$$\overline{\lim_{|s|\to\infty}} |s|^{-2} \log |e^{\rho s}\varphi^*(s)| \leqq 0 \qquad \text{Rl } s \leqq c;$$

that is, $e^{\rho s}\varphi^*(s)$ is at most of order two minimal type in the half plane Rl $s \leqq c$. It follows from (6) that

$$\left| e^{\rho s}\varphi^*(s) \right| \leqq O(1) \prod_1^\infty \left| \left(1 - \frac{s}{a_k}\right)\left(1 - \frac{\sigma}{a_k}\right)^{-1} \right|,$$

$$\leqq O(1) \prod_1^\infty \left[1 + \left(\frac{\tau}{a_k'}\right)^2\right]^{1/2} \qquad \text{Rl } s \leqq c,$$

where

$$a_k' = a_k - c.$$

We have

$$\sum_{k=1}^\infty \left(\frac{1}{a_k'}\right)^2 < \infty.$$

Therefore given $\epsilon > 0$ we can choose n so large that

$$\sum_{n+1}^\infty \left(\frac{1}{a_k'}\right)^2 < 2\epsilon.$$

Making use of the elementary inequality $(1 + x) \leqq e^x$ for $x \geqq 0$ we find that

$$\left| e^{\rho s}\varphi^*(s) \right| \leqq 0(1) \prod_1^n \left[1 + \left(\frac{\tau}{a_k'}\right)^2\right]^{1/2} e^{\epsilon\tau^2}.$$

From this it follows that

$$\overline{\lim_{|s|\to\infty}} |s|^{-2} \log |e^{\rho s}\varphi^*(s)| \leqq \epsilon \qquad \text{Rl } s \leqq c.$$

Since ϵ is arbitrary our assertion is proved.

We now know that $e^{\rho s}\phi^*(s)$ is bounded on the lines ($\sigma = c$, $-\infty < \tau < \infty$) and ($-\infty < \sigma \leqq c$, $\tau = 0$) and is at most of order two minimal type for $\sigma \leqq c$. Applying the Phragmen-Lindelöf principle, see Titchmarsh [1939; 178], to each of the two quadrants of this configuration we find that $e^{\rho s}\varphi^*(s)$ is bounded for $\sigma \leqq c$, and thus in the entire plane. By Liouville's theorem $e^{\rho s}\varphi^*(s)$, being entire and bounded, is a constant. See Titchmarsh [1939; 85]. Since

$$e^{i\rho\tau}\varphi^*(c + i\tau) = \int_0^\infty \varphi(t + \rho)e^{-it\tau}\, dt$$

$$= o(1) \qquad (\tau \to \pm\infty),$$

it follows that $\varphi^*(s) \equiv 0$. This implies that $\varphi(t) = 0$ almost everywhere $(\rho \leqq t < \infty)$; q.e.d.

THEOREM 3.2a. *If*

1. $G(t) \in$ class II,
2. $f(x) = \int_{-\infty}^{\infty} G(x-t)\, d\alpha(t)$ *has abscissa of convergence* γ_c,
3. $f(x) = O[G(x-\rho)] \qquad (x \to +\infty)$,

then $\alpha(t)$ *is constant for* $\rho < t < \infty$.

Let $\mu_2 + 1$ be the multiplicity of $s - \alpha_2$ as a zero $E(s)$, and let

$$\int_{-\infty}^{\infty} H_1(t)e^{-st}\, dt = \left[1 - \frac{s}{\alpha_2}\right]^{-\mu_2-1},$$

$$\int_{-\infty}^{\infty} H(t)e^{-st}\, dt = \left[1 - \frac{s}{\alpha_2}\right]^{\mu_2+1} / E(s),$$

so that

$$G = H_1 * H.$$

By Theorem 8.1b of Chapter VI

$$f(x) = \int_{-\infty}^{\infty} H(x-t)A(t)\, dt \qquad x > \gamma_c,$$

where

$$A(t) = \int_{-\infty}^{\infty} H_1(t-u)\, d\alpha(u).$$

Since

$$H(t) = \left(1 - \frac{D}{\alpha_2}\right)^{\mu_2+1} G(t)$$

it follows as in § 5.2 of Chapter VI that if $\epsilon > 0$ then

$$G(t-\rho)/H(t-\rho-\epsilon) = O(1) \qquad (t \to +\infty).$$

To see this let $L(t)$ and $L_1(t)$ be defined by the relations

$$t = b + \sum_{k=1}^{\infty} \frac{L}{a_k(a_k + L)},$$

$$t = b + \sum_{k=1}^{\infty} \frac{L_1}{a_k(a_k + L_1)} + \frac{\mu_2 + 1}{\alpha_2} - \frac{(\mu_2+1)L_1}{\alpha_2(\alpha_2 + L_1)}.$$

By Theorem 3.4 of V

$$(d/dt) \log G(t-\rho)/H(t-\rho-\epsilon)$$
$$= -L(t - \rho + o(1)) + L_1(t - \rho - \epsilon + o(1)).$$

The relation defining L_1 may be written in the form

$$L^{-1}[L_1(t)] = t - \frac{\mu_2 + 1}{\alpha_2} + \frac{(\mu_2+1)L_1}{\alpha_2(\alpha_2 + L_1)},$$

and thus

$$L_1(t) = L\left(t - \frac{\mu_2 + 1}{\alpha_2} + \frac{(\mu_2+1)L_1}{\alpha_2(\alpha_2 + L_1)}\right),$$

$$L_1(t) = L(t + o(1)).$$

Consequently

$$(d/dt) \log G(t-\rho)/H(t-\rho-\epsilon)$$
$$= -L(t-\rho+o(1)) + L(t-\rho-\epsilon+o(1)).$$

Since $L(t)$ is increasing we have, if t is sufficiently large,

$$(d/dt) \log G(t-\rho)/H(t-\rho-\epsilon) \leqq 0,$$

or equivalently $G(t-\rho)/H(t-\rho-\epsilon) \in \downarrow$. Our assertion now follows. Thus assumption 3 implies that

$$f(x) = O[H(x-\rho-\epsilon)] \qquad x \to +\infty. \tag{7}$$

Choose $x_0 > \gamma_c$. We have

$$\int_{-\infty}^{\rho+\epsilon} H(x-t)A(t)\,dt = \int_{-\infty}^{\rho+\epsilon} \frac{H(x-t)}{H(x_0-t)} H(x_0-t)A(t)\,dt.$$

If $x > x_0$ then, by Theorem 3.1 of IV, $H(x-t)/H(x_0-t) \in \uparrow$. By the mean value theorem

$$\int_{-\infty}^{\rho+\epsilon} H(x-t)A(t)\,dt$$
$$= \frac{H(x-\rho-\epsilon)}{H(x_0-\rho-\epsilon)} \int_{\xi}^{\rho+\epsilon} H(x_0-t)A(t)\,dt \qquad (-\infty < \xi < \rho+\epsilon).$$

The number ξ depends upon x. However there is a constant M such that

$$\left| \int_{\xi}^{\rho+\epsilon} H(x_0-t)A(t)\,dt \right| \leqq M$$

for all ξ, so that

$$\int_{-\infty}^{\rho+\epsilon} H(x-t)A(t)\,dt = O[H(x-\rho-\epsilon)] \qquad x \to +\infty. \tag{8}$$

Theorem 2.1 of Chapter VII implies that $A(t) = o(e^{\alpha_2 t})$ as $t \to +\infty$. Let β_2 be the smallest pole of $(1 - s\alpha_2^{-1})^{\mu_2+1}/E(s)$, and choose c, $\alpha_2 < c < \beta_2$. Then if $\varphi(t) = e^{-ct}A(t)$ we have

$$\varphi(t) \in L^1(\rho+\epsilon, \infty), \tag{9}$$

and (7) and (8) combined give

$$\int_{\rho+\epsilon}^{\infty} H(x-t)e^{ct}\varphi(t) = O[H(x-\rho-\epsilon)]. \tag{10}$$

It follows from Lemma 3.2 that $\varphi(t) = 0$ for $\rho+\epsilon < t < \infty$. From this it is easily deduced that $\alpha(t)$ is constant for $\rho+\epsilon < t < \infty$, and since ϵ is arbitrary for $\rho < t < \infty$.

THEOREM 3.2b. *If*

1. $G(t) \in$ class II,
2. $f(x) = \int_{-\infty}^{\infty} G(x-t)\,d\alpha(t)$ *has abscissa of convergence* γ_c,
3. $f(x) = O[G(x-\rho)]$ $\qquad (x \to +\infty$, for every $\rho)$,

then $f(x) \equiv 0$.

This is an immediate consequence of Theorem 3.2a. Applied to the Laplace transform Theorem 3.2b gives the result mentioned in § 3.1.

3.3. Let $G(t) \in$ class I be defined as in 3.1(2) and 3.1(4). We associate with $G(t)$ a class II kernel $\mathscr{G}(t)$ as follows. We set

$$\mathscr{E}(s) = e^{bs} \prod_{k=1}^{\infty} \left(1 - \frac{s}{|a_k|}\right) e^{s/|a_k|}, \tag{1}$$

$$\mathscr{G}(t) = (2\pi i)^{-1} \int_{-i\infty}^{i\infty} [\mathscr{E}(s)]^{-1} e^{st}\, ds. \tag{2}$$

We shall show that if

$$f(x) = \int_{-\infty}^{\infty} G(x-t)\, d\alpha(t)$$

and if

$$f(x) = O[\mathscr{G}(x-\rho)] \qquad (x \to +\infty, \text{for every } \rho), \tag{3}$$

or

$$f(x) = O[\mathscr{G}(\rho - x)] \qquad (x \to -\infty, \text{for every } \rho), \tag{3'}$$

then

$$f(x) \equiv 0.$$

In order to illustrate the essential point of the argument which follows let us consider a very special case. Let

$$f(x) = \int_{-\infty}^{\infty} G(x-t)\varphi(t)\, dt \tag{4}$$

where $\varphi(t) \in L^2(-\infty, \infty)$. We denote by $\mathscr{F}$ the Fourier transformation

$$\mathscr{F}_t\psi = (2\pi)^{-1/2} \underset{T\to\infty}{\overset{(2)}{\text{l.i.m.}}} \int_{-T}^{T} \psi(u)e^{itu}\, du,$$

and by $\mathscr{F}^{-1}$ its inverse

$$\mathscr{F}_t^{-1}\psi = (2\pi)^{-1/2} \underset{T\to\infty}{\overset{(2)}{\text{l.i.m.}}} \int_{-T}^{T} \psi(u)e^{-itu}\, du.$$

We assert that if $f(x)$ is given by equation (4) and if

$$\psi(t) = \mathscr{F}_t^{-1} E(-iu)\mathscr{E}(-iu)^{-1} \mathscr{F}_u \varphi(t) \tag{5}$$

then

$$f(x) = \int_{-\infty}^{\infty} \mathscr{G}(x-t)\psi(t)\, dt. \tag{6}$$

The transformation $\varphi(t) \to \psi(t)$ defined by (5) is unitary; that is $||\varphi||_2 = ||\psi||_2$. This follows from the relation

$$|E(-iu)/\mathscr{E}(-iu)| = 1 \qquad (-\infty < u < \infty)$$

and Plancherel's theorem. Transformations of this type are called Watson transformations, see Titchmarsh [1948; 212 et seq.]. To establish (6) we note that as a consequence of (4) we have

$$\mathscr{F}_u f = E(-iu)\mathscr{F}_u\varphi.$$

Equation (6) is equivalent to

$$\mathscr{F}_u f = \mathscr{E}(-iu)\mathscr{F}_u \psi$$
$$= \mathscr{E}(-iu)E(-iu)\mathscr{E}(-iu)^{-1}\mathscr{F}_u \varphi = E(-iu)\mathscr{F}_u \varphi,$$

which we have just seen to be true. Thus $f(x)$ is represented as a transform with a class II kernel $\mathscr{G}(t)$. If (3) holds then Theorem 3.2b shows that $f(x)$ must be identically zero. (A similar argument shows that if (3′) holds then $f(x)$ is again identically zero.)

THEOREM 3.3. *If*

1. *$G(t)$ is defined by* 3.1 (2) *and* 3.1 (4) *and belongs to class I,*

2. $f(x) = \int_{-\infty}^{\infty} G(x - t)\, d\alpha(t)$ *converges,*

3. $f(x) = O[\mathscr{G}(x - \rho)]$ *as* $(x \to +\infty$, *for every* $\rho)$, *or*

3′. $\quad = O[\mathscr{G}(\rho - x)]$ *as* $(x \to -\infty$, *for every* $\rho)$,

then $f(x) \equiv 0$.

Using Theorem 8.1a of Chapter VI and Theorem 2.1 of Chapter VII and arguing as in the proof of Theorem 3.2a, it is easy to see that it is no restriction to assume that

$$f(x) = \int_{-\infty}^{\infty} G(x - t)\varphi(t)\, dt \tag{7}$$

where

$$\varphi(t) = O(e^{\beta_2 t}) \text{ as } t \to +\infty, \quad \varphi(t) = O(e^{\beta_1 t}) \text{ as } t \to -\infty, \tag{8}$$

β_1 and β_2 being such that

$$\beta_1 > \alpha_1, \qquad \beta_2 < \alpha_2.$$

Suppose that 3 holds. Let

$$c \geqq 2\alpha_2$$

and let

$$A_k = |a_k| + c,$$

$$\mathscr{E}_1(s) = e^{bs} \prod_{k=3}^{\infty} \left(1 - \frac{s}{A_k}\right) e^{s/A_k},$$

$$\mathscr{G}_1(t) = (2\pi i)^{-1} \int_{-i\infty}^{i\infty} [\mathscr{E}_1(s)]^{-1} e^{st}\, ds.$$

$\mathscr{G}_1(t)$ is a class II kernel closely related to $\mathscr{G}(t)$. We will show that $f(x)$ can be expressed in the form

$$f(x) = \int_{-\infty}^{\infty} \mathscr{G}_1(x - u)\psi(u)\, du.$$

This is analogous to the formula (6) and serves the same purpose. We note that if $\eta > 0$ then

$$
\begin{aligned}
\mathscr{G}_1(t) &= O(e^{(\alpha_1+\eta)t}) && t \to +\infty \\
&= O(e^{(\alpha_2-\eta)t}) && t \to -\infty.
\end{aligned} \tag{9}
$$

(Actually much more is true.)

It is evident that

(a) $\mathscr{E}_1(s)/E(s)$ is analytic for $\alpha_1 < \text{Rl}\, s < \alpha_2$.

Because of our choice of c

$$\left|1 - \frac{\sigma + i\tau}{A_k}\right| \left|1 - \frac{\sigma + i\tau}{a_k}\right|^{-1} \qquad k = 3, 4, \cdots$$

decreases as $|\tau|$ increases if $\alpha_1 < \sigma < \alpha_2$. It follows that if $\alpha_1 < \sigma_1 \leqq \sigma_2 < \alpha_2$ then there exists a constant A such that

(b) $|\mathscr{E}_1(s)/E(s)| \leqq A/(1 + \tau^2)$ for $\sigma_1 \leqq \text{Rl}\, s \leqq \sigma_2$.

A simple application of a theorem of Hamburger, see Widder [1946; 265], now shows that if

$$J(t) = (2\pi i)^{-1} \int_{-i\infty}^{i\infty} [\mathscr{E}_1(s)/E(s)]e^{st}\, ds$$

then

$$\int_{-\infty}^{\infty} J(t)e^{-st}\, dt = \mathscr{E}_1(s)/E(s)$$

the bilateral Laplace transform converging absolutely for $\alpha_1 < \text{Rl}\, s < \alpha_2$. Because of the order relation (b) we may deform the line of integration of the integral defining $J(t)$ to $\text{Rl}\, s = \alpha_2 - \eta$ or to $\text{Rl}\, s = \alpha_1 + \eta$. We obtain

$$
\begin{aligned}
J(t) &= (2\pi i)^{-1} \int_{\alpha_2-\eta-i\infty}^{\alpha_2-\eta+i\infty} [\mathscr{E}_1(s)/E(s)]e^{st}\, ds \\
&= (2\pi i)^{-1} \int_{\alpha_1+\eta-i\infty}^{\alpha_1+\eta+i\infty} [\mathscr{E}_1(s)/E(s)]e^{st}\, ds
\end{aligned}
$$

and these together with (b) imply that

$$
\begin{aligned}
J(t) &= O(e^{(\alpha_2-\eta)t}) && t \to -\infty, \\
&= O(e^{(\alpha_1+\eta)t}) && t \to +\infty.
\end{aligned} \tag{10}
$$

Finally the convolution theorem for the bilateral Laplace transform, D. V. Widder [1946; 257], gives

$$\int_{-\infty}^{\infty} \mathscr{G}_1(x - t)J(t)\, dt = G(x).$$

Thus

$$f(x) = \int_{-\infty}^{\infty} \varphi(t)\, dt \int_{-\infty}^{\infty} \mathscr{G}_1(x - u) J(u - t)\, du.$$

Because of the conditions (8), (9), and (10) this iterated integral is absolutely convergent. Consequently the order of the integrations may be inverted to give

$$f(x) = \int_{-\infty}^{\infty} \mathscr{G}_1(x - u)\psi(u)\, du, \tag{11}$$

where

$$\psi(u) = \int_{-\infty}^{\infty} J(u - t)\varphi(t)\, dt.$$

We next assert that if

$$r = \frac{1}{|a_1|} + \frac{1}{|a_2|} + \sum_{k=3}^{\infty} \frac{c}{|a_k| A_k}$$

then

$$\mathscr{G}(t)/\mathscr{G}_1(t - r - \epsilon) = O(1) \qquad \text{as } t \to +\infty$$

for any $\epsilon > 0$. To prove this we introduce

$$\mathscr{E}_2(s) = e^{bs} \prod_{k=3}^{\infty} \left(1 - \frac{s}{|a_k|}\right) e^{s/|a_k|},$$

$$\mathscr{G}_2(t) = (2\pi i)^{-1} \int_{-i\infty}^{i\infty} [\mathscr{E}_2(s)]^{-1} e^{st}\, ds.$$

Proceeding as in § 5.2 of Chapter VI one may show that

$$\mathscr{G}(t)/\mathscr{G}_2(t - |a_1|^{-1} - |a_2|^{-1} - \epsilon) = O(1) \qquad \text{as } t \to +\infty.$$

We have, see § 10.2 of III,

$$\mathscr{E}_2(-c) e^{ct} \mathscr{G}_2(t) = \mathscr{G}_1\left(t - \sum_{k=3}^{\infty} \frac{c}{|a_k| A_k}\right)$$

from which it follows that

$$\mathscr{G}_2(t - |a_1|^{-1} - |a_2|^{-1} - \epsilon)/\mathscr{G}_1(t - r - \epsilon) = O(1) \qquad \text{as } t \to \infty,$$

thus proving our assertion. Assumption 3 now implies that

$$f(x) = O[\mathscr{G}_1(x - \rho)] \qquad (x \to \infty, \text{ for every } \rho). \tag{12}$$

The relations (11) and (12) and Theorem 3.2b show that $f(x)$ must vanish identically.

3.4. The following result may be proved exactly as Theorem 3.2a was proved. The analogue of Lemma 3.2 which is needed here reduces to a special case of a well known result, see Titchmarsh [1948; 322–327].

THEOREM 3.4. *If*

1. $G(t)$ *defined by* 3.1 (2) *belongs to* class III,
2. $\alpha(t)$ *is of bounded variation for* $T < t_1 \leqq t \leqq t_2 < \infty$,
3. $f(x) = \int_{-\infty}^{\infty} G(x - t)\, d\alpha(t)$ *converges for* $x > T + b + \sum_1^\infty a_k^{-1}$,
4. $f(x) = 0$ *for* $x > \rho + b + \sum_1^\infty a_k^{-1}$, $\rho \geqq T$,

then $\alpha(t)$ *is constant for* $(\rho < t < \infty)$.

4. THE ANALYTIC CHARACTER OF KERNELS OF CLASSES I AND II

4.1. The best known example of a class II kernel is

$$e^{-e^t}e^t = (2\pi i)^{-1} \int_{-i\infty}^{i\infty} \Gamma(1 - s)e^{st}\, ds$$

which is an entire function. This suggests that conceivably all class II kernels may be entire functions. We shall prove that this is true.

THEOREM 4.1. *If*

1. $E(s) = e^{bs} \prod_1^\infty \left(1 - \frac{s}{a_k}\right) e^{s/a_k}$,
2. $a_k > 0$ $k = 1, 2, \cdots$, $\sum_1^\infty a_k^{-1} = \infty$,
3. $G(t) = (2\pi i)^{-1} \int_{-i\infty}^{i\infty} [E(s)]^{-1} e^{st}\, ds$,

then $G(t)$ *is the restriction to the real axis of a function* $G(w)$ *analytic in the entire plane.*

We may assume without loss of generality that b is zero. It is easily verified that if $a \geqq 0$, then

$$\left| \left(1 - \frac{s}{a}\right) e^{s/a} \right| \geqq 1 \qquad \pi/4 \leqq |\arg s| \leqq \pi/2.$$

It follows that for s in these sectors we have

$$\left| \prod_{k=1}^\infty \left(1 - \frac{s}{a_k}\right) e^{s/a_k} \right| \geqq \left| \prod_{k=1}^n \left(1 - \frac{s}{a_k}\right) e^{s/a_k} \right|. \tag{1}$$

From inequality (1) we see that for each n there exists a constant $B > 0$ such that

$$\left| \prod_{k=1}^\infty \left(1 - \frac{s}{a_k}\right) e^{s/a_k} \right| \geqq B\, |s|^n \qquad \pi/4 \leqq |\arg s| \leqq \pi/2. \tag{2}$$

Let C be the contour below where $\pi/4 \leqq \theta_0 \leqq \pi/2$.

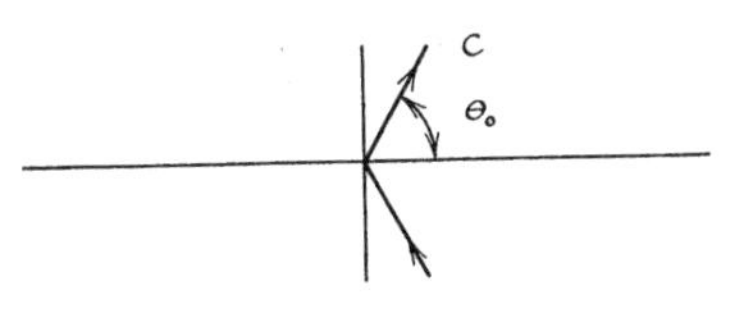

FIG. III

Inequality (2) immediately enables us to deduce by a standard argument that

$$G(t) = (2\pi i)^{-1} \int_C [E(s)]^{-1} e^{st}\, ds. \tag{3}$$

Again using inequality (1) we see that for each n, there exists a constant $B' > 0$, such that

$$\left| \prod_{k=1}^{\infty} \left(1 - \frac{s}{a_k}\right) e^{s/a_k} \right| \geqq B' \left| e^{s\left(\sum_1^n a_k^{-1}\right)} \right| \qquad \pi/4 \leqq |\arg s| \leqq \pi/2. \tag{4}$$

It follows that the integral (3), with t replaced by w, converges uniformly for w in any compact subset of the open sector shaded in the figure below.

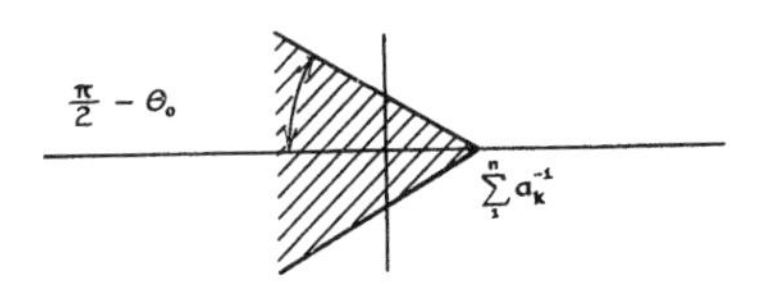

FIG. IV

Since n may be chosen arbitrarily large and since $\lim_{n\to\infty} \sum_1^n a_k^{-1} = \infty$ the integral (3) (with t replaced by w) converges uniformly for w in any compact set, and our theorem is proved.

4.2. In section 9.9 of Chapter III it was shown that if

$$G(t) = (2\pi i)^{-1} \int_{-i\infty}^{i\infty} \left[\prod_1^{\infty} \left(1 - \frac{s^2}{k^2}\right)\right]^{-1} e^{st}\, ds,$$

then $G(t) = \vartheta_3'(\frac{1}{2}, t)$ for $\infty < t < 0$. Thus $G(t)$ is, for $-\infty < t < 0$, the restriction to the real axis of a function $G(w)$ analytic for $\mathrm{Rl}\, w \leqq 0$. This is a special case of the following result.

THEOREM 4.2. *If*

1. $E(s) = e^{bs} \prod_1^{\infty} \left(1 - \dfrac{s}{a_k}\right) e^{s/a_k}$,

2. $a_k > 0$ $k = 1, 2, \cdots$, $\sum_1^{\infty} a_k^{-1} < \infty$,

3. $G(t) = (2\pi i)^{-1} \displaystyle\int_{-i\infty}^{i\infty} [E(s)]^{-1} e^{st}\, ds$,

then $G(t)$ *is, for* $-\infty < t < b + \sum_1^\infty a_k^{-1}$, *the restriction to the real axis of a function* $G(w)$ *analytic in the half-plane* $\text{Rl}\, w < b + \sum_1^\infty \frac{1}{a_k}$.

It is no restriction to assume that b is equal to zero. It is easily verified that given $\theta > 0$, there exists a constant A such that

$$\left|\left(1 - \frac{s}{a}\right)e^{As/a}\right| \geqq 1 \qquad \theta \leqq |\arg s| \leqq \pi/2.$$

It follows that for s in these sectors we have

$$\left|\prod_{k=1}^{\infty}\left(1 - \frac{s}{a_k}\right)e^{sA/a_k}\right| \geqq \left|\prod_{k=1}^{n}\left(1 - \frac{s}{a_k}\right)e^{As/a_k}\right|. \tag{1}$$

From inequality (1) we see that for each n there exists a constant $B > 0$ such that

$$\left|\prod_{k=1}^{\infty}\left(1 - \frac{s}{a_k}\right)e^{As/a_k}\right| \geqq B|s|^n \qquad \theta \leqq |\arg s| \leqq \pi/2. \tag{2}$$

Let C be the contour below.

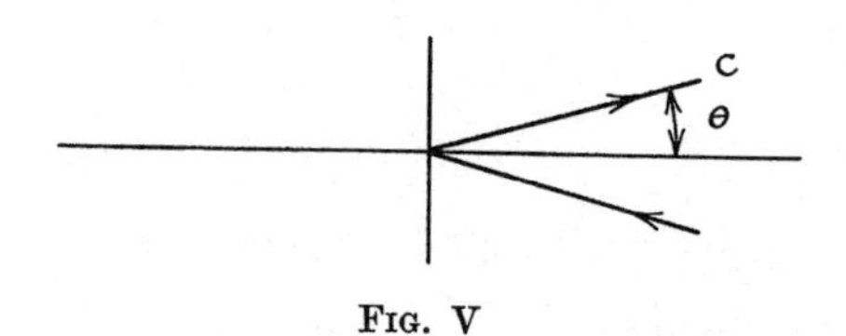

FIG. V

We have

$$G(t) = (2\pi i)^{-1}\int_{-i\infty}^{i\infty}\left[\prod_{k=1}^{\infty}\left(1 - \frac{s}{a_k}\right)e^{As/a_k}\right]^{-1} e^{s[t+(A-1)\sum_1^\infty a_k^{-1}]}\, ds.$$

Inequality (2) enables us to deduce by a standard argument that

$$G(t) = (2\pi i)^{-1}\int_C\left[\prod_{k=1}^{\infty}\left(1 - \frac{s}{a_k}\right)e^{As/a_k}\right]^{-1} e^{s[t+(A-1)\sum_1^\infty a_k^{-1}]}\, ds. \tag{3}$$

From inequality (1) we see that for each n there exists a constant B' such that

$$\left|\prod_{k=1}^{\infty}\left(1 - \frac{s}{a_k}\right)e^{As/a_k}\right| \geqq B'\left|e^{As\sum_{k=1}^{n}\frac{1}{a_k}}\right| \qquad \theta \leqq |\arg s| \leqq \pi/2.$$

It follows that the integral (3) with t replaced by w converges uniformly for w in any compact subset of the open sector shaded below.

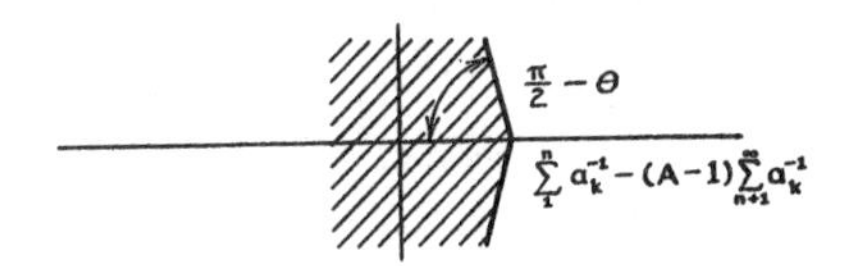

FIG. VI

Since θ may be taken arbitrarily small and since

$$\lim_{n\to\infty}\left[\sum_{k=1}^{n}\frac{1}{a_k}-(A-1)\sum_{k=n+1}^{\infty}\frac{1}{a_k}\right]=\sum_{k=1}^{\infty}\frac{1}{a_k},$$

we see that the integral (3) converges for w in any compact subset of the half plane $\text{Rl } w < \sum_{k=1}^{\infty}\frac{1}{a_k}$.

It can be further shown that under the assumptions of Theorem 4.2 the line $\text{Rl } w = b + \sum_{1}^{\infty} a_k^{-1}$ is a cut for $G(w)$. Since the demonstration of this makes use of results from the theory of general Dirichlet series we omit it here.

5. QUASI-ANALYTICITY

5.1. Let M_n, $n = 0, 1, \cdots$, be a sequence of positive constants, normalized by the condition $M_0 = 1$. The sequence $\{M_n\}_0^\infty$ is said to be non-quasi-analytic if there exists a function $f(x) \not\equiv 0$ such that

$$|f^{(n)}(x)| \leqq Ak^nM_n \qquad (-\infty < x < \infty;\ n = 0, 1, \cdots)$$

(here A and k are constants which may depend upon $f(x)$) and such that for some x_0

$$|f^{(n)}(x_0)| = 0 \qquad (n = 0, 1, \cdots).$$

Using our theory of convolution transforms we can prove the following result which is of fundamental importance in the theory of quasi-analyticity. See also I. I. Hirschman and D. V. Widder [1952].

THEOREM 5.1. *If $\{M_n\}_0^\infty$ is a normalized sequence of positive constants and if*

$$\sum_{n=1}^{\infty} M_{n-1}/M_n < \infty$$

then M_n is non-quasi-analytic.

Let $a_k = M_k/M_{k-1}$ $k = 1, 2, \cdots$, and let us define

$$E(s) = \prod_{k=1}^{\infty}\left(1 - \frac{s}{a_k}\right),$$

$$G(t) = (2\pi i)^{-1}\int_{-i\infty}^{i\infty}[E(s)]^{-1}e^{st}\,ds.$$

$G(t)$ is a kernel belonging to class III. We assert that if $\varphi(t) \in L^\infty(-\infty, \infty)$ and if

(1) $$f(x) = \int_{-\infty}^{\infty} G(x-t)\varphi(t)\,dt$$

then

(2) $$|f^{(n)}(x)| \leqq 2^n \,||\varphi||_\infty M_n \qquad (-\infty < x < \infty;\quad n = 0, 1, \cdots).$$

We know from Theorem 4.2a of Chapter VII that if $\{A_1, \cdots, A_m\}$ is any selection from $\{a_1, a_2, \cdots\}$ then

(3) $$\left|\prod_{k=1}^{m}\left(1 - \frac{D}{A_k}\right)f(x)\right| \leqq ||\varphi||_\infty.$$

Since

$$\frac{D}{a_k} = 1 - \left(1 - \frac{D}{a_k}\right)$$

$D^n/a_1 \cdots a_n$ can be expressed as a sum of 2^n terms of the form

$$\prod_{k=1}^{m}(1 - D/A_k).$$

Thus

$$\begin{aligned}|D^n f(x)| &\leqq 2^n a_1 \cdots a_n \,||\varphi||_\infty,\\ &\leqq 2^n\,||\varphi||_\infty M_n.\end{aligned}$$

If $\varphi(t) = 0$ for $t \geqq T$ then $f(x) = 0$ for $t \geqq T$, which implies that $f^{(n)}(x_0) = 0$ for $T \leqq x_0 < \infty$. It is obvious that f cannot be identically zero unless $\varphi(t) = 0$ almost everywhere.

For an extensive treatment of the theory of quasi-analytic functions see S. Mandelbrojt [1942].

Bibliography

AISSEN, M., EDREI, A., SCHOENBERG, I. J., and WHITNEY, A.
1951. On the generating functions of totally positive sequences. *Proceedings of the National Academy of Sciences*, vol. 37, pp. 303–307.

AISSEN, M., SCHOENBERG, I. J., and WHITNEY, A.
1953. On the generating functions of totally positive sequences. *Journal d'Analyse Mathématique*, vol. 2, pp. 93–103.

AKUTOWICZ, E. J.
1948. The third iterate of the Laplace transform. *Duke Mathematical Journal*, vol. 15, pp. 1093–1132.

BANACH, S.
1932. *Théorie des opérations linéaires.* Warsaw.

BARRUCAND, P.
1950. Généralization de la transformation de Stieltjes itérée: transformation d'ordre quelconque. *Comptes Rendus des Séances de l'Académie des Sciences*, vol. 231, pp. 748–750.

BERNSTEIN, V.
1933. *Leçons sur les progrès reçents de la théorie des series de Dirichlet.* Paris.

BOAS, R. P., JR.
1937. Asymptotic relations for derivatives. *Duke Mathematical Journal*, vol. 3, pp. 637–646.
1942. Inversion of a generalized Laplace integral. *Proceedings of the National Academy of Sciences*, vol. 28, pp. 21–24.

BOAS, R. P., JR., and WIDDER, D. V.
1939. The iterated Stieltjes transform. *Transactions of the American Mathematical Society*, vol. 45, pp. 1–72.

BÔCHER, M.
1917. *Leçons sur les méthodes de Sturm.* Paris.

BOCHNER, S., and CHANDRASEKHARAN, K.
1949. *Fourier transforms.* Princeton.

BOCHNER, S., and WIDDER, D. V.
1948. A homogeneous differential system of infinite order with non-vanishing solution. *Bulletin of the American Mathematical Society*, vol. 54, pp. 409–415.

CHANDRASEKHARAN, K. *See* BOCHNER, S.

EDREI, A.
1953. On the generating functions of totally positive sequences. *Journal d'Analyse Mathematique*, vol. 2, pp. 104–109.
See also AISSEN, M.

GANTMAKHER, F., and KREIN, M.
1937. Sur les matrices complétement non-negatives et oscillatoires. *Compositio Mathematica*, vol. 4, pp. 445–476.

GARDER, A. O.
1954. Topics in the theory of convolution transforms. Thesis, Washington University.

GELFOND, A. O.
1950. On the generalized polynomials of S. N. Bernstein (in Russian). *Izv. Akad. Nauk S.S.S.R.*, ser. math., vol. 14, pp. 413–420.

HERGLOTZ, A.
1911. Über Potenzreihen mit positivem reelen Teil im Einheitskreis. *Berichte über die Verhandlungen der koniglich Sachsischen Gesellschaft der Wissenschaften zu Leipzig, mathematisch-physische Klasse*, vol. 63 [6–9], pp. 501–511.

HILLE, E.
1948. Functional analysis and semi-groups. *American Mathematical Society Colloquium Publication*, vol. 31.

HIRSCHMAN, I. I., JR.
1951. The behaviour at infinity of certain convolution transforms. *Transactions of the American Mathematical Society*, vol. 70, pp. 1–14.
1953. Systems of partial differential equations which generalize the heat equation. *Canadian Journal of Mathematics*, vol. 5, pp. 118–128.

HIRSCHMAN, I. I., JR., and WIDDER, D. V.
1948a. Generalized inversion formulas for convolution transforms. *Duke Mathematical Journal*, vol. 15, pp. 659–696.
1948b. An inversion and representation theory for convolution transforms with totally positive kernels. *Proceedings of the National Academy of Sciences*, vol. 34, pp. 152–156.
1949a. Generalized Bernstein polynomials. *Duke Mathematical Journal*, vol. 16, pp. 433–438.
1949b. The inversion of a general class of convolution transforms. *Transactions of the American Mathematical Society*, vol. 66, pp. 135–201.
1949c. A representation theory for a general class of convolution transforms. *Transactions of the American Mathematical Society*, vol. 67, pp. 69–97.
1950a. Generalized inversion formulas for convolution transforms, II. *Duke Mathematical Journal*, vol. 17, pp. 391–402.
1950b. A miniature theory in illustration of the convolution transform. *American Mathematical Monthly*, vol. 57, pp. 667–674.
1951a. Convolution transforms with complex kernels. *Pacific Journal of Mathematics*, vol. 1, pp. 211–225.
1951b. On the products of functions represented as convolution transforms. *Proceedings of the American Mathematical Society*, vol. 2, pp. 97–99.
1952. A note on quasi-analytic functions. *Publications de l'Institut Mathématique de l'Académie Serbe des Sciences*, vol. 4, pp. 57–60.

HOBSON, E. W.
1926. *The theory of functions of a real variable*, vol. 2, ed. 2, Cambridge.

KREIN, M. *See* GANTMAKHER, F.

LAGUERRE, E.
1882. Sur les fonctions du genre zéro et du genre un. *Comptes Rendus des Séances de l'Académie des Sciences*, vol. 98, pp. 828–831.

LÉVY, P.
1925. *Calcul des probabilités.* Paris.

LORENTZ, G. G.
1953. *Bernstein Polynomials.* Toronto.

MAGNUS, W., and OBERHETTINGER, F.
1948. *Formeln und Sätze für die speziellen Funktionen der mathematischen Physik.* Berlin.

MANDELBROJT, S.
1942. Analytic functions and classes of infinitely differentiable functions. *Rice Institute Pamphlet*, vol. 29, pp. 1–142.

MOTZKIN, T.
1936. *Beitrage zur Theorie der linearen Ungleichungen.* Jerusalem.

OBERHETTINGER, F. *See* MAGNUS, W.

POLLARD, H.
1946a. The representation of e^{-x^λ} as a Laplace integral. *Bulletin of the American Mathematical Society*, vol. 52, pp. 908–910.
1946b. Integral transforms. *Duke Mathematical Journal*, vol. 13, pp. 307–330.
1947a. The inversion of the transforms with reiterated Stieltjes kernels. *Duke Mathematical Journal*, vol. 14, pp. 129–142.
1947b. The integral transforms with iterated Laplace kernels. *Duke Mathematical Journal*, vol. 14, pp. 659–675.

PÓLYA, G.
1913. Über Annäherung durch Polynome mit lauter reelen Wurzeln. *Rendiconti del Circolo Matematico del Palermo*, vol. 36, pp. 279–295.

PÓLYA, G., and SZEGÖ, G.
1925. *Aufgaben und Lehrsätze aus der Analysis*, vol. 1.

SCHOENBERG, I. J.
1930. Über variationsvermindernde lineare Transformationen. *Mathematische Zeitschrift*, vol. 32, pp. 321–328.
1947. On totally positive functions, Laplace integrals and entire functions of the Laguerre-Pólya-Schur type. *Proceedings of the National Academy of Sciences*, vol. 33, pp. 11–17.
1948a. Some analytical aspects of the problem of smoothing. *Studies and Essays Presented to R. Courant on his 60th Birthday*, January 8, 1948, pp. 351–370. Interscience Publishers, Inc., New York.
1948b. On variation diminishing integral operators of the convolution type. *Proceedings of the National Academy of Sciences*, vol. 34, pp. 164–169.
1950. On Pólya frequency functions. II. Variation-diminishing integral operators of the convolution type. *Acta Scientiarum Mathematicarum Szeged*, vol. 12, Leopoldo Fejér et Frederico Riesz LXX annos natis dedicatus, Pars B, pp. 97–106.
1951. On Pólya frequency functions I. The totally positive functions and their Laplace transforms. *Journal d'Analyse Mathématique*, vol. 1, pp. 331–374.

SCHOENBERG, I. J., and WHITNEY, A.
1949. Sur la positivité des determinants de translations de functions de fréquence de Pólya, avec une application à une problème d'interpolation. *Comptes Rendus des Séances de l'Académie des Sciences*, vol. 228, pp. 1996–1998.
1951. A theorem on polygons in n dimensions with applications to variation diminishing and cyclic variation diminishing linear transformations. *Compositio Mathematica*, vol. 9, pp. 141–160.
See also AISSEN, M.

SHOHAT, J. A., and TAMARKIN, J. C.
1943. *The problem of moments.* New York.

SIERPINSKI, W.
1920. Sur les fonctions convexes mesurables. *Fundamenta Mathematicae*, vol. 1, pp. 125–129.

SUMNER, D. B.
1949. An inversion formula for the Stieltjes transform. *Bulletin of the American Mathematical Society*, vol. 55, pp. 174–183.

SZEGÖ, G. *See* PÓLYA, G.

TAMARKIN, J. D. *See* SHOHAT, J. A.

TITCHMARSH, E. C.
1939. *The theory of functions*, 2 ed. Oxford.
1948. *Introduction to the theory of Fourier integrals*, 2 ed. Oxford.
1951. *The theory of the Riemann zeta-function.* Oxford.

TYCHONOFF, A.
1935. Théorèmes d'unicité pour l'équation de la chaleur. *Matematiceskii Sbornik*, vol. 42, pp. 199–215.

DE LA VALLEE POUSSIN, C. J.
1914. *Cours d'analyse infinitésimale.* Paris.

WHITNEY, A.
1953. A reduction theorem for totally positive matrices. *Journal d'Analyse Mathematique*, vol. 2, pp. 88–92.
See also AISSEN, M., and SCHOENBERG, I. J.

WIDDER, D. V.
1940. The Green's function for a differential system of infinite order. *Proceedings of the National Academy of Sciences*, vol. 26, pp. 213–215.
1944. Positive temperatures on an infinite rod. *Transactions of the American Mathematical Society*, vol. 55, pp. 85–95.
1946. *The Laplace transform*, 2 ed. Princeton.
1947a. *Advanced calculus*. New York.
1947b. Green's functions for linear differential systems of infinite order. *Proceedings of the National Academy of Sciences*, vol. 33, pp. 31–34.
1947c. Inversion formulas for convolution transforms. *Duke Mathematical Journal*, vol. 14, pp. 217–249.
1947d. The inversion of a generalized Laplace transform. *Proceedings of the National Academy of Sciences*, vol. 33, pp. 295–297.
1948. A symbolic form of an inversion formula for a Laplace transform. *American Mathematical Monthly*, vol. 55, pp. 489–491.
1950a. An inversion of the Lambert transform. *Mathematics Magazine*, vol. 23, pp. 171–182.
1950b. Symbolic inversions of the Fourier sine transform and of related transforms. *Journal of the Indian Mathematical Society*, vol. 14, pp. 119–128.
1951a. Necessary and sufficient conditions for the representation of a function by a Weierstrass transform. *Transactions of the American Mathematical Society*, vol. 71, pp. 430–439.
1951b. A symbolic form of the classical complex inversion formula for a Laplace transform. *American Mathematical Monthly*, vol. 58, pp. 179–181.
1951c. Weierstrass transforms of positive functions. *Proceedings of the National Academy of Sciences*, vol. 37, pp. 315–317.
See also BOAS, R. P., JR., BOCHNER, S., and HIRSCHMAN, I. I., JR.

WINTNER, A.
1938. *Asymptotic Distributions and Infinite Convolutions*. Edwards Brothers, Ann Arbor.

Symbols and Notations

Index

A CATALOG OF SELECTED

DOVER BOOKS

IN SCIENCE AND MATHEMATICS

Math–Geometry and Topology

ELEMENTARY CONCEPTS OF TOPOLOGY, Paul Alexandroff. Elegant, intuitive approach to topology from set-theoretic topology to Betti groups; how concepts of topology are useful in math and physics. 25 figures. 57pp. 5⅜ x 8½. 60747-X

COMBINATORIAL TOPOLOGY, P. S. Alexandrov. Clearly written, well-organized, three-part text begins by dealing with certain classic problems without using the formal techniques of homology theory and advances to the central concept, the Betti groups. Numerous detailed examples. 654pp. 5⅜ x 8½. 40179-0

EXPERIMENTS IN TOPOLOGY, Stephen Barr. Classic, lively explanation of one of the byways of mathematics. Klein bottles, Moebius strips, projective planes, map coloring, problem of the Koenigsberg bridges, much more, described with clarity and wit. 43 figures. 210pp. 5⅜ x 8½. 25933-1

CONFORMAL MAPPING ON RIEMANN SURFACES, Harvey Cohn. Lucid, insightful book presents ideal coverage of subject. 334 exercises make book perfect for self-study. 55 figures. 352pp. 5⅝ x 8¼. 64025-6

THE GEOMETRY OF RENÉ DESCARTES, René Descartes. The great work founded analytical geometry. Original French text, Descartes's own diagrams, together with definitive Smith-Latham translation. 244pp. 5⅜ x 8½. 60068-8

PRACTICAL CONIC SECTIONS: The Geometric Properties of Ellipses, Parabolas and Hyperbolas, J. W. Downs. This text shows how to create ellipses, parabolas, and hyperbolas. It also presents historical background on their ancient origins and describes the reflective properties and roles of curves in design applications. 1993 ed. 98 figures. xii+100pp. 6½ x 9¼. 42876-1

THE THIRTEEN BOOKS OF EUCLID'S ELEMENTS, translated with introduction and commentary by Thomas L. Heath. Definitive edition. Textual and linguistic notes, mathematical analysis. 2,500 years of critical commentary. Unabridged. 1,414pp. 5⅜ x 8½. Three-vol. set. Vol. I: 60088-2 Vol. II: 60089-0 Vol. III: 60090-4

GEOMETRY OF COMPLEX NUMBERS, Hans Schwerdtfeger. Illuminating, widely praised book on analytic geometry of circles, the Moebius transformation, and two-dimensional non-Euclidean geometries. 200pp. 5⅝ x 8¼. 63830-8

DIFFERENTIAL GEOMETRY, Heinrich W. Guggenheimer. Local differential geometry as an application of advanced calculus and linear algebra. Curvature, transformation groups, surfaces, more. Exercises. 62 figures. 378pp. 5⅜ x 8½. 63433-7

CURVATURE AND HOMOLOGY: Enlarged Edition, Samuel I. Goldberg. Revised edition examines topology of differentiable manifolds; curvature, homology of Riemannian manifolds; compact Lie groups; complex manifolds; curvature, homology of Kaehler manifolds. New Preface. Four new appendixes. 416pp. 5⅜ x 8½. 40207-X

History of Math

THE WORKS OF ARCHIMEDES, Archimedes (T. L. Heath, ed.). Topics include the famous problems of the ratio of the areas of a cylinder and an inscribed sphere; the measurement of a circle; the properties of conoids, spheroids, and spirals; and the quadrature of the parabola. Informative introduction. clxxxvi+326pp; supplement, 52pp. 5⅜ x 8½. 42084-1

A SHORT ACCOUNT OF THE HISTORY OF MATHEMATICS, W. W. Rouse Ball. One of clearest, most authoritative surveys from the Egyptians and Phoenicians through 19th-century figures such as Grassman, Galois, Riemann. Fourth edition. 522pp. 5⅜ x 8½. 20630-0

THE HISTORY OF THE CALCULUS AND ITS CONCEPTUAL DEVELOPMENT, Carl B. Boyer. Origins in antiquity, medieval contributions, work of Newton, Leibniz, rigorous formulation. Treatment is verbal. 346pp. 5⅜ x 8½. 60509-4

THE HISTORICAL ROOTS OF ELEMENTARY MATHEMATICS, Lucas N. H. Bunt, Phillip S. Jones, and Jack D. Bedient. Fundamental underpinnings of modern arithmetic, algebra, geometry, and number systems derived from ancient civilizations. 320pp. 5⅜ x 8½. 25563-8

A HISTORY OF MATHEMATICAL NOTATIONS, Florian Cajori. This classic study notes the first appearance of a mathematical symbol and its origin, the competition it encountered, its spread among writers in different countries, its rise to popularity, its eventual decline or ultimate survival. Original 1929 two-volume edition presented here in one volume. xxviii+820pp. 5⅜ x 8½. 67766-4

GAMES, GODS & GAMBLING: A History of Probability and Statistical Ideas, F. N. David. Episodes from the lives of Galileo, Fermat, Pascal, and others illustrate this fascinating account of the roots of mathematics. Features thought-provoking references to classics, archaeology, biography, poetry. 1962 edition. 304pp. 5⅜ x 8½. (Available in U.S. only.) 40023-9

OF MEN AND NUMBERS: The Story of the Great Mathematicians, Jane Muir. Fascinating accounts of the lives and accomplishments of history's greatest mathematical minds–Pythagoras, Descartes, Euler, Pascal, Cantor, many more. Anecdotal, illuminating. 30 diagrams. Bibliography. 256pp. 5⅜ x 8½. 28973-7

HISTORY OF MATHEMATICS, David E. Smith. Nontechnical survey from ancient Greece and Orient to late 19th century; evolution of arithmetic, geometry, trigonometry, calculating devices, algebra, the calculus. 362 illustrations. 1,355pp. 5⅜ x 8½. Two-vol. set. Vol. I: 20429-4 Vol. II: 20430-8

A CONCISE HISTORY OF MATHEMATICS, Dirk J. Struik. The best brief history of mathematics. Stresses origins and covers every major figure from ancient Near East to 19th century. 41 illustrations. 195pp. 5⅜ x 8½. 60255-9

Mathematics

FUNCTIONAL ANALYSIS (Second Corrected Edition), George Bachman and Lawrence Narici. Excellent treatment of subject geared toward students with background in linear algebra, advanced calculus, physics, and engineering. Text covers introduction to inner-product spaces, normed, metric spaces, and topological spaces; complete orthonormal sets, the Hahn-Banach Theorem and its consequences, and many other related subjects. 1966 ed. 544pp. 6⅛ x 9¼. 40251-7

ASYMPTOTIC EXPANSIONS OF INTEGRALS, Norman Bleistein & Richard A. Handelsman. Best introduction to important field with applications in a variety of scientific disciplines. New preface. Problems. Diagrams. Tables. Bibliography. Index. 448pp. 5⅜ x 8½. 65082-0

VECTOR AND TENSOR ANALYSIS WITH APPLICATIONS, A. I. Borisenko and I. E. Tarapov. Concise introduction. Worked-out problems, solutions, exercises. 257pp. 5⅝ x 8¼. 63833-2

THE ABSOLUTE DIFFERENTIAL CALCULUS (CALCULUS OF TENSORS), Tullio Levi-Civita. Great 20th-century mathematician's classic work on material necessary for mathematical grasp of theory of relativity. 452pp. 5⅝ x 8¼. 63401-9

AN INTRODUCTION TO ORDINARY DIFFERENTIAL EQUATIONS, Earl A. Coddington. A thorough and systematic first course in elementary differential equations for undergraduates in mathematics and science, with many exercises and problems (with answers). Index. 304pp. 5⅜ x 8½. 65942-9

FOURIER SERIES AND ORTHOGONAL FUNCTIONS, Harry F. Davis. An incisive text combining theory and practical example to introduce Fourier series, orthogonal functions and applications of the Fourier method to boundary-value problems. 570 exercises. Answers and notes. 416pp. 5⅜ x 8½. 65973-9

COMPUTABILITY AND UNSOLVABILITY, Martin Davis. Classic graduate-level introduction to theory of computability, usually referred to as theory of recurrent functions. New preface and appendix. 288pp. 5⅜ x 8½. 61471-9

ASYMPTOTIC METHODS IN ANALYSIS, N. G. de Bruijn. An inexpensive, comprehensive guide to asymptotic methods–the pioneering work that teaches by explaining worked examples in detail. Index. 224pp. 5⅜ x 8½ 64221-6

APPLIED COMPLEX VARIABLES, John W. Dettman. Step-by-step coverage of fundamentals of analytic function theory–plus lucid exposition of five important applications: Potential Theory; Ordinary Differential Equations; Fourier Transforms; Laplace Transforms; Asymptotic Expansions. 66 figures. Exercises at chapter ends. 512pp. 5⅜ x 8½. 64670-X

INTRODUCTION TO LINEAR ALGEBRA AND DIFFERENTIAL EQUATIONS, John W. Dettman. Excellent text covers complex numbers, determinants, orthonormal bases, Laplace transforms, much more. Exercises with solutions. Undergraduate level. 416pp. 5⅜ x 8½. 65191-6